KU-690-963

IEE TELECOMMUNICATIONS SERIES 27

Series Editors: Professor J. E. Flood
Professor C. J. Hughes
Professor J. D. Parsons

Transmission systems

Other volumes in this series:

Transmission systems

Edited by
J. E. Flood and P. Cochrane

Peter Peregrinus Ltd. on behalf of the Institution of Electrical Engineers

Published by: Peter Peregrinus Ltd., London, United Kingdom

© 1991: Peter Peregrinus Ltd.

Apart from any fair dealing for the purposes of research or private study, or criticism or review, as permitted under the Copyright, Designs and Patents Act, 1988, this publication may be reproduced, stored or transmitted, in any forms or by any means, only with the prior permission in writing of the publishers, or in the case of reprographic reproduction in accordance with the terms of licences issued by the Copyright Licensing Agency. Inquiries concerning reproduction outside those terms should be sent to the publishers at the undermentioned address:

Peter Peregrinus Ltd.,
Michael Faraday House,
Six Hills Way, Stevenage,
Herts. SG1 2AY, United Kingdom

While the editors and the publishers believe that the information and guidance given in this work is correct, all parties must rely upon their own skill and judgment when making use of it. Neither the editors nor the publishers assume any liability to anyone for any loss or damage caused by any error or omission in the work, whether such error or omission is the result of negligence or any other cause. Any and all such liability is disclaimed.

The right of the editors to be identified as editors of this work has been asserted by them in accordance with the Copyright, Designs and Patents Act 1988.

British Library Cataloguing in Publication Data

A CIP Record for this book
is available from the British Library

ISBN 0 86341 148 7

Printed in England by Short Run Press Ltd., Exeter

Contents

List of contributors

P.F. Adams BSc CEng MIEE
BT Laboratories

M.J. de Belin BEng PhD
GEC Plessey Telecommunications

L. Bickers MSc CEng MIEE
BT Laboratories

R.J. Catchpole MA AMIEE
BNR Europe Ltd

Professor P. Cochrane MSc PhD
  CEng FIEE
BT Laboratories

R.M. Dorward MA CEng MIEE
GEC Plessey Telecommunications
PO Box 53
Coventry CV3 1HJ

C.M. Earnshaw BSc CEng MIEE
BT

S.P. Ferguson MSc
GEC Plessey Telecommunications

Professor J.E. Flood OBE DSc(Eng)
  PhD CEng FIEE
Aston University

P.J. Howard BSc MIEE
BNR Europe Ltd

L.D. Humphrey BSc AMIEE and
  A.D. Wallace BA
BNR Europe Ltd

D.J. Kingdom
BT

R.W. McLintock BSc CEng MIEE
BT

A.W. Muir CEng MIEE
BT

M Nouri PhD CEng MIEE
Marconi Communication Systems

P.A. Rosher BSc MSc
BT Laboratories

M.J. Sexton BSc(Eng) MIEE
STC

S. Welch OBE MEng CEng MIEE
Ministry of Defence

Preface

In earliest times, unless somebody was prepared to travel, man could only communicate as far as he could shout. Complex pressure waves, propagated through the medium of the air, reproduced the sounds of the speaker's voice at the ear of the listener. This was an analogue transmission system. Methods were then developed to enable communication to extend over longer distances. For example, a trumpet was blown, a beacon fire was lit or a flag was waved. These systems used only a very limited variety of elementary signals. Nowadays, they would be called digital systems.

Electrical communications began with the invention of the telegraph by Wheatstone and by Morse in 1837. Long-distance communications remained digital, but began to develop rapidly. When Bell invented the telephone in 1876, his receiver produced continuous current variations which corresponded to the sound waves at his transmitter. The era of analogue transmission had begun. Although telegraphy continued to be important for transmitting 'hard-copy' texts, the subsequent history of telecommunications is dominated by telephony.

The invention of the thermionic triode valve by de Forest in 1907 led to signal processing. Its use as an amplifier made telephony possible over longer distances and enabled underground cables to replace open-wire overhead lines. Its use as an oscillator and as a modulator made radio telephony possible. It also enabled a large number of voice signals to be multiplexed onto a single pair of conductors by modulating different carrier frequencies. This art of frequency-division-multiplex (FDM) carrier telephony developed over the years until it became possible to convey over 10 000 voice channels on a single coaxial-cable pair. The cost per kilometre of telephone circuits thus fell dramatically over the years. Extensive national trunk networks developed and international and inter-continental telephony became commercial realities.

The invention of pulse-code modulation (PCM) by Reeves in 1938 enabled analogue voice signals to be converted into digital signals. These signals can be regenerated at repeaters and so can be practically immune to noise, regardless of the distance over which they are transmitted. Time-division multiplexing (TDM), which had been introduced into telegraphy by Baudot in 1874, made multi-channel PCM systems possible. However, the complexity of the electronic circuits required made PCM uneconomic compared with FDM carrier systems until after the invention of the transistor by Bardeen and Brattain in 1948. The subsequent introduction of solid-state integrated circuits finally made PCM equipment much cheaper than FDM carrier systems and the era of digital telephony commenced.

The past fifteen years have seen very rapid progress in the technology of telecommunications and the development of services which have totally eclipsed the evolution of the previous 100 years. Transmission-system development has played a key role in this progress, with the migration from analogue carrier systems operating over twisted-pair, coaxial and radio bearers to digital systems which are now concentrated on optical fibre. These changes have themselves been fuelled by the dramatic progress made in semiconductor technology, which opened the door to an unprecedented revolution in electronic equipment and information technology. The direct result has been a very-rapid fall in the cost of transmitting information over long distances. Numerous new service opportunities have thus arisen with the realisation of low-cost transmission and the availability of highly-integrated terminal equipment. Now, there is no facet of life which is not influenced by telecommunications. Transmission networks form an essential part of the infrastructure of all modern industrialised nations, and a key contributor to their economic success and social stability.

The dramatic advances made in transmission technology were, in some respects, mirrored by advances in switching, but not to the same extent. The cost of information transport has been falling by over an order of magnitude per decade, whilst switching has achieved much less than half this rate during the same 15-year period. However, the advent of digital exchanges has enabled circuits conveyed over digital transmission links to be interconnected without the need for multiplexing equipment at the ends of these links. The use of stored-program control in these exchanges led to common-channel signalling between the exchange processors, thus eliminating the need for signalling apparatus associated with every individual speech circuit and giving further substantial cost savings. Thus, the combination of digital switching with digital transmission has led to the introduction of integrated digital networks (IDNs). Digital exchanges also enable digital transmission to extend over customers' lines. This can enable a wide range of services to be provided using common telecommunication plant, leading ultimately to an integrated-services digital network (ISDN).

In the early 1970s, several countries formulated plans for the rapid modernisation in their telecommunications infrastructure which were to be based on a combination of digital transmission and switching. The extent of these plans in the UK was revealed at the IEE Conference on Telecommunications Transmission in 1976 and this prompted an enthusiastic response from industry and education. A series of IEE Vacation Schools was organised to promote the understanding of these technology changes and in 1979 the first School dedicated to Telecommunications Transmission was held at the University of Aston in Birmingham. This School has been maintained by the rapid migration of transmission technology and has continued to be held at two-year intervals. Contributors to the School have been drawn from industry and academia to give up-to-date views of the state of the art on each occasion. After the sixth School in July 1989, a clear pattern had begun to emerge with the migration from wholly-analogue operation over copper cables and microwave radio to the dominance of digital operation on optical fibre and radio, but with a significant proportion of mobile systems peripheral to the fixed terrestrial network.

The possibility is now emerging of migration from wholly-digital operation back to a near-analogue mode with optical amplifiers based on laser structures and doped fibre, plus wavelength-division multiplexing that closely resembles its FDM predecessor. This latest cycle of change is yet another of those from *analogue to digital and back to analogue* that have occurred with man's desire to communicate at a distance, but is only the 'second cycle' during the period where he has used electromagnetic waves.

It therefore seems timely to bring together the combined expertise, exposed during the six Vacation Schools so far completed, into a single volume to address telecommunications transmission over the modern period, using both analogue and digital technology. In compiling this book, we therefore asked the authors to attempt to embody a broad view of their individual topics which would reflect the migration of the technology over the 12-year history of the Vacation Schools but with due deference to the earlier history.

The book is organised to form a natural progression from the fundamental need for telecommunications, the transmission media available, systems options and design, performance specification and requirements, the network planning process and, finally, future trends. It can be read as a reasonably-continuous text by those visiting the topic for the first time. Some repetition and duplication has purposely been retained to reinforce the learning process. This follows the style of the individual schools themselves. For example, Chapter 3 gives a general overview of transmission media, which is then substantially expanded in Chapters 10, 11 and 12 in respect to their practical application and limitations in a systems context. This has been augmented by the use of practical examples with real design data and condensed mathematical descriptions to give direct insights into the design process.

Our prime job as both authors and editors has been to try and provide a balance across the full scope of the book to reflect fully the evolution of transmission technology, past, present and future. This we have found both an interesting and rewarding task, inasmuch as the result represents the distillation of the experience, views and thoughts of many practising engineers and leading authorities in the field. To each of them we afford a vote of thanks for their preparation of the original lectures, and for their authorship of the individual chapters, as well as their active contributions to the Vacation Schools themselves. The students have also played an important part in the refinement process by asking all of those questions you wished you'd thought of before you had actually presented the lecture! Their individual companies also deserve a mention for sponsoring their attendance and that of the authors who have no doubt leaned on their kindness in the preparation of their source material for each School. We are also indebted to our secretaries, Mrs Esther Vennell and Mrs Helen Turner, as well as the Secretariat of the IEE and the staff of Peter Peregrinus for the preparation and marshalling of the material during many editorial phases.

Acknowledgment is made to the following organisations for permission to publish much information contained in the book: BT, Marconi Communication Systems, GPT Ltd. and STC plc.

PETER COCHRANE   JOHN FLOOD
BT Laboratories   Aston University
Martlesham Heath   Aston Triangle
Ipswich   Birmingham
Suffolk IP5 7RE   B4 7ET
England   England

May 1991

Abbreviations

A/D	Analogue-to-Digital
ACF	Autocorrelation Function
ACSSB	Amplitude-Companded Single-Side-Band
ADC	Analogue-to-Digital Convertor
ADPCM	Adaptive Differential Pulse-Code Modulation
AGC	Automatic Gain Control
AIS	Alarm Indication Signal
AM	Amplitude Modulation
AMI	Alternate Mark Inversion
ANSI	American National Standards Institute
APD	Avalanche Photodiode
ARQ	Automatic Repeat Request
ASK	Amplitude Shift Keying
ATM	Asynchronous Transfer Mode
AU	Administrative Unit
BECM	Band-Edge Component Maximisation
BER	Bit Error Ratio
BIB	Backward Indicator Bit
BIDS	Broadband Integrated Distributed Star
BISDN	Broadband Integrated Services Network
BPON	Broadband Passive Optical Network
BPSK	Bipolar Phase Shift Keying
BRL	Bell Research Labs/Balance Return Loss
BSB	British Sky Broadcasting (Previously British Satellite Broadcasting)
BSN	Backward Sequence No.
BZS	Bipolar Zero-Substituted
C/N	Carrier to Noise Ratio
CAI	Common Air Interface
CATV	Cable Television
CCIR	Comité Consultatif International des Radio Communications
CCITT	Comité Consultatif International Télégraphique et Téléphonique
CDM	Code Division Multiplexing
CDMA	Code Division Multiple Access
CDS	Cumulative Digital Sum
CEPT	Conference of European Post and Telegraphs Administrations

CMI	Coded-Mark Inversion
CR	Carrier Recovery
CVD	Continuously-Variable-Delta Modulation/Continuous Vapour-phase Deposition
CVSD	Continuously-Variable-Slope Delta Modulator
CW	Carrier Wave
D/A	Digital-to-Analogue
DAC	Digital-to-Analogue Convertor
DASS	Digital Access Signalling System
DBS	Direct Broadcast Satellite
DCCE	Digital Cell Centre Exchange
DCS	Digital Communications System/Digital Cross-Connect System
DDSSC	Digital Derived Services Switching Centre
DFB	Distributed Feedback
DFE	Decision Feedback Equaliser
DIV	Data In Voice
DMSU	Digital Main Switching Unit
DOV	Data Over-Voice
DPC	Data Processing Centre
DPCM	Differential Pulse-Code Modulation
DPSK	Differential Phase-Shift Keying
DSBSC	Double-sideband Suppressed-carrier
DSI	Digital Speech Interpolation
DSP	Digital Signal Processing
DSSU	Digital Derived Switching Unit
DSV	Digital Sum Variation
DUP	Data User Part
DXS	Digital Cross-Connect System
ECL	Emitter-Coupled Logic
EHF	Extra-High Frequencies
EIRP	Effective Isotropic Radiated Power
ELED	Edge-Emitting Diode
ELF	Extra Low Frequency
EMI	Electromagnetic Interference
ETSI	European Telecommunications Standards Institute
FAS	Flexible Access System
FAW	Frame-Alignment Word
FCC	Federal Communications Commission
FCP	Frequency Comparison Pilot
FDD	Frequency Division Duplex
FDM	Frequency-Division Multiplex
FDMA	Frequency Division Multiple Access
FEC	Forward Error Correction
FET	Field Effect Transistor
FEXT	Far-End Crosstalk
FIB	Forward Indicator Bit
FIFO	First-In First-Out

FM	Frequency Modulation
FSK	Frequency-Shift Keying
FSN	Forward Sequence Number
FT	Fractional Tap
FTTC	Fibre to the Curb
GSC	Group Switching Centre
GSM	Groupe Speciale Mobile/Group Services Mobile
GSO	Geo-Stationary Orbit
HDB	High-Density Bipolar
HDLC	High Level Data Link Control
HDTV	High Definition Television
HF	High Frequencies
HPA	Hypothetical Path Attenuation
HRC	Hypothetical Reference Circuit
HRX	Hypothetical Reference Connection
IBS	Intelsat Business Services
IDA	Integrated Digital Access
IDN	Integrated Digital Network
IDR	Intermediate Data Rate
IF	Intermediate Frequency
IFRB	International Frequency Registration Board
IN	Intelligent Network
IND	Intelligent Network Database
INDB	Intelligent Network Data Interbase
ISDN	Integrated Services Digital Network
ISI	Intersymbol Interference
ISO	International Standards Organisation
ISU	Initial Signal Unit
IT	Information Technology
ITU	International Telecommunications Union
LAN	Local Area Network
LD	Laser Diode
LED	Light Emitting Diode
LF	Low Frequencies
LNA	Low-Noise Amplifier
LPC	Linear Predictive Coding
LRE	Low Rate Encoding
LSI	Large-Scale Integrated
LSU	Lone Signal Unit
MAN	Metropolitan-Area Network
MBHL	Mean Busy-Hour Loading
MCVFT	Multichannel Voice-Frequency Telegraph
MF	Medium Frequencies/Multi Frequency
MIAC	Multi-Point Interactive Audio-Visual Conferencing
MM–SM	Multi Mode–Single Mode
MMOF	Multi Mode Optical Fibre
MSC	Main Switching Centre
MSOH	Multiplex Section Overhead

MTTF	Mean Time to Failure
MUF	Maximum Usable Frequency
MUM	Multi-Unit Message
NA	Numerical Aperture
NEXT	Near-End Crosstalk
NFB	Negative Feedback
NIC	Near-Instantaneous Companding
NMC	Network Management Centre
NNI	Network Node Interface
NPR	Noise Power Ratio
NRB	Non-Redundant Binary
NRZ	Non Return to Zero
NTSC	National Television System Committee
O & M	Operation & Maintenance
OC	Optical Carriers
OFTEL	Office of Telecommunications
OLR	Overall Loudness Rating
OMC	Operations & Management Centre
OSI	Open Systems Interconnection
OW	Order Wire
P & T	Posts & Telegraph
PABX	Private Automatic Branch Exchange
PAL	Phase Alternate Line
PAM	Pulse-Amplitude Modulation
PCM	Pulse-Code Modulation
PCN	Personal Computer Network
PDH	Plesiochronous Digital Hierarchy
PFM	Pulse Frequency Modulation
PIN	P type Insulator N type
PLE	Principal Local Exchange
PLL	Phase-Locked Loop
PLM	Pulse-Length Modulation
PM	Phase Modulation
PN	Pseudo Noise
PON	Passive Optical Network
POTS	Plain Old Telephone Service
PRBS	Pseudo-Random Binary Sequence
PSK	Phase-Shift Keying
PSTN	Public Switched Telephone Network
PTT	Post, Telegraph & Telephone
PWM	Pulse-Width Modulation
QAM	Quadrature Amplitude Modulation
QPSK	Quarternary Phase Shift Keying
RADA	Random-Access Discrete-Address
RARC 83	Regional Administrative Conference
RBER	Residual Bit Error Ratio
RBQPSK	Reduced-Bandwidth Quaternary Phase-Shift Keying
RCU	Remote Concentrator Unit

RF	Radio Frequency
RIN	Relative Intensity Noise
RLR	Receive Loudness Rating
RMS	Root Mean Square
RPE–LTP	Regular-Pulse-Excitation Long-Term Prediction
RSCH	Regenerator Section Overhead
RSU	Remote Switching Unit
RX	Receiving
SCPC	Single Channel per Carrier
SDH	Synchronous Digital Hierarchy
SDLC	Synchronous Data Link Control
SECAM	Sequential Colour and Memory
SGASS	Super Group Assembly
SHF	Super-High Frequencies
SIF	Signalling Information Field
SIO	Service Information Octet
SLEDS	Surface Light Emitting Diodes
SLR	Send Loudness Rating
SMOF	Single Mode Optical Fibre
SMS	Satellite Multiservice System
SOH	Section Overhead
SONET	Synchronous Optical Network
SQNR	Signal/Quantisation-Noise Ratio
SSB	Single-Sideband
SSBSC	Single-Sideband Suppressed-Carrier
SSU	Subsequent Signal Unit
STATMUX	Statistical Multiplexer
STM	Synchronous Transport Module
STS	Synchronous Transport Signals
SYNTRAN	Synchronous Transmission
TASI	Time-Assignment Speech Interpolation
TAT8	TransAtlantic Telephone Cable No 8
TCM	Time-Compression Multiplexing
TDD	Time Division Duplex
TDMA	Time-Division Multiple Access
TMUX	Transmultiplexers
TN	Tributary Network
TPON	Telephony on a Passive Optical Network
TR	Timing Recovery
TSC	Transit Switching Centre
TTL	Transistor Transistor Logic
TUG	Tributary Unit Group
TUP	Telephone User Part
TVE	Transversal Equaliser
TVRO	Television Receive-only
TWT	Travelling-Wave-Tube
TX	Transmitting
TX/RX	Transmitting/Receiving

UHF	Ultra-High Frequencies
UI	Unit Interval
VC	Virtual Container
VF	Voice Frequency
VHF	Very-High Frequencies
VLF	Very-Low Frequencies
VLSI	Very Large Scale Integration
VSAT	Very Small Aperture Terminal
VSB	Vestigial Sideband
VT	Virtual Tributary
WAN	Wide Area Network
WARC-77	1977 World Administrative Radio Conference
WDM	Wavelength Division Multiplexing
WDMA	Wavelength Division Multiple Access
XPD	Cross Polar Discrimination

Chapter 1
Introduction
C. M. Earnshaw
Director Network, BT, United Kingdom

1.1 General

Telecommunications is assuming a progressively more important role in the infrastructure of society as a whole. The growth of business and the social structure of any country relies on and interacts with telecommunications, the development of which is thus a key element in the economic health of a nation [1].

The telephone network comprises transmission paths and switching nodes disposed to enable users, on a world-wide basis, to make satisfactory calls to any other users of the system, on demand and in an economic manner. Customers connected to the network must be uniquely identified; the network must indicate to the caller the availability or otherwise of the required number, must protect each connection from interference from other connections, must generate information to allow revenue to be collected and must be dimensioned in such a way as to be economic and capable of expansion. It must also be flexible enough to cater for a wide and expanding range of non-voice and visual services and be able to absorb new technology and cater for unexpected variations in growth [2].

There are over 670 million telephones in over 129 countries around the world, all of which must be capable of being interconnected through national and international telecommunication networks [3]. It is clearly impracticable to provide direct connections between all combinations of telephones, or even groups of telephones, world-wide. The optimisation of switching, signalling and transmission techniques and the adoption of global standards are therefore fundamental requirements of designing and operating telecommunication networks which offer customers ease of operation and good reliability at a cost-effective tariff.

During recent years telecommunication networks have gone through a process of radical re-structuring as a result of the availability of high-speed digital transmission and switching systems, together with increasing customer demand for new services and facilities which become available when an integrated digital network is realised. This explosion in both the demand and availability of services is indicated in Fig. 1.1.

This chapter will consider the constituent parts of an integrated digital network, and discuss some of the trends and pressures which are shaping telecommunication networks of the future. However, as the networks of

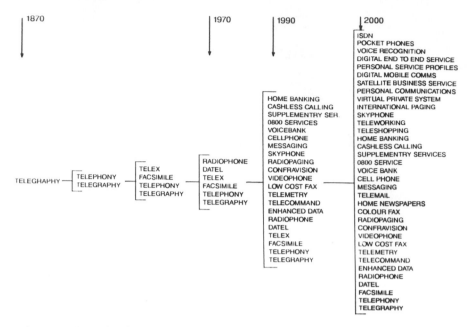

Fig. 1.1   Growth of telecommunications services (past, present and predicted)

today and of the future are based on many years' experience with analogue networks, the chapter will begin with a brief review of the topology of an analogue telecommunication network.

1.2   Analogue networks

Just as it is impracticable to interconnect all telephones in a network, so it is impractical to connect all local lines to a telephone exchange, or even to interconnect all local telephone exchanges of a realistic size. A hierarchical network must therefore be adopted [4]. An example is shown in Fig. 1.2.

Historically, call switching in analogue networks was performed using mechanical step-by-step switches (Strowger), which were then augmented and partially superseded by crossbar switching and then later by electronically-controlled exchanges which evolved to adopt new switching techniques employing reed relays [5]. The trend towards the introduction of semiconductor technology brought about a reduction in manufacturing costs, together with significantly-lower operation and maintenance costs [6].

It would have been uneconomic to design a switched network which provided satisfactory performance between every possible combination of customers. Network design was therefore carried out on a statistical basis to ensure that there were sufficient paths available for a defined percentage of calls to be connected on their first attempt [7]. Analogue transmission

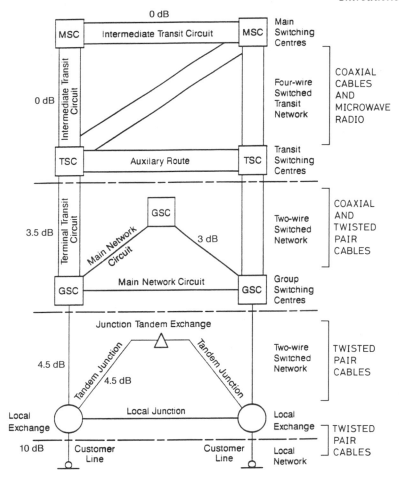

Fig. 1.2 A switched analogue network

design and planning has been based on ensuring that, for any connection between telephones, the impairments are within certain limits. Transmission performance requirements are discussed in Chapter 4. In implementing a national transmission plan, it was necessary to allocate the permissible overall loss in such a way as to obtain the most economic design. For example, customers lines to local exchanges are very significant for overall costs because of their predominance in the network. Junction circuits between local exchanges and GSCs are less numerous, so it is economic to provide better-quality cables. Trunk circuits could be provided at higher cost and use amplifiers to minimise circuit attenuation [8].

Local line and junction traffic capacity was provided via dedicated copper pairs and employed simple DC signalling systems. Most of the local line network consists of external plant, i.e. buried duct, manholes, cabinets, poles and overhead cables, whilst the junction network is usually carried on cables in buried duct work.

Analogue trunk transmission capacity was provided using frequency-division multiplexing (FDM) techniques to combine a large number of audio channels into groups, supergroups and hypergroups, which are then transmitted on co-axial cables or on microwave radio systems [9]. These systems are described in Chapter 5. The complexity of such systems varied with required traffic capacity but some 'backbone' routes required the use of single co-axial cables which carried 10 800 audio channels multiplexed within a 60 MHz bandwidth [10, 11]. Such systems required a vast quantity of terminal equipment to amplify, translate and filter the individual audio channels, together with large quantities of buried repeater equipment along each cable. In addition to the co-axial cable trunk network, microwave radio systems were used in the 4 GHz, and 6 GHz frequency bands, each band supporting a number of radio-frequency bearers each of which carried 960 or 1800 analogue FDM channels. The combined network thus provided both alternative traffic routing and a choice of transmission media. The microwave network was also used to distribute television signals from studios to their broadcast transmitter sites. Line transmission systems are discussed in Chapter 10 and microwave radio systems in Chapter 12. Signalling within an analogue transmission network is performed by the use of audio tone frequencies associated with each telephony circuit in the trunk and junction network. Signalling systems are discussed in Chapter 14.

1.3 Network modernisation

Analogue telecommunication networks can provide a relatively high-quality basic voice service; however, their customer-to-customer performance is dominated by electromechanical switching equipment with limited inter-exchange signalling facilities. As a result, analogue networks are relatively slow, noisy and expensive to maintain. They are also limited in their ability to provide the customer and administrative facilities required in the future. During the 1960s and 1970s, major network operators world-wide began to study changing customer and network requirements, and to consider the implications of the introduction of new technologies which were starting to become both technically feasible and economic. Such studies concluded that the combination of digital transmission and digital switching to create an integrated digital network (IDN) could result in significant network economies. As a consequence, most major networks are now undergoing a transformation from analogue to digital operation [12].

1.4 Digital networks

1.4.1 General
Digital networks can be considered to consist at the highest level of a Transmission Bearer Network which supports a number of specialised

networks in addition to the well-established public switched telephone network (PSTN). The bearer network consists of a dense network of high-capacity digital transmission systems which interconnect a number of switching centres.

The Transmission Bearer Network can be thought of as the basic building block of any telecommunication network. It provides the basic transportation medium for a number of separate functional or service networks, each in turn carrying a number of network services, as shown in Fig. 1.3. The transmission bearer network consists of a large number of high-capacity transmission systems which are used to interconnect the switching nodes of the various functional networks. The systems are interconnected at flexibility nodes, or repeater stations. A highly-meshed network of systems comprises the network core which then interconnects with smaller-capacity systems fanning out at the periphery. The planning of transmission networks is discussed in Chapter 16.

The functional networks vary in use from the public switched telephone network (PSTN) to broadband visual networks, private circuits, and a synchronisation network, as shown in Fig. 1.3. However, this interim structure is very much a product of today's customer requirements based on an existing network infrastructure. It can be foreseen that its topology will need to evolve further in order to provide a full Integrated Services Digital Network (ISDN).

The modernisation of a PSTN to create an integrated digital network (IDN) is characterised by the use of stored-program-controlled digital switching, digital transmission systems and the use of common-channel inter-processor signalling. Extending the 64 kbit/s digital path together with the high-capability signalling direct to the customer premises creates an ISDN and brings an increased range of network service capabilities direct to the customer [13].

The IDN has a number of conceptual layers, supported on the transmission network. It consists of both switched and logical layers together with a number of service-specific or functional layers, as shown in Fig. 1.4.

The switched layer contains the network switching centres which are interconnected by transmission capacity. The logical layer of the IDN is

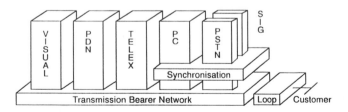

Fig. 1.3 Functional networks

PSTN = public switched telephone network
SIG = signalling network
PC = private circuits
PDN = public data network

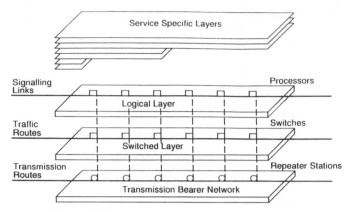

Fig. 1.4   Layers of the PSTN

analogous to a network of interlinked computers each having an operating system with specific application software packages to provide particular network services. The logical layer describes the application software in the switching nodes in terms of generic and service-specific attributes. Generic attributes are common to all nodes, such as switching a call between the input and output ports of an exchange. Service-specific attributes are those necessary for the network to carry particular services. The establishment of calls requiring specific features may require an interchange of information and instructions between exchange processors; this is carried out via the common-channel signalling network.

The addition of centrally-located software, detached from exchange processors, in Intelligent Network Databases can be used to provide advanced network management, administration and maintenance infrastructures. Today, major telecoms operators are looking to introduce this new technology to create the 'Intelligent Network' of the 1990s. The technology also opens new possibibilities for customer control of services and the integration of private and public network capabilities [14].

1.4.2   The core network

Whilst the topology of digital networks is still broadly based on the historical evolution of the analogue network, there are a number of significant changes which arise out of a need to minimise operating costs whilst providing a high-quality and flexible network. For example, the British Telecom network has a three-tier structure consisting of Digital Derived Services Switching Centres (DDSSC), Digital Main Switching Units (DMSU) and Digital Cell Centre Exchanges (DCCE) with associated Remote Concentrator Units (RCU), as shown in Fig. 1.5. Three main changes are taking place as the network is converted to digital operation: a reduction in the number of main switching centres, a trend towards co-siting of main and local pro-

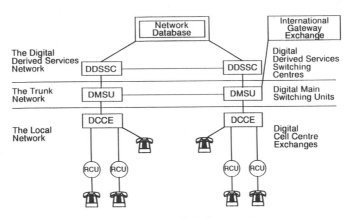

Fig. 1.5 The British Telecom switched digital network

cessors, and a trend towards removing small local exchanges by the introduction of Remote Concentrator Units [15]. One of the greatest changes that has taken place in terms of network topology is at the trunk switching level. The new digital network consists of a total of about 60 main network switching centres, whilst the equivalent analogue system contained over 400 Group Switching Centres. An equivalent change is also taking place in the junction and local networks. Over 6000 local exchanges are being parented on about 100 processor sites, with the original exchanges being replaced with RCUs. These changes have been made possible by the inherently high quality of optical-fibre-based digital transmission equipment and by a recognition of the higher cost of providing switching equipment compared to that for transmission.

Reducing the number of exchanges has three principal effects. First, exchanges will be larger and therefore more economical. A modern digital exchange has a high fixed cost associated with the central processing, but the incremental costs associated with increasing size to cope with more traffic and customer connections are lower than for analogue technologies. Secondly, most calls are local to the area in which they are generated. Therefore, increasing the geographical coverage of an exchange to concentrate more traffic into one switch will not only eliminate junction-network transmission equipment costs, but also reduce the consequential costs of double switching. Thirdly, concentrating exchanges will also result in a lower whole-life cost for exchange operation and maintenance (O&M).

At the top tier of the network, all the exchanges can be fully interconnected by a mesh network of high-capacity optical-fibre and microwave radio transmission systems. These systems are considered in depth in Chapters 11 and 12, respectively. Inter-exchange signalling can be performed using CCITT No. 7 protocols, a system whereby all signalling information is concentrated into a 64 kbit/s data link. This allows tremendous flexibility and has already resulted in a significant reduction in the cost of providing signalling equipment, as common–channel CCITT No. 7 signalling costs are only about 5% of that required for signalling with equivalent analogue equipment.

To protect the trunk transmission network from equipment failures and from failures caused by human intervention, e.g. road works, the digital network can be provided with a network of automatically-switched 140 Mbit/s digital service protection links [16]. Typically, about 10% of the available total transmission capacity, is allocated exclusively for circuit restoration, for manual traffic off-loads to enable maintenance to be undertaken, and for the transmission of traffic under special circumstances or special events. Automatic protection switching allows a number of different restoration paths to be pre-programmed to enable restoration within a very-short period of time [17].

1.4.3 The local network

The local distribution network lies between the local exchange and the customer and is sometimes referred to as the 'Access Network' or 'Local Loop'. This network represents a very substantial proportion of the assets of a telecommunications network operator and, as all customer services ultimately rely upon this part of the network, the management and development of it are of fundamental importance. Customer access networks are discussed in Chapter 15. In most countries the physical network structure was laid down over 50 years ago and still dominates today's local network. Whilst the majority of the Access Network is currently provided using copper-based analogue technology, there will be a steady evolution towards a network having significant proportions of digital electronic circuits [18]. The demand for this change is being led by business customers, but in the long term it may well be dominated by residential demand for visual and inter-active services. Apart from the need to develop suitable system components, this evolution has implications for the network structure itself as well as for planning, installation and maintenance practices.

Until quite recently there has been a relatively low penetration of digital transmission in the local network. However, this is growing with the development of ISDN (Integrated Services Digital Network) access and the provisions of digital private circuits operating at 64 kbit/s or 1·5 or 2 Mbit/s.

Optical fibre has been introduced into local networks to serve major business customers [19]. In the longer term, it is expected that optical fibre will also be used in the local network for small business, and residential customers by locating electronics in 'street furniture' and using simple traditional copper for the final link to the customers' premises. A further development, known as Telephony Over Passive Optical Networks (TPON), could take fibre direct to customers' premises using passive networks comprising single-mode fibres fed from the local exchange and fanned out via optical splitters at cabinets and distribution points to serve a number of customers [20]. Such a system could also carry television signals and broadband ISDN services as well as telephony services by employing wavelength-division multiplexing techniques; it is termed Broadband Passive Optical Networks (BPON) [21].

1.5 Network services and management

1.5.1 Network services

Telecommunications offers customers access to a wide range of services which are known as 'network services'. The following paragraphs highlight some of the features of typical services.

Derived services network: A derived services network employs dedicated switching capacity to offer customers special services such as 'freephone' and message services. The control information for enhanced features is programmed directly into an Intelligent Network Database (INDB) which provides call-routing information, charging information and network-management and maintenance facilities, together with call-pattern statistics for both the network operator and the service provider.

Private services network: This provides the business community with private circuits, both analogue and digital. Analogue private circuits support telephony, data and such services as music circuits for broadcast radio services, and wide-bandwidth customer circuits. In addition, terrestrial broadcast services are provided by video circuits. Digital private circuits are available to customers at bit rates of between 2·4 and 64 kbit/s, and at 1·5, 2 and 8 Mbit/s.

Packet switching: This allows the use of CCITT X.25 packet-switching techniques to support a variety of data services, including electronic fund transfer and credit-card transaction validation. The network requires dedicated switching equipment which is configured in a tiered structure.

Telex: This provides a low-speed (50 baud) telegraphy service that is widely used within the business world. The network provides automatic direct dialling to more than 200 countries throughout the world via dedicated switching centres. The telex network allows direct connection of personal computers, which enables the customer to make use of facilities such as store-and-forward, multi-address messages and pre-recorded address lists. The network can also accommodate gateway services to provide access between private digital networks and electronic message-handling services.

Integrated services digital network (ISDN): The modern business environment is rapidly moving into widespread use of text creation, storage and manipulation, retrieval of information, and new methods of communication. As a consequence, such offices now utilise machines which combine the power of computers with telecommunications to provide a single paper-less integrated system. Networks therefore need to handle a wide variety of voice and data services and the traditional functions of switching, transmission and signalling need to be supplemented by information recognition, processing and retrieval to create a service-providing 'active' network. ISDN is intended to provide a variety of services such as, Closed User Groups, Data-Message Store and Retrieval, Voice-Message Storage and Retrieval,

Data Conversion between different terminal types, and concentration of services for low-bit-rate telemetry usage.

1.5.2 Management of telecommunication networks

A data network that interconnects the intelligence embedded in the switching and transmission systems to the real-time network-control systems and data-processing centres is known generically as the Administration or Network-Management Network. It performs a vitally important role in maximising the efficient usage of the network to the mutual benefit of both the network operator and the customer [22]. Fig. 1.6 shows the different physical elements of this network, which are as follows:

(*a*)   Network Data Collectors, responsible for the bulk collection of data (statistics, billing etc.) from switching centres and transmission surveillance points.

(*b*)   Operations and Maintenance Centres (O&M), for remote man–machine communication to switching centres, which allow manipulation of switching software, change of customer routing/service, and receive fault reports.

(*c*)   Network Management Centres, which receive and analyse real-time traffic flows, and allow traffic re-routing to overcome congestion.

Management Systems can be sub-divided into a number of levels to facilitate complete supervision and control. Five are shown in Fig. 1.7, which are as follows:

(*a*)   *Plant and plant management*: This provides access to network plant by reporting alarms, and allowing access to equipment control systems.

(*b*)   *Network control*: This allows event management, alarm correlation, traffic management and allocation of resources. It also provides correlation and co-ordination of events in different parts of the network.

(*c*)   *Service management*: This provides correlation between network faults and service availability and collation of service statistics.

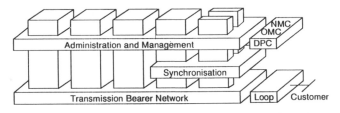

NMC - Network Management Centre
OMC - Operations and Maintenance Centre
DPC - Data Processing Centre

Fig. 1.6   Network management

NMC = Network management centre
OMC = Operations and maintenance centre
DPC = Data processing centre

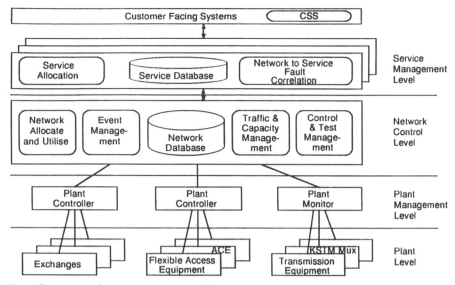

Fig. 1.7 Network management architecture

(*d*) *Customer facing*: This provides a single interface between the customer and the network operator for such items as order processing, fault progressing and billing. Increasingly, major customers are seeking to have direct real-time access and control of their telecommunications facilities.

1.6 The changing telecommunications environment

1.6.1 General
Telecommunication networks are under increasing pressure from a number of differing forces, many of which will result in future telecommunication networks which look quite different from those of today [23]. Hence network operators world-wide are having to address these factors when planning for services in the future. Some of the main issues which will impact on the development of networks are considered in the following sections. However, it should be recognised that many of these issues are of a volatile nature; hence the emphasis on network flexibility.

1.6.2 Customer expectation
Consider the expectations of a customer, whether a single residential customer or a commmercial organisation of considerable size. The customer quite simply expects each year a lot more for a lot less, and his or her expectations are continuously fuelled by developments in the mass consumer market. For example, customers will expect not only a better quality of connection once made, but will also expect fewer call-attempt failures (either

due to switching equipment or to insufficient transmission plant), faster repair of faults and a declining incidence of faults. So the network of the future will have to offer good reliability and a far greater variety of services tailored specifically to the requirements of the customers.

Networks of the future must therefore be extremely flexible to allow on-line re-configuration on a flexible basis to adjust to the requirements of each individual customer. As a result, the administration, maintenance and management of future networks will play a vital role in determining the level of customer satisfaction and will require the implementation of a comprehensive administration network and management systems. Customers will also expect a network operator to be capable of providing services at very short notice, ultimately on demand.

1.6.3 Customer mobility

During recent times there has been a rapidly increasing demand for services which allow the customer to move location easily without losing the ability to access a wide range of telecommunication facilities. Such services used to be provided to a very small number of customers with the aid of VHF radio systems. More recently, service has been offered on a large scale using physically compact equipment via cellular radio networks. The introduction of these services has been accompanied by unprecedented demand for mobile/portable facilities which, in turn, has put pressure on the available frequency spectrum. The pressure is likely to mean that large portions of the UHF and low SHF frequency bands are dedicated exclusively to mobile services. These will shortly include the next generation of pan-European digital cellular network, due to start operation during the early 1990s [24].

The present generation of cellular equipment allows both outgoing and incoming calls to be made. This requires considerable inherent intelligence within the network to switch calls between customers. Whilst this gives the customer a highly versatile 'wire-less' communication medium, the real task for the network operator is in giving the telecommunication network the capability to locate each mobile customer.

A new generation of low-cost mobile services has recently been implemented and are known as Telepoint. These offer outgoing calls only to be made from up to around 200 metres from base stations, which are sited at prominent locations and provide a personal alternative to telephone boxes. For these services to become fully two-way, enabling incoming calls to be received, would require the development of the locating network intelligence discussed in the previous paragraph. Handsets are CT2 Common Air Interface (CAI) standard and can be used from a domestic base station or a CAI wireless PABX, as well as from a Telepoint. These handsets are truly portable and are 'pocket' sized.

Personal Communications Networks (PCNs) have also been licensed recently and use a new mobile standard called Digital Communications System 1800 (DCS1800) which is based on the Group Services Mobile (GSM) standard. These networks will offer full mobility and use GSM technology in their own intelligent networks. Access will be provided in the

1·71–1·88 GHz region. Handsets will be smaller than the current generation of cellular handsets, but the PCNs will require many more base stations and a much more complex network infrastructure. Initially, the PCN services will be aimed at the current cellular market; however, it is expected that the longer-term aim will be to provide an alternative mass-market service, competing directly with the fixed network operators.

In the longer term, customers may demand a truly mobile telecommunications environment. From a single personalised communicator, which is carried by each customer, they should be able to make and receive calls irrespective of their locations. Similarly, the fixed network will need to cater for customers on the move who wish to take personal numbers with them [25]. Demand along these lines is having significant implications for the design of telecommunication networks.

1.6.4 Advancing technology

In terms of recent history, advances in the availability and capability of technology have created strong pressure for change. The introduction of new technology, such as digital switching and common-channel signalling, has revolutionised the telecommunication network and enabled operators to realise many new customer requirements. However, this in itself creates further customer demand and increases the pressure for new services within the network.

The introduction of large-scale integration on a single silicon substrate is enabling a huge reduction in physical size to be accompanied by an equally large increase in complexity and computational power. More interesting perhaps is the trend towards standardised silicon, a process whereby a highly-complex gate array is designed with a number of tasks in mind, so that large-volume production can be used to keep the hardware costs down. The equivalent process is happening within the transmission and switching equipment markets, where the trend is towards standard items of hardware that are customised via the addition of specific software. Such standardisation can offer economies to the network operator. As a purchaser of compatible telecommunication equipment on a world market, the network operator can obtain better value for money, whilst ensuring a well tested and supported product for its customers.

At the same time, software is becoming an increasingly significant component and cost in the provision of telecommunication networks. Apart from its ability to provide diverse facilities that can be readily upgraded, software also allows a greater efficiency of hardware utilisation to be realised. Beyond call control, network monitoring and management, software is currently being applied to a wide range of customer services within networks. Voice synthesis, analysis, recognition, verification and control, language translation and man–machine conversation are among a number of the present and future applications of modern software.

1.6.5 Satellite technology

Traditionally, satellites have been used for high-capacity international telephony routes, and for broadcast television purposes. However, low-capacity systems are now available which allow services such as electronic news gathering, data services, telex and paging facilities, many of which were once offered solely by terrestrial telecommunication networks. Satellite networks, particularly data networks, are also ideally suited for use by new network operators who wish to introduce their service to a large number of widely-dispersed customers. New opportunities and services will also arise as low-cost direct broadcast television becomes more popular. Although the initial cost of building and launching satellites is high, currently available technology can offer instantaneous wide-area coverage with international coverage if allowed [25, 26]. Satellite communication systems are discussed in Chapter 13.

1.6.6 Network flexibility

As highlighted earlier, networks in the future will need the flexibility to respond to a much wider range of service demands, both technically and in a timely manner. In addition, customers will require the network to offer 'intelligent' services, to provide 'feedback' to customers and to assist customers in their use of network services. In the business community, many customers will require greater control over the facilities at their disposal and will expect to have visibility of what is gong on. Although they will expect the network operator to provide the services, they will want more and more opportunity to use the intelligence of the network on a selective basis to the benefit of their company. From the network operator's viewpoint the network must be capable of being 'managed' by providing remote re-structuring facilities, sophisticated fault and maintenance management and on-line performance assessment. All these factors have created pressures which are leading to the development of new network technologies.

One solution which has been proposed and which potentially offers great flexibility is known as Asynchronous Transfer Mode (ATM). The system is similar to packet switching in that traffic is transmitted in bursts which are of short length, and contain a short header for identification and routing purposes. Studies have shown that such a system may offer advantages over the existing circuit-switched system by offering a single transport medium for all services, by offering flexibility in the allocation of bandwidth, and greater flexibility for the introduction of new services [27]. Such a system is therefore very much in line with the ideals of an integrated digital network. It is anticipated that a move towards an ATM-based network would have greatest effect on switching and transmission equipment and the topography of the access network, whilst the high-capacity core network could be retained in a similar form to that which exists today.

1.6.7 Network intelligence

Networks of the future are certain to require a high degree of intelligence if they are to satisfy customer demand for new services. The separation of service-control from basic switch-control software provides a more flexible structure for the deployment of complex network services. Work is being carried out to define the interface standards and protocols to enable data bases to be developed which are capable of access from any switching centre and enable the network operator to respond rapidly to market demands for new services. In addition, the network intelligence will be able to assist the customer in use of the network, either by offering advanced user-selectable services, or by offering more sophisticated communication by way of a dialogue between customer and network.

1.7 Telecommunication liberalisation

In most parts of the world, the telecommunications service is the responsibility of a single administration. Usually, this is a government posts and telegraphs (P&T) department or a public corporation. In the UK, the former public monopoly corporation became a private company (British Telecom) in 1984 and another company (Mercury) was licensed to compete with it. There are also two competing operators of cellular radio services. Similar developments have taken place in Japan. In most European countries the telecommunications service still remains a public monopoly. However, the European Community is requiring these public network operators to permit the attachment of privately-owned customers' terminal equipment.

The privatisation of telecommunications has had a number of consequences. Since competition remains imperfect, direct government control through ownership is replaced by supervision by an official regulatory body. In the USA, this is the Federal Communications Commission (FCC) and in the UK it is the Office of Telecommunications (OFTEL). Where there are competing networks, interconnection must be provided between them to enable customers of one network to make calls to those on another network. This requires the network operators to comply with common transmission standards and with a common national numbering scheme for their customers. Liberalisation of the market for terminal apparatus means that this apparatus must operate to agreed standards to ensure that it is suitable to interwork with the network.

1.8 Telecommunication standards

National telecommunication networks need to interface with the networks of other countries for international calls. The interconnection is via international transmission systems and switching centres (Gateways). Such connections may be achieved via submarine cables and satellite links from a number of ground stations. Whilst these international links historically made use of analogue technology, modern submarine links employ fibre optics and digital transmission technology (for example, TAT 8 connecting the UK, France and the USA). Satellite systems also now deploy digital technology to provide world-wide coverage for PSTN and private-circuit usage, and localised coverage for special services such as European television distribution or private closed-user-group communications.

From the necessity for national and international telecommunication networks to inter-work on a world-wide scale it follows that common standards are essential. These are of major interest to network operators and equipment suppliers alike. Owing to the scale of the communications business, it is now most important that equipment standards are internationally agreed, to reduce the increasing cost of equipment development (avoiding one-off, or small-scale variants), to ensure inter-operability and to allow manufacturers to take advantage of the world market place.

The development of standards which make an efficient international network possible is carried out through the auspices of the International Telecommunications Union (ITU), which was founded in 1865. The ITU has 124 member countries and has its headquarters in Geneva. The work of the ITU is carried out through two main committees:

The Comité Consultatif International Télégraphique et Téléphonique (CCITT), formed in 1956, is responsible for the study of technical questions, operating methods and tariffs for telephony, telegraphy, and data transmission.

The Comité Consultatif International des Radio Communications (CCIR), formed in 1906, it responsible for the study of technical matters and operating methods relating to radio communications. Associated with the CCIR is the International Frequency Registration Board (IFRB), which regularises the assignment of radio frequencies to prevent undesirable interference between different transmissions.

The CCITT and the CCIR are composed of representatives of telecommunication operators and industrial organisations. Both committees have a large number of active study groups, which meet frequently and carry out thorough studies over the whole range of telecommunications. The recommendations of the study groups are reported to plenary sessions of the CCITT and CCIR which meet every four years. The results of the plenary sessions are published in a series of volumes which provide recommended standards. In theory, the recommendations of these consultative committees apply only to international communications. However, international traffic must pass over part of the national networks of at least two countries in addition to the international circuits concerned. Consequently, national network standards are inevitably affected.

In addition to the ITU, there are a number of international bodies established to co-ordinate the provision of telecommunication facilities in different regions of the world. In Europe there are a number of bodies set up to deal with operating conditions for telecommunications. The two main ones are CEPT (Conference of European Post and Telegraphs Administrations), which is constituted from individual national P&T operators, and ETSI (European Telecommunications Standards Institute), which is a much wider based organisation constituted of members from most of the telecommunications suppliers and user groups as well as the P&T operators. The latter has only recently been established and the technical work previously undertaken by CEPT transferred into ETSI, leaving CEPT dealing with the commercial issues facing network operators. ETSI is recognised as the European telecommunications standards body by the European Commission and it is a condition of membership that all national variants are to be withdrawn. This has the effect of making some standards mandatory when supported by the Commission via an enabling Directive.

One result of this rather complex system is that standards are slow to evolve, and to react, to increasing world-wide liberalisation of telecommunication services. The challenge for the future is therefore to ensure that international standards are agreed in a flexible way which does not restrict the development of modern cost-effective networks.

1.9 Conclusions

Telecommunications is assuming increasing importance in the health of the nation and individual life styles. The combined influences of customer expectation, technology trends and the regulatory environment all interact to shape the way telecommunication networks need to evolve.

The successful design and operation of telecommunication networks requires an increasing awareness of all these factors and their inter-relationship if we are to achieve the goal of cost-effective high-quality global communications.

1.10 References

1 SAUNDERS, R.J. *et al.*: 'Telecommunications and economic growth'. World Bank Publication 1983
2 Various Papers: 'Telecom at 150 years', *IEEE COMSOC–MAG*, August 1989, **27**, (8)
3 MARTIN, J.: 'The wired society' (Prentice Hall, 1978)
4 FLOOD, J.E.: 'Telecommunication networks' (Peter Peregrinus, 1975)
5 HUGHES, C.J.: 'Switching—State-of-the-Art: Part 1', *BTTJ* 1986, **4**, (1)
6 HUGHES, C.J.: '*ibid* Part 2', *BTTJ.*, 1986, **4**, (2)
7 BEAR, D.: 'Principle of telecommunication traffic engineering' (Peter Peregrinus, 1972)
8 HILLS, M. T., and EVANS, B.C.: 'Transmission systems—Telecommunications systems design: Vol 1'. (George Allen and Unwin, 1973)
9 'Transmission systems for communications' (Bell Telephone Laboratories, 1970, 4th edn.)
10 KELCOURSE, L.C., and HILL, F.J.: 'L5 System: Overall description and system design', *Bell Syst. Tech. J.* 1974, **53**, pp. 1901–1933
11 'The 60 MHz FDM transmission system,' Special Issue *POEEJ.* 1973, **66**, Pt. 3
12 MUIR, A., and HART, G.: 'The conversion of a telecommunications network from the analogue to digital operation'. First IEE National Conference on UK Telecommunications Networks — Present and Future, 1987
13 FOGARTY, K.D.: 'ISDN services and network recommendations', *BTTJ.*, 1986, **4**, pp. 50–57

14 LEWIS, D.: 'Telecommunications services in the 1990s', *BTEJ.*, 1989, **7**, pp. 226–232
15 Various Papers: Second National Conference on Telecommunications, York, April 1989. IEE Conf. Publ. 300
16 SHICKNER, M.J.: 'Service protection in the trunk network', *BTEJ*, 1988, **7**, pp. 89–95
17 DAVIDSON, J., *et al.*: 'The evolution of service protection in the BT Network'. Globecom 89, Houston, Texas, November 1989
18 Special Issue on Fibre Optic Systems for Terrestrial Applications, *IEEE J.*, 1986, **SAC-4**, (9)
19 DUFOUR, I.G.: 'Flexible access systems', *BTEJ.*, 1989, **3**, pp. 233–237
20 STERN, J.R. *et al.*: 'Passive optical local networks for telephony applications and beyond', *Electron. Lett.*, 1987, **23**, pp. 1255–1257
21 'The local network', *Special Issue, BTTJ*, 1989, **7**, (2)
22 Various Papers: 'Telecommunications network operations and management', *IEEE J.*, 1988, **SAC-6**, (4)
23 Various Papers: 'Telecommunication deregulation', *IEEE COMSOC–MAG*, 1989, **27**, (1)
24 BALSTAN, D.M.: 'Pan-European cellular radio or 1991 and all that', *IEE Electron. & Commun. Engng. J.*, Jan/Feb 1989
25 Various Papers: Special Series 'VSAT communication networks', *IEEE COMSOC-MAG*, 1988, **26**, (7, 8, 9)
26 Various Papers: 'VSATS views on future trends', *IEEE COMSOC-MAG*, 1989, **27**, pp. 43–63
27 GETCHER, J., and O'Reilly, P.: 'Conceptual issues for ATM.', *IEEE Networks*, 1989, **3**, pp. 14–16

Chapter 2
Transmission principles
J.E. Flood

Aston University

2.1 Introduction

Transmission systems exist to provide circuits for transmitting speech and other signals between the nodes of a telecommunication network. A *circuit* provides for the transmission of these signals in both directions. If the circuit uses a separate transmission path for each direction, then each of these unidirectional paths is called a *channel*. In general, a complete channel consists of sending equipment at a *terminal station*, a *transmission link*, which may contain repeaters at *intermediate stations*, and receiving equipment at another terminal station.* Present-day transmission systems [1, 2, 3] range in complexity from simple unamplified audio-frequency lines to satellite radiocommunication systems.

Both transmission channels and the signals they convey may be classified in two broad classes: *analogue* and *digital*. An analogue signal is a continuous function of time; at any instant it may have any value between limits set by the maximum power that can be transmitted. Speech signals are an obvious example. A digital signal can only have discrete values. The commonest digital signal is a binary signal, having only two values (e.g. 'mark' and 'space' or '1' and '0'). Telegraph signals and outputs of binary-coded data from computers are thus digital signals. A television waveform is a mixture of analogue and digital signals, since it transmits both the picture contents and synchronising pulses. Some analogue and digital signals that are transmitted in telecommunication networks are listed in Table 2.1.

A signal consisting of a single sinusoidal waveform is completely predictable; thus it conveys no information. A useful analogue signal must therefore contain a range of frequencies; this is known as its *bandwidth*. For a digital signal, the number of signal elements transmitted per second is called the signalling rate in *bauds*. If a non-redundant binary code is used, the rate of transmission of information (in bits per second) equals the signalling rate in bauds. If the coding contains redundancy, the bit rate is less than the number of bauds. If a multilevel signal is used (e.g. ternary or quarternary), each element conveys more than one bit of information; the bit rate is thus greater than the number of bauds.

---

* In practice, a building that houses any kind of transmission equipment is usually called a *repeater station*.

Table 2.1   Typical signals transmitted in telecommunication networks

(*a*)   Analogue signals

Type of signal	Bandwidth, Hz
Telephone speech	300 Hz to 3·4 Hz
Facsimile Group 2*	0·3 kHz to 2·7 kHz
Broadcast programmes (e.g. music)	50 Hz to 15 kHz
Colour television (625 lines)	0 to 5·5 MHz

(*b*)   Digital signals

Type of signal	Digit rate
Teleprinter	50 bauds
Data	200, 600, 1200, 2400, 4800, 9600 bit/s and 48 000 and 64 000 bit/s
PCM telephony (per channel)	64 kbit/s
ADPCM telephony	32 kbit/s
Video conferencing	2 Mbit/s and lower

* Vestigial sideband transmission on 2·1 kHz carrier.

To transmit an analogue signal without distortion, the channel must be a linear system. Cable systems and radio-relay systems equipped with linear amplifiers are examples of analogue channels. A digital channel does not require to be linear, since its output provides a number of discrete conditions corresponding to the input signal. An example of a digital channel is a telegraph circuit, whose output signal is provided by the operation of a relay. It does not follow that analogue signals must always be transmitted over analogue channels and digital signals over digital channels. Data communication and voice-frequency telegraphy over telephone lines are examples of transmitting digital signals over analogue channels. Analogue signals may be coded for transmission over digital channels by means of analogue-to-digital convertors. An example is the transmission of speech by means of pulse-code modulation over lines equipped with regenerators.

An advantage of digital transmission over analogue transmission is its relative immunity to interference. For example, error-free transmission of a binary signal only requires detection of the presence or absence of each pulse, and this can be done correctly in the presence of a high level of noise. It is also possible to employ *regeneration*. Provided that a received signal is not so corrupted that it is detected erroneously, it can cause the generation of an almost perfect signal for retransmission. The use of regenerative repeaters enables the transmission performance of digital circuits to be almost independent of their length, whereas analogue signals deteriorate progressively with distance.

If a link can provide adequate transmission over a band of frequencies which is wider than that of the signals to be sent, it can be used to provide a number of channels. At the sending terminal, the signals of different channels are combined to form a composite signal of wider bandwidth. At the receiving terminal, the signals are separated and retransmitted over separate channels. This process is known as *multiplexing*. The separate channels that enter and leave the terminal stations are called *baseband channels* and the transmission link, which carries the multiplex signal, is called a *broadband channel* or a *bearer channel*.

The principal multiplexing methods are *frequency-division multiplexing* (FDM) and *time-division multiplexing* (TDM). In FDM transmission, each baseband channel uses the bearer channel for all of the time, but it is allocated only a fraction of the bandwidth. In TDM transmission, each base band channel uses the entire bandwidth of the bearer channel, but only for a fraction of the time.

2.2 Power levels

A wide range of power levels is encountered in telecommunication transmission systems. It is therefore convenient to use a logarithmic unit for powers. This is the *decibel* (dB), which is defined as follows:

If the output power P_2 is greater than the input power P_1, then the gain G in decibels is

$$G = 10 \log_{10} \frac{P_2}{P_1} \, \text{dB} \tag{2.1a}$$

If, however, $P_2 < P_1$, then the loss or attenuation in decibels is

$$L = 10 \log_{10} \frac{P_1}{P_2} \, \text{dB} \tag{2.1b}$$

If the input and output circuits have the same impedance, then $P_2/P_1 = (V_2/V_1)^2 = (I_2/I_1)^2$ and

$$G = 20 \log_{10} \frac{V_2}{V_1} = 20 \log_{10} \frac{I_2}{I_1} \, \text{dB} \tag{2.2}$$

In some countries, the unit employed is the *neper* (N), defined as follows:

$$\text{Gain in nepers} = \log_e \frac{I_2}{I_1} \, \text{N} \tag{2.3}$$

Thus, if the input and output circuits have the same impedance, a gain of 1 N corresponds to 8·69 dB.

A logarithmic unit of power is convenient when a number of circuits having gain or loss are connected in tandem. The overall gain or loss of a number of circuits in tandem is simply the algebraic sum of their individual gains and losses measured in decibels or nepers.

If a passive network, such as an attenuator pad or a filter, is inserted in a circuit between its generator and load, the increase in the total loss of the circuit is called the *insertion loss* of the network. If an active network, such as an amplifier, is inserted, the power received by the load may increase. There is thus an *insertion gain*.

The decibel, as defined above, is a unit of relative power level. To measure absolute power level in decibels, it is necessary to specify a *reference level*. This is usually taken to be 1 mW and the symbol dBm is used to indicate power levels relative to 1 mW. For example, 1 W = +30 dBm and 1 μW = −30 dBm. Sometimes (e.g. in satellite systems) the reference level is taken to be 1 W. The symbol used is then dBW.

Since a transmission system contains gains and losses, a signal will have different levels at different points in the system. It is therefore convenient to express levels at different points in the system in relation to a chosen point called the *zero reference point*. The *relative level* of a signal at any other point in the system with respect to its level at the reference point is denoted by dBr. This is, of course, equal to the algebraic sum of the gains and losses betweeen that point and the reference point, as shown in Fig. 2.1. For a 4-wire circuit (see Section 2.4), the zero-level reference point is usually taken to be the 2-wire input to the hybrid transformer.

It is often convenient to express a signal level in terms of the corresponding level at the reference point; this is denoted by dBm0. Consequently,

$$dBm0 = dBm - dBr$$

For example, if a signal has an absolute level of −6 dBm at a point where

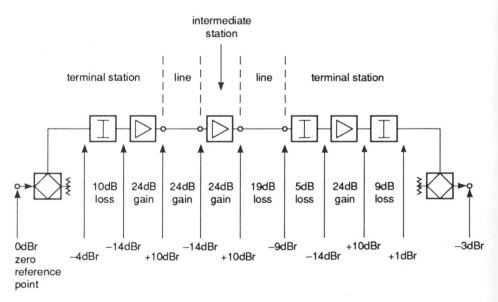

Fig. 2.1   Example of relative levels in a transmission system.

the relative level is -10 dBr, the signal level referred back to the zero reference point is $+4$ dBm0.

2.3 Distortion

2.3.1 Nonlinear distortion

When a transmission path is nonlinear, the amplitude of the output signal is no longer directly proportional to the amplitude of the input signal. The effect is sometimes called *amplitude distortion*. The output signal also contains frequency components that are not present in the input signal. If a sinusoidal signal $v_i = V \cos \omega t$ is applied to a non-linear transmission path whose transfer characteristic is given by

$$v_0 = a_0 + a_1 v_i + a_2 v_i^2 + a_3 v_i^3 + \dots$$

the output signal is

$$v_0 = [a_0 + \tfrac{1}{2}a_2 V^2] + [a_1 V + \tfrac{3}{4}a_3 V^3] \cos \omega t$$
$$+ \tfrac{1}{2}a_2 V^2 \cos 2\omega t + \tfrac{1}{4}a_3 V^3 \cos 3\omega t + \dots$$

Thus, the nonlinearity generates harmonics of the input frequency. The effect is therefore called *harmonic distortion*.

If the input signal contains two or more frequencies, the output signal does not only contain the fundamental frequencies and their harmonics; it contains sum and difference frequency components. The effect is called *intermodulation distortion*.

For example, if the input signal is

$$v_i = A \cos \omega_1 t + B \cos \omega_2 t$$

the output signal is

$$v_0 = \{a_0 + \tfrac{1}{2}a_2(A^2 + B^2) + a_1[A \cos \omega_1 t + B \cos \omega_2 t]\}$$
$$+ a_2\{\tfrac{1}{2}A^2 \cos 2\omega_1 t + \tfrac{1}{2}B^2 \cos 2\omega_2 t$$
$$+ AB \cos(\omega_1 + \omega_2)t + AB \cos(\omega_1 - \omega_2)t\}$$
$$+ a_3\{\tfrac{3}{4}A(A^2 + 2B^2) \cos \omega_1 t + \tfrac{3}{4}B(B^2 + 2A^2) \cos \omega_2 t$$
$$+ \tfrac{1}{4}A^3 \cos 3\omega_1 t + \tfrac{1}{4}B^3 \cos 3\omega_2 t$$
$$+ \tfrac{3}{4}A^2 B[\cos(2\omega_1 + \omega_2)t + \cos(2\omega_2 - \omega_1)t]$$
$$+ \tfrac{3}{4}B^2 A[\cos(2\omega_2 + \omega_1)t + \cos(2\omega_2 - \omega_1)t]\}$$
$$+ \dots$$

If the input signal contains more than two frequencies, the output signal contains many more intermodulation components. If the input signal occupies a wide frequency band, the spectrum of the intermodulation products is extremely complex. The result is usually called *intermodulation noise*.

2.3.2 Attenuation and delay distortion

When a signal is transmitted from one point to another, there is inevitably attenuation due to energy losses in the transmission medium and delay due to the distance travelled at a finite velocity of propagation. The effect of the channel may be represented by its gain/frequency response, $H(\omega)$, and its phase/frequency response, $\phi(\omega)$. If the output signal is to be undistorted, then the channel must have a constant gain g and delay T for every frequency component contained in the input signal S_i. If the spectrum of S_i extends from ω_1 to ω_2, it is clearly required that

$$H(\omega) = g \quad \text{for } \omega_1 \leqq \omega \leqq \omega_2 \qquad (2.4)$$

Any departure from this ideal condition is termed *attenuation distortion*.

Attenuation distortion may be corrected by inserting an *attenuation equaliser* [1]. This is a network designed to have a gain/frequency response, $E(\omega)$, which is the inverse of that of the channel,

$$\text{i.e.} \quad E(\omega) \cdot H(\omega) = g \quad \text{for } \omega_1 \leqq \omega \leqq \omega_2$$

A flat overall gain/frequency response is thus obtained.

If a signal component $V \cos \omega t$ is delayed by a time T, the output is

$$gV \cos \omega(t - T) = gV \cos[\omega t - \phi(\omega)]$$

where

$$\phi(\omega) = \omega T \qquad (2.5)$$

Thus, if the delay T is to be the same for all frequencies in the signal, we require $\phi \propto \omega$ for $\omega_1 \leqq \omega \leqq \omega_2$.

Any departure of the phase shift from the linear law is termed *phase distortion* or *delay distortion*. To obtain distortionless transmission, the phase/frequency characteristic should be a straight line through the origin (i.e. $\phi(0) = 0$). This is not possible when the circuit contains transformers or coupling capacitors. Fortunately, it is not necessary for speech transmission, since the ear is insensitive to differences in phase. However, for a signal whose waveshape must be preserved (e.g. television and data signals), phase distortion must be minimised. It is then necessary to use a *phase equaliser*. This is a network designed to have a phase/frequency response $\theta(\omega)$ such that

$$\theta(\omega) + \phi(\omega) = \omega T \quad \text{for } \omega_1 \leqq \omega \leqq \omega_2$$

The quantity $d\phi/d\omega$, which is the slope of the phase/frequency characteristic, is called the *group delay* of the channel. A convenient measure of delay distortion is the difference between the maximum and minimum values of group delay occurring within the band ω_1 to ω_2. This is called the *differential delay* of the channel.

An equaliser may also be designed in the time domain. The equaliser is designed so that the combination of channel and equaliser in tandem have an impulse response which is within the limits specified. Now, the transfer function, $H(\omega)\angle\phi(\omega)$, is the Fourier transform of the impulse response. Consequently, improving the impulse response is equivalent to improving the gain/frequency and phase/frequency responses.

2.3.3 Multipath tranmission

One cause of severe attenuation and delay distortion is the *multipath effect*, shown in Fig. 2.2*a*. This effect arises when the signal arrives at the receiving end of a channel over two or more paths having different delays. Examples of this are as follows:

(*a*) In a cable, there may be multiple reflections due to impedance irregularities.

(*b*) In a 4-wire circuit there may be echoes due to imperfect balances at the 4-wire/2-wire terminations.

(*c*) In long-distance HF radio transmission, the signal reflected from the ionosphere may be received both over a single-hop path and a multi-hop path.

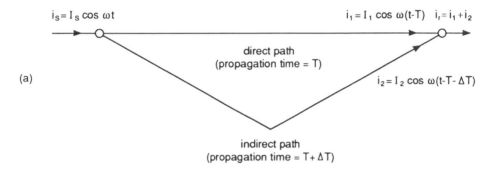

(a)

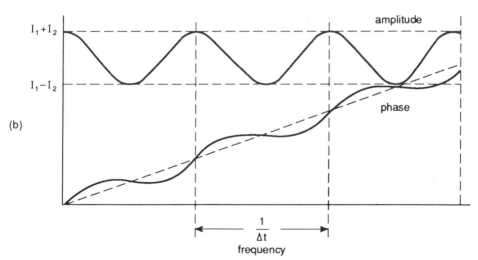

(b)

Fig. 2.2 Distortion caused by multipath transmission

a Interference between signals received over direct and indirect paths
b Variation with frequency of amplitude and phase of received signal

(*d*)   In VHF and UHF radio transmission, the signal may be received both by direct transmission and by reflection from the ground or from an obstacle.

If the difference between the lengths of the two paths is an exact multiple of the wavelength λ, the signals arive in phase and add. If the path difference is an odd number of half wavelengths, the signals arrive in antiphase and cancel. However, $\lambda = v/f$ (where v is the velocity of propagation and f the frequency), so the phase difference varies with frequency. Interference between the two signals thus causes the amplitude of the received signal to vary with frequency, as shown in Fig. 2.2*b*. The phase/frequency characteristic contains ripples with the same periodicity.

2.4 Four-wire circuits

2.4.1 Principle of operation

It is frequently necessary to use amplifiers to compensate for the attenuation of a transmission path. Since most amplifiers are unidirectional, it is usually necessary to provide separate channels for the 'go' and 'return' directions of transmission. This results in a 4-*wire circuit*, as shown in Fig. 2.3. Each pair of amplifiers (one for each direction of transmission) is known as a 4-*wire repeater*. Each section of the line together with its associated repeater is called a *repeater section*. Although Fig. 2.3 shows a 4-wire audio-frequency

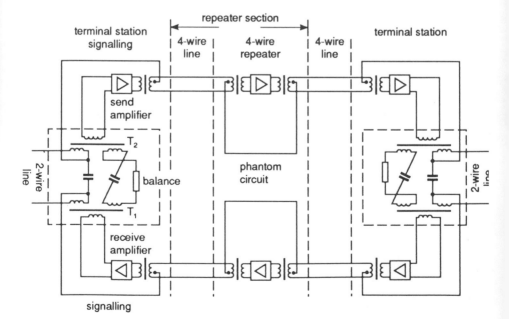

Fig. 2.3   four-wire amplified circuit

line, in practice, the go and return paths may be provided by channels in a high-frequency transmission system instead of on physical cable pairs.

At each end, the 4-wire circuit must be connected to a 2-wire circuit leading to an exchange. If both paths of the 4-wire circuit were connected directly to the 2-wire circuit at each end, a signal could circulate round the complete loop thus created. This would result in continuous oscillation or *singing*, unless the sum of the gains in the two directions was less than zero. To avoid this possibility, the 2-wire line at each end is connected to the 4-wire circuit by means of a *4-wire/2-wire terminating set*. This contains a *hybrid transformer** (consisting of two cross-connected transformers, as shown in Fig. 2.3), and a *line-balance network* whose impedance is similar to that of the 2-wire circuit over the required frequency band.

The output signal from the 'receive' amplifier causes equal voltages to be induced in the secondary windings of transformer T_1. If the impedances of the 2-wire line and the line balance are equal, then equal currents flow in the primary windings of transformer T_2. These windings are connected in antiphase; thus, no EMF is induced in the secondary winding of T_2 and no signal is applied to the 'send' amplifier. It should be noted that the output power from the receive amplifier divides equally between the 2-wire line and the line balance. When a signal is applied from the 2-wire line, the cross-connection between the transformer windings results in zero current in the line-balance impedance. The power thus divides equally between the input of the send amplifier and the output of the receive amplifier, where it has no effect. The price paid for avoiding singing is thus 3 dB loss in each direction of transmission, together with any losses in the transformers (typically 0·5 to 1·0 dB).

An alternative form of 4-wire/2-wire termination, which uses a bridge circuit, is shown in Fig. 2.4. If the balance impedance equals the impedance of the 2-wire line, the bridge is balanced. The output of the receive amplifier then produces zero voltage at the input terminals of the transmit amplifier.

The impedance of the 2-wire line varies with frequency. To achieve correct operation of 4-wire/2-wire terminations, it would be necessary to measure the impedance/frequency characteristic of each line and to design a balance network to match it closely. In practice, the balance usually consists of resistor of value equal to the nominal impedance of the line (usually 600 Ω). This is known as a *compromise balance*. Thus, a small fraction of the power received from one side of the 4-wire circuit will pass through the hybrid transformer and be retransmitted in the other direction.

If the go and return channels are each provided by means of a balanced line terminated with transformers at the ends of each repeater section, a signal applied to the centre tap of an output transformer secondary causes equal currents to flow in the two wires. These currents flow in opposite directions in the two halves of the primary winding of the next input transformer and so produce zero output voltage from its secondary. Thus,

---

* The anti-sidetone transformer (induction coil) in a telephone also acts as a hybrid transformer, connecting the 2-wire subscriber's line to the 4-wire circuit consisting of the microphone and receiver.

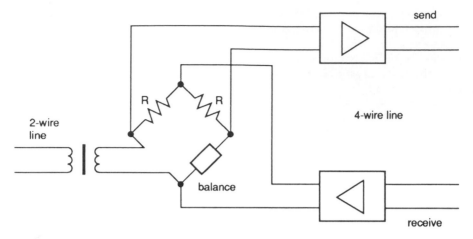

Fig. 2.4   4-wire/2-wire termination using bridge circuit

an additional *longitudinal* signal voltage can be transmitted which does not interfere with the *transverse* signal voltage across the transformer winding. By connecting the centre taps of the input and output transformers at each repeater, as shown in Fig. 2.3, a complete 2-wire circuit can be provided between the transformers at each end of the 4-wire circuit. Such a circuit is called a *phantom circuit,* because it does not require additional physical conductors.

Telephone transmission systems must cater for the transmission of inter-exchange signals in addition to speech. On a 4-wire audio-frequency circuit, it is therefore usual to connect the phantom circuit to the 2-wire circuits at the hybrid transformers, as shown in Fig. 2.3. This provides DC connections between the conductors of the 2-wire circuits to transmit the necessary signalling conditions.

2.4.2 Echoes

In a 4-wire circuit, such as that shown in Fig. 2.3, an imperfect line balance causes part of the signal energy transmitted in one direction to return in the other. Both the talker and the listener may be able to hear the reflected signal and the effect is termed *echo.* The signal reflected to the speaker's end of the circuit is called *talker echo* and that at the listener's end is called *listener echo.* The paths traversed by these echoes are shown in Fig. 2.5.

The attenuation between the 4-wire line and 2-wire line or between 2-wire line and 4-wire line at each hybrid coil has been shown in Section 2.4.1 to be 3 dB. Thus, the total attenuation from one 2-wire circuit to the other is

$$L_2 = 6 - G_4 \text{ dB} \tag{2.6}$$

where G_4 is the net gain of one side of the 4-wire circuit (i.e. total amplifier gain minus total line loss).

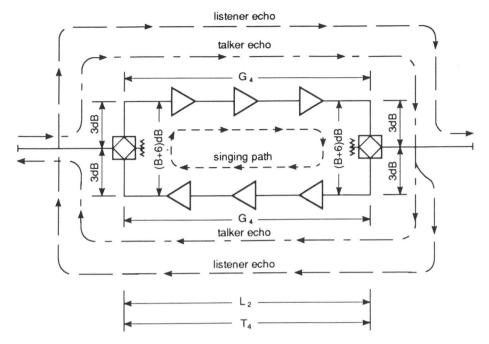

Fig. 2.5 Echo and singing paths in a 4-wire circuit

The attenuation through the hybrid transformer from one side of the 4-wire circuit to the other is called the *transhybrid loss*. It is equal to $6+B$ dB [2] where

$$B = 20 \log_{10} \left| \frac{N+Z}{N-Z} \right| \text{ dB} \qquad (2.7)$$

where Z = impedance of the 2-wire line
        N = impedance of the balance network

The loss B represents that part of the trans-hybrid loss which is due to the impedance mismatch between the 2-wire line and the balance network (see Section 3.2.2). It is known as the *balance-return loss* (BRL).

The attenuation, L_t, of the echo that reaches the talker's 2-wire line, round the path shown in Fig. 2.5, is

$$L_t = 3 - G_4 + (B+6) - G_4 + 3 \text{ dB}$$

$$= 2L_2 + B \text{ dB}$$

This echo is delayed by a time

$$D_t = 2T_4$$

where T_4 is the delay of the 4-wire circuit (between its 2-wire terminations). The attenuation L_1 of the echo that reaches the listener's 2-wire line (relative to the signal received directly) is

$$L_1 = (B+6) - G_4 + (B+6) - G_4 \text{ dB}$$
$$= 2L_2 + 2B \text{ dB}$$

and this is delayed by a time $2T_4$ relative to the signal received directly.

The effect of an echo is different for the speaker and the listener. For the speaker it interrupts his conversation and for the listener it reduces the intelligibility of the received speech. Talker echo is the more troublesome because it is louder (by an amount equal to the BRL). The annoying effect of echo increases with its magnitude and the delay. The longer the circuit, the greater is the echo attenuation L_t required. This can be achieved by making the overall loss L_2 increase with the length of the circuit.

There is a limit to which the loss of connections can be increased to control echo. This is usually reached when the *round-trip delay*, $2T_4$, is about 40 ms. This delay is exceeded on very-long transcontinental and intercontinental circuits, so it is impossible to obtain both an adequately-low transmission loss and an adequately-high echo attenuation. On such circuits, it is necessary to control echo by fitting devices called *echo suppressors*.

An echo suppressor consists of a voice-operated attenuator, which is fitted in one path of the 4-wire circuit and operated by speech signals on the other path. Whenever speech is being transmitted in one direction, transmission on the opposite direction is attenuated, thus interrupting the echo path. There is one such suppressor (called a 'half echo suppressor') at each end of the circuit.

A number of difficulties arise with simple echo suppressors of this type. In a very-long-distance switched connection, it is possible to have a number of circuits fitted with echo suppressors connected in tandem; if these operate independently, 'lock-out' conditions can arise. It is therefore necessary to disable the echo suppressors on intermediate links in the connection. It is also necessary to disable echo suppressors during data transmission, since data-transmission systems often use a return channel to request retransmission of blocks of information when errors are detected. More-sophisticated echo suppressors have been designed [4] both to provide these facilities and to cater for the very-long propagation times (270 ms each way) encountered on geosynchronous satellite links.

Echo cancellers are now also used [5]. The echo is cancelled by subtracting a replica of it from the received signal. This replica is synthesised by means of a filter, controlled by a feedback loop, which adapts to the transmission characteristic of the echo path and tracks any variations in it that may occur during a conversation.

2.4.3 Stability

If the balance-return losses of the terminations of a 4-wire circuit are sufficiently small and the gains of its amplifiers are sufficiently high, the net gain round the loop may exceed zero and singing will occur. The net loss L_s of the singing path shown in Fig. 2.5 is

$$L_s = 2(B + 6 - G_4) \text{ dB} \tag{2.8}$$

Substituting from eqn. 2.6 into eqn. 2.8 gives

$$L_s = 2(B + L_2) \, \text{dB} \qquad (2.9)$$

Thus, the loss of the singing path equals the sum of the 2-wire to 2-wire losses in the two directions of transmission and the BRLs at each end. The necessary conditions for stability is $L_s > 0$. This requires that $L_2 + B > 0$, i.e.

$$G_2 < B \quad (\text{where } G_2 = -L_2) \qquad (2.10)$$

The gain G_2 which can be obtained over a 4-wire circuit is thus limited by the BRL. Eqn. 2.7 shows that if $N = Z$, the balance return loss is infinite. In the limiting cases when either Z or N is zero or infinite, the balance return loss is zero. The loss between the return and go channels is then only 6 dB (plus any loss due to transformer inefficiencies).

In practice, the 2-wire line may be open-circuited for long periods because it is waiting for a call to be connected to it at the exchange. It may be short-circuited by a line or equipment fault. Although the circuit is not carrying traffic under these conditions, it is still undesirable for singing to occur. Because the magnitude of the oscillation increases until it is limited by the overloading of an amplifier, the power level of a channel that is singing greatly exceeds that for which transmission systems are designed. In a frequency-division multiplex system, this increase in loading can cause intereference to every other circuit routed over the system (see Section 2.5). 4-wire circuits are therefore usually set up to be stable even when the 2-wire lines at each end are open-circuited or short-circuited ($B = 0$). This requires operation with an overall net loss ($G_2 < 0$).

In practice, the attenuation of the singing path is deliberately made greater than zero. This provides a safety margin and avoids the attenuation distortion caused by echoes when the circuit is operating close to its singing point, (see Section 2.3.3). The *singing point* of a circuit is defined as the maximum gain S that can be obtained (from 2-wire to 2-wire line) without producing singing. Thus, from eqn. 2.10, $S = B$; i.e. the singing point is given by the BRL (or the average of the two BRLs if these are different at the two ends of the circuit). The *stability margin* is defined as the maximum amount of additional gain M that can be introduced (equally and simultaneously) in each direction of transmission without causing singing, i.e. $L_s - 2M = 0$. Hence, from eqn. 2.9,

$$M = B + L_2 \, \text{dB} \qquad (2.11)$$

Thus, the stability margin is the sum of the 2-wire to 2-wire loss and the BRL.

In practice, a stability margin of 3 dB is found to be adequate* (i.e. $L_s = 6$ dB). If the circuit is to cater for zero BRL, the overall loss from 2-wire circuit to 2-wire circuit is then 3 dB.

In setting up long-distance switched connections, it is often necessary to connect a number of 4-wire circuits in tandem. It is advantageous to eliminate terminating sets from the interfaces between the 4-wire lines rather

---

* If the standard deviation of G_4 is $1 \cdot 0$ dB and a normal probability distribution is assumed, then the probability of the gain exceeding the 3 dB stability margin is 1 in 1000.

than interconnect them on a 2-wire basis. The complete connection therefore consists of a number of 4-wire circuits in tandem with a 4-wire/2-wire termination at each end of the connection. It is necessary to ensure that this complete circuit has adequate stability.

The loss of a 4-wire circuit may depart from its nominal value for a number of reasons. These include:

(a)   Variation of line losses and amplifier gains with time and temperature.
(b)   Gain at other frequencies being different from that measured at the test frequency (usually 800 Hz or 1600 Hz).
(c)   Errors in making measurements and lining up circuits.

The effect of these deviations is to cause the losses of circuits to have a statistical distribution about their mean (nominal) values. The standard deviation σ is usually between 1 dB (for modern systems) and 1·5 dB (for older systems). As a result of these variations a connection may become unstable, although its nominal loss provides an adequate stability margin. The probability of this happening must be very small (e.g. 1 in 1000).

If a number of circuits are connected in tandem, then:

(a)   Total loss = sum of losses of the individual circuits
(b)   Variance σ^2 of total loss = sum of variances of losses of the individual circuits.

If it is assumed that these losses and the BRLs have normal probability distributions, the probability of a connection becoming unstable can be readily calculated [2].

Since the standard deviation σ increases with the number of circuits in tandem, so must the overall loss. A simple rule that has been adopted by operating administrations in a number of countries is

$$L_2 = 4·0 + 0·5n \text{ dB}$$

where L_2 is overall loss (2-wire to 2-wire) and n is number of 4-wire circuits in tandem. When a circuit incorporates an HF radio link, variations in ionospheric conditions cause fading which can result in short-term variations of 20 dB in the receiver output level. Although the output level of the circuit can be controlled by a constant-volume amplifier at each radio terminal, it is impossible to ensure that the circuit always operates below its singing point. To prevent singing, a device known as a *singing suppressor* is fitted at each end of the 4-wire circuit. A singing suppressor contains a voice-operated attenuator in both the 'go' and 'return' channel. In the quiescent condition, the suppressor introduces minimum loss in the return path and maximum loss in the go path. When incoming speech from the 2-wire circuit is detected, the loss is removed from the go direction and introduced in the return direction. A singing suppressor thus also acts as an echo suppressor.

2.5 Crosstalk

Any sound heard in the telephone receiver associated with one channel resulting from the transmission of a signal over another channel is called *crosstalk.* Even low-level crosstalk can be intelligible and give the user the impression of a lack of secrecy. This interference can be caused by couplings between the wires of different lines due to capacitance, leakage or mutual inductance.

In multipair cables the greatest cause of interference is capacitance. It can be shown that the interference between two pairs is zero if each wire of a pair has an equal capacitance to earth and has equal capacitance to each wire of the other pair. Care is therefore taken in the manufacture of multipair cables to make the capacitance unbalances as small as possible. Since any induced voltage due to capacitance unbalance or mutual inductance is directly proportional to frequency, crosstalk increases with frequency and determines the upper frequency limit for use of the cable. Crosstalk due to impedance unbalance may also arise when the two wires of a line have slightly different resistances (e.g. due to poor jointing) or if the impedance of the equipment terminating the line is not balanced with respect to earth.

An additional source of crosstalk which is present in frequency-division multiplex systems (carrier systems) is produced by amplifier nonlinearities. If an amplifier is nonlinear, then two signals of frequencies f_1 and f_2 at its input produce at the output not only f_1 and f_2 but also intermodulation products of frequencies $f_1 \pm f_2$, $f_1 \pm 2f_2$ etc., as shown in Section 2.3.1. If these components lie within the frequency bands of other channels, then inter-channel crosstalk results. Amplifiers for carrier systems therefore have large amounts of negative feedback to obtain very low intermodulation levels in addition to gain stability. It is also necessary for amplifiers to have a sufficiently high overload level to handle the total power of a multichannel signal and to line up circuits carefully to ensure that signal levels do not exceed their nominal values by more than a small tolerance.

When crosstalk energy is transferred from one channel to another, it can usually be detected at each end of the disturbed channel, as shown in Fig. 2.6. When the crosstalk is propagated over the disturbed channel in the same direction as its own signal, the crosstalk is called *far-end crosstalk* (FEXT) or distant-end crosstalk. When the crosstalk is propagated over the disturbed channel in the opposite direction to its own signal, it is called *near-end crosstalk* (NEXT).

If the power of the signal entering the disturbing channel is P_1 and the power of the crosstalk received by the disturbed channel (at its near or far end) is P_2, then

$$\text{Crosstalk attenuation} = 10 \log_{10} \frac{P_1}{P_2} \, \text{dB}$$

If the power of the disturbed channel's own signal is P_3, then

$$\text{Crosstalk ratio} = 10 \log_{10} \frac{P_3}{P_2} \, \text{dB}$$

The crosstalk ratio is usually measured when test signals of the same level are applied to both the disturbing and disturbed channels.

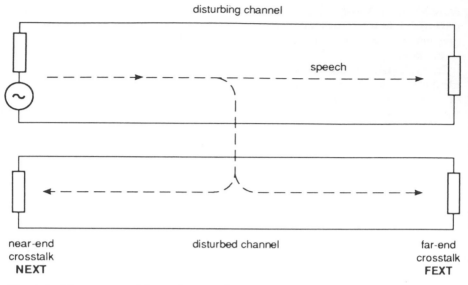

disturbing channel

speech

near-end
crosstalk
NEXT

disturbed channel

far-end
crosstalk
FEXT

Fig. 2.6   Near-end and far-end crosstalk

In a multipair audio cable, power is transferred from the disturbing pair to the disturbed pair along the whole length of the cable. For FEXT, the ratio P_2/P_3 is therefore proportional to the length of the cable. In the case of NEXT, the power transferred at a distance x from the sending end is attenuated by travelling distance x over the disturbing pair and distance x back again over the disturbed pair, a total distance of $2x$. Thus, most of the received crosstalk power is due to only a short length of the cable near the sending end and P_2 is almost independent of the total length. It can be shown [1] that for FEXT the crosstalk ratio decreases with frequency at a rate of 6 dB per octave and for NEXT it decreases by 4·5 dB per octave.

Crosstalk is most serious when it is heard by a subscriber during a silent interval in the conversation. The risk of crosstalk being audible is negligible if the crosstalk ratio exceeds 65 dB. A crosstalk ratio of 55 dB can often be tolerated when the crosstalk, although audible, is not intelligible.

2.6 Noise

2.6.1 General
Noise in a transmission channel may be generated internally or may be the result of interference from external sources. Interference enters a channel through unwanted electromagnetic coupling between a circuit and a source of interference such as a power line (i.e. the same mechanism as crosstalk). Internal line noise may be due to faulty joints whose contact resistance is varied by vibration caused, for example, by passing traffic. This is called *microphonic noise*. There is also an inherent source of noise due to the thermal

agitation of electrons. This produces energy which is uniformly distributed over the frequency spectrum and is therefore often called *white noise*. In telephony, this is heard as a hiss. However, noise also occurs in short bursts; in telephony these are heard as clicks. This is called *impulse noise*. It is caused by electrical discharges, switching transients in electromechanical telephone exchanges and intermodulation in FDM systems.

2.6.2 Thermal noise

If the thermal noise has a power density of n watts/Hz, the noise power N in an ideal channel of bandwidth W hertz is therefore given by $N = nW$ watts. If the gain/frequency characteristic, $H(f)$, of the channel is not flat then:

$$N = \int_0^\infty nH(f)^2 \, df$$

The same noise power would be obtained from an ideal channel of bandwidth W whose gain is equal to the maximum value, H_m, of $H(f)$ if

$$W = \frac{1}{H_m^2} \int_0^\infty H(f)^2 \, df$$

W is then the *noise bandwidth* of the channel.

The noise voltage, v_n, is the result of the thermal agitation of a very large number of electrons. It therefore varies randomly with time and has a normal or Gaussian probability-density distribution, given by

$$p(v_n) = \frac{1}{\sigma\sqrt{2\pi}} \exp(-v_n^2/2\sigma^2) \tag{2.12a}$$

where σ is the standard deviation of v_n which is thus the RMS noise voltage.

The probability of the noise voltage being greater than $+V$ is

$$P(v_n > +V) = \int_{+V}^\infty p(v_n) \, dv = \tfrac{1}{2} - \int_0^V p(v_n) \, dv \tag{2.12b}$$

and the probability of it being less than $-V$ is

$$P(v_n < -V) = \int_{-\infty}^{-V} p(v_n) \, dv = p(v_n > +V) \tag{2.12c}$$

These probabilities can therefore by obtained from standard statistical tables.

It has been shown [6, 7] that the RMS noise voltage V_n produced by thermal agitation in a resistance of R ohms is given by

$$V_n^2 = 4kTWR \text{ volts}^2$$

where

W = bandwidth, Hz
T = absolute temperature, K
k = Boltzmann's constant $(1 \cdot 37 \times 10^{-23} \text{ J/K})$.

If this resistance is connected to a matched load R, the noise power delivered to it is

$$P_n = \frac{(V_n/2)^2}{R} = kTW \text{ watts} \qquad (2.13a)$$

The *available noise power* is thus proportional both to bandwidth and temperature, but is independent of the resistance.

From eqn. 2.13a, the available noise power expressed in dBW is:

$$N = 10 \log_{10} k + 10 \log_{10} T + 10 \log_{10} W$$

$$= -228 \cdot 6 + 10 \log_{10} T + 10 \log_{10} W \text{ dBW} \qquad (2.13b)$$

The available noise power is thus $-228 \cdot 6$ dBW per Hz per K. The noise power delivered into a circuit at a temperature of 290 K (i.e. 17°C) is 4×10^{-21} W per Hz of bandwidth, i.e. -174 dBm/Hz. For any specified signal-to-noise ratio, the minimum signal level that may be used is thus ultimately determined by thermal agitation noise.

When this resistive source is connected to a load, the load is also generating noise. If thermal agitation is the only source of noise, the total noise power in the matched load is the sum of the available noise powers of the source and the load (since each delivers half its total noise power to the other and dissipates half internally).

2.6.3 Noise factor and noise temperature

In an amplifier, in addition to thermal noise, there is also noise due to random fluctuations in the currents of transistors or other devices, i.e. shot noise and flicker, [8, 9]. The total noise present therefore exceeds that due to thermal agitation. The *noise factor* f_n of the amplifier is given by:

$$f_n = \frac{\text{signal-to-noise ratio at input}}{\text{signal-to-noise ratio at output}}$$

$$= \frac{\text{total noise power at output}}{\text{noise power due to thermal agitation}}$$

$$= \frac{N_0}{gkT_0 W}$$

where

N_0 = total output noise power
g = power gain of the amplifier
T_0 = reference temperature, i.e. 290 K

Since the output noise power from the amplifier is greater than that due to thermal noise alone, it is equal to the amount of thermal noise that would be present at a higher temperature than that actually present. This is known as the *noise temperature* T_n. Thus, $T_n = f_n T_0$. The difference between the noise temperature and the standard room temperature T_0 corresponds to the internally-generated noise. It is called the *excess noise temperature* T_x. Thus:

$$T_x = T_n - T_0 = T_0(f_n - 1)$$

and

$$f_n = 1 + T_x / T_0$$

The noise power at the output of an amplifier can thus be expressed in terms of an equivalent noise temperature. The noise appearing at the output terminals of a receiving antenna can also be expressed in terms of its noise temperature. In this case, the noise is due to radiation received from objects on earth, from objects in space and from interstellar gases.

If the noise factor expressed in decibels is F_n (where $F_n = 10 \log_{10} f_n$) the output noise power, N_0, from an amplifier is given by

$$N_0 = F_n + G - 174 + 10 \log_{10} W \text{ dBm}$$

where

$$G = \text{power gain in decibels}$$

If $W = 4$ kHz, then $10 \log_{10} W = 36$ dB and

$$N_0 = F_n + G - 138 \text{ dBm}$$

2.6.4 Noise in analogue transmission systems

If an analogue transmission system contains n identical repeater sections, the total noise output power N_T is n times that of a single amplifier, i.e.

$$N_T = F_n + G - 138 + 10 \log_{10} n \text{ dBm} \qquad (2.14)$$

If the relative output level of each repeater is L dBr, then the total output noise referred to the zero reference point is

$$N_{RO} = N_T - L \text{ dBm0}$$

Analogue transmission systems normally obtain an adequately low noise level by having repeaters at sufficiently short intervals along each route. As an example, consider the transmission of a 10 mW signal over 120 km of line having an attenuation of 1·5 dB/km to produce a signal at the receiving end also of 10 mW. The noise level of the line is -138 dBm. There might appear to be three possible methods of overcoming the 180 dB attenuation of the line:

(a)   Amplify the 10 mW input signal by 180 dB before sending it to line
(b)   Amplify the attenuated signal by 180 dB at the receiving end of the line
(c)   Amplify the signal at various points along the line.

Method (a) is obviously impracticable, because it would require an amplifier with an output power of 10^{10} MW. With method (b), the signal level falls to -170 dBm at the receiving end and this is 32 dB below the noise level. Clearly, only method (c) is possible. If an amplifier is available which has 60 dB gain, three amplifiers are required. If the noise factor is 8 dB, the output noise power, from eqn. 2.14, is -65 dBm. The output signal level is $+10$ dBm, so the output signal/noise ratio is 75 dB.

Because the sensitivities of the ear and a telephone receiver vary with frequency, total noise power is not an accurate measure of its subjective effect. Therefore, noise is usually measured by a meter having a frequency-weighting network which has been standardised by the CCITT [10]. Such instruments [11] are called *psophometers*. Noise power is thus measured in units of pWp (picowatts psophometrically weighted) or dBmp (decibels relative to 1 mW psophometrically weighted). If the noise power is referred to the zero-level reference point in a system, then the units are pWOp and dBmOp. In the USA, noise power is usually referred to −90 dBm instead of 0 dBm (to make all measurements positive) and weighted noise is measured in dBrnc (where 'rn' stands for reference noise and 'c' for C message weighting). A noise level of 0 dBrnc is thus equivalent to −90 dBmp. For white noise in the band 0—4 kHz, the effect of psophometric weighting is to reduce the noise level by 3·6 dB. Thus, −N dBm corresponds to −(N + 3·6) dBmp.

2.6.5 Companding

On some very long routes it is not possible to achieve an adequate signal/noise ratio using normal transmission techniques. For these circuits a method known as *companding* is used to reduce the effects of noise and interference [1]. At the sending end of the channel a *compressor* is used. This unit consists of an amplifier whose gain is controlled by the incoming speech signal so that low-level signals are amplified more than peak-level signals. The range of output levels (in dB) is thus a fraction, say half, of the range of input levels. The signal/noise ratio for high-level input signals is unaffected; however, low-level input signals are transmitted at a relatively higher level, so they obtain an improvement in signal/noise ratio. At the receiving end of the channel an *expander* is used; this introduces a loss equal to the gain introduced by the expander, thus restoring the range of signal levels. For example, if the range of speech levels is from −30 dBm to 0 dbm and the noise level is −45 dBm, for a non-companded system the signal/noise ratio for the weakest signals would be only 15 dB. If a compressor raises the lowest signal level to −15 dBm, the signal/noise ratio is increased to 30 dB. In the absence of speech, the expander reduces the noise level from −45 dBm to −90 dBm. This heavy attenuation of noise during quiet periods provides the major benefit of using companders. During speech, companding increases the range of levels over which a minimum signal/noise ratio is exceeded. This increase (in dB) is referred to as the *companding advantage*.

The control signals applied to the compressor and expander are required to correspond to the average power of the input speech signal. They are therefore derived by a rectifier with a time constant approximating to the duration of the shortest elements of speech. This type of compandor is therefore called a *syllabic compandor* [12]. The action of an expander, in increasing the level range of output signals, magnifies the effect of any variation in the gain of the transmission channel. This can lead to instability, so syllabic compandors are only used when they are essential to make a long-distance circuit usable. For use on HF radio circuits, an improved form of compandor has been developed in which a separate control signal is transmitted over the radio link to adjust the gain of the expander to suit the loss of the compressor. This system is known as Lincompex [13] (Linked

Compressor and Expander). This enables HF radio circuits to have much improved performance and removes the need for operators to make frequent adjustments.

2.7 Modulation

2.7.1 General
The processing of a signal to make it suitable for sending over a transmission medium is called *modulation* [14]. Reasons for using modulation are:

(a) Frequency translating (e.g. when an audio-frequency baseband signal modulates a radio-frequency carrier)
(b) Improving signal/noise ratio by increasing the bandwidth (e.g. using frequency modulation)
(c) Multiplexing (see Section 2.8)

Modulation is performed by causing the baseband modulating signal to vary a parameter of a carrier wave. A sinusoidal carrier, $v_c = A \cos(\omega t + \phi)$, is defined by three parameters: amplitude A, frequency $\omega/2\pi$ and phase ϕ. Thus there are three basic modulation methods: *amplitude modulation* (AM), *frequency modulation* (FM) and *phase modulation* (PM).

When modulation is employed, a modulator is needed at the sending end of a channel and a demodulator at the receiving end recovers the baseband signal from the modulated carrier. The combination of modulator and demodulator at a terminal is often referred to as a *modem*.

2.7.2 Amplitude modulation
The simplest form of modulation is amplitude modulation. The modulator causes the envelope of the carrier wave to follow the waveform of the modulating signal and the demodulator recovers it from this envelope.

If a carrier, $v_c = V_c \cos \omega_c t$ is modulated to a depth m by a sinusoidal modulating signal, $v_m = V_m \cos \omega_m t$, the resulting AM signal is

$$v = (1 + m \cos \omega_m t)v_c$$
$$= (1 + m \cos \omega_m t) V_c \cos \omega_c t$$
$$= V_c[\cos \omega_c t + \tfrac{1}{2}m \cos(\omega_c + \omega_m)t + \tfrac{1}{2}m \cos(\omega_c - \omega_m)t]$$

If the modulating signal contains several components, $f_1, f_2, \ldots$, etc., then the modulated signal contains $f_c - f_1, f_c - f_2, \ldots$, etc., and $f_c + f_1, f_c + f_2, \ldots$, etc. in addition to f_c. If the modulating signal consists of a band of frequencies, as shown in Fig. 2.7a, the modulated signal consists of two *sidebands*, each occupying the same bandwidth as the baseband signal, as shown in Fig. 2.7b. In the *upper sideband*, the highest frequency corresponds to the highest frequency in the baseband; this is therefore known as an *erect sideband*. In the *lower sideband*, the highest frequency corresponds to the lowest frequency in the baseband; this is known as an *inverted sideband*. Simple amplitude modulation makes inefficient use of the transmitted power, as information is transmitted only in the sidebands but the majority

(a)

0   F~m~

(b)

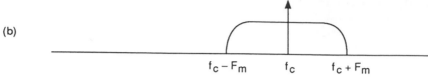

f~c~ – F~m~   f~c~   f~c~ + F~m~

(c)

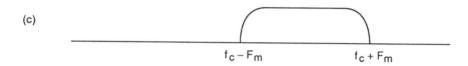

f~c~ – F~m~   f~c~ + F~m~

(d)

f~c~   f~c~ + F~m~

(e)

f~c~   f~c~ + F~m~

Fig. 2.7   Frequency spectra for amplitude modulation

a Baseband signal
b Simple amplitude modulation (AM)
c Double-sideband suppressed-carrier (DSBSC) modulation
d Single sideband suppressed-carrier (SSBSC) modulation
e Vestigial-sideband (VSB) modulation

of the power is contained in the carrier. However, AM permits the use of a relatively simple receiver using an envelope demodulator. It is therefore used for services, such as broadcasting and mobile radio, where this is an important advantage.

It is possible, by using a balanced modulator [15], to eliminate the carrier and generate only the sidebands, as shown in Fig. 2.7c. This is known as *double-sideband suppressed-carrier modulation* (DSBSC). To demodulate a DSBSC signal, it is necessary to use a *coherent demodulator* (consisting of a balanced modulator supplied with a locally-generated carrier) instead of the envelope demodulator used with simple AM.

If the incoming DSBSC signal is

$$v_i = \tfrac{1}{2}mV_c[(\cos(\omega_c + \omega_m)t + \cos(\omega_c - \omega_m)t]$$

and the coherent demodulator multiplies this with a local carrier $v_c = (\cos \omega_c t + \theta)$, its output voltage is

$$v = \tfrac{1}{4}mV_c\{\cos[(2\omega_c + \omega_m)t + \theta] + \cos[(2\omega_c - \omega_m)t + \theta]$$
$$+ \cos(\theta + \omega_m)t + \cos(\theta - \omega_m)t\}$$

The components at frequencies $(2\omega_c \pm \omega_m)$ are removed by a low-pass filter and the baseband output signal is

$$v_0 = \tfrac{1}{4}mV_c[\cos(\theta + \omega_m t) + \cos(\theta - \omega_m t)] \tag{2.15a}$$

$$= \tfrac{1}{2}mV_c \cos\theta \cos\omega_m t \tag{2.15b}$$

Thus v_0 represents the original baseband signal, provided that the phase θ of the local carrier is stable.

A further economy in power, and a halving in bandwidth, can be obtained by producing a *single-sideband suppressed-carrier* (SSBSC) signal, as shown in Fig. 2.7d. If the upper sideband is used, the effect of the modulator is simply to produce a frequency translation of the baseband signal to a position in the frequency spectrum determined by the carrier frequency. If the lower sideband is used, the band is inverted as well as translated.

A coherent demodulator is required for demodulating a SSBSC signal. Since only one sideband is present, the demodulated output signal, from eqn. 2.15a, is

$$v_0 = \tfrac{1}{4}mV_c \cos(\omega_m t \pm \theta)$$

The SSBSC signal requires the minimum possible bandwidth for transmission. Consequently, the method is used whenever its complexity is justified by the saving in bandwidth. An important example is the use of SSBSC for multichannel carrier results in a corresponding shift in the frequencies of the components in the baseband output signal. For speech transmission, frequency shifts of the order of ±10 Hz are not noticeable, but the errors that can be tolerated for telegraph and data transmission are less. The CCITT specifies [16] that the frequency shift should be less than ±2 Hz.

By using SSBSC, it is possible to transmit two channels through the bandwidth needed by simple AM for a single channel; one uses the upper sideband of the carrier and the other uses the lower sideband. This is known as *independent-sideband modulation* and is used in HF radio communication [2]. However, it is also possible to do this using DSBSC. The transmitter uses two modulators whose carriers are in quadrature. The receiver uses two coherent demodulators whose local carriers are in quadrature. Eqn. 2.15b shows that each demodulator produces a full output from the signal whose carrier is in phase (since $\cos 0 = 1$) and zero output from the signal whose carrier is in quadrature (since $\cos \pi/2 = 0$). This is called *quadrature amplitude modulation* (QAM). In practice, the method is not used for analogue baseband signals since small errors in the phases of the local carriers cause a fraction of the signal of each channel to appear as crosstalk in the output from the other. However, the method is used for transmitting digital signals (see Section 2.7.4).

If the baseband signal extends down to very low frequencies, as in television, it is almost impossible to suppress the whole of the unwanted sideband without affecting low-frequency components in the wanted sideband. Use is then made of *vestigial sideband* (VSB) transmission instead of SSBSC. A conventional AM signal (as shown in Fig. 2.7b) is first generated and this is then applied to a filter having a transition between its pass and stop band that is skew symmetric about the carrier frequency. This results in an output signal having the spectrum shown in Fig. 2.7e. If a coherent demodulator is used, the original baseband signal can be recovered without distortion. It is also possible to use a simple envelope demodulator for VSB, but some nonlinear distortion then results [14]. VSB transmission does, of course, require a greater channel bandwidth than SSB. However, for a wideband signal such as television, the bandwidth saving compared with DSB is considerable.

2.7.3 Frequency and phase modulation

The instantaneous angular frequency of an alternating voltage is $\omega = d\phi/dt$ radians/s. This relationship between frequency and phase means that frequency modulation (FM) and phase modulation (PM) are both forms of *angle modulation.*

A sinusoidal carrier modulated by a sinusoidal baseband signal may be represented by

$$v = V_c \cos[\omega_c t + \beta \sin \omega_m t]$$

where $\beta = modulation\ index$

The maximum phase deviation is $\Delta\phi = \pm\beta$ and, since $d\phi/dt = \omega_m\beta \cos \omega_m t$, the maximum frequency deviation is $\Delta F = \pm\beta f_m$. In PM, the phase deviation is proportional to the modulating voltage; therefore β is independent of its frequency and the frequency deviation is proportional to it. In FM, the frequency deviation is proportional to the modulating voltage; therefore the deviation frequency is independent of its frequency and β is inversely proportional to it. Thus, for FM, the modulation index may be defined as

$$\beta = \frac{\text{maximum frequency deviation of carrier}}{\text{maximum baseband frequency}} = \frac{\Delta F}{F_m}$$

In angle modulation, the information is conveyed by the instantaneous phase of the signal. Consequently, phase distortion in the transmission path causes attenuation distortion of the received signal. The differential delay of the transmission path must therefore be closely controlled over the bandwidth required to transmit the signal.

The frequency spectrum of the transmitted signal contains higher-order sideband components [14] at frequencies $f_c \pm nf_m$ (where $n = 1, 2, 3, \ldots$). The bandwidth required is thus greater than for AM. However, there is a little energy outside the band $(1 \pm \beta)f_c$. To a first approximation the bandwidth W required [1] is

$$W = 2F_m(1 + \beta) \quad \text{(Carson's rule)}$$

Thus, for FM:

$$W = 2(F_m + \Delta F)$$

where ΔF = maximum frequency deviation

For low-index modulation ($\beta < 1$) the bandwidth needed is little more than that for AM, but for large values of β it is much more.

In FM, the amplitude of the signal conveys no information. Thus, amplitude variations due to fading have no effect on the output signal level. Amplitude variations caused by noise also have no effect.* However, noise also perturbs the phase of the signal and thus its instantaneous frequency. For white noise, it can be shown [14] that the improvement in signal/noise ratio compared with fully-modulated AM, for the same transmitter carrier power and same noise-power density at the receiver, is given by:

$$\frac{\text{output signal/noise power ratio for FM}}{\text{output signal/noise power ratio for AM}} = 3\beta^2$$

The improvement obtained is thus proportional to the square of the frequency deviation used.

The amplitude of the disturbance produced in the output of an FM receiver by an interfering signal is proportional to the frequency difference between the interference and the unmodulated carrier frequency. White noise thus produces a 'triangular' output-noise spectrum in which the RMS noise voltage (per Hz) is proportional to frequency [1]. It is therefore common practice to insert in front of the modulator at the transmitter a *pre-emphasis* network which has a rising gain/frequency characteristic across the baseband. The receiver contains a *de-emphasis* network with the inverse gain/frequency characteristic, to obtain a channel having a flat overall gain/frequency characteristic and a uniform noise spectrum. For single-channel telephony or broadcasting this provides a better output signal/noise ratio. When FM is used to transmit a wideband signal consisting of a block of telephone channels assembled by frequency-division multiplexing (see Section 2.8.1), the use of pre-emphasis is essential [1]. Otherwise, there

---

* For this assumption to be valid, the input signal/noise ratio much exceed a *threshold* [1, 14] of approximately 12 dB. When the input signal/noise ratio decreases below this, peaks of noise voltage begin to obliterate the carrier and the output signal/noise ratio deteriorates rapidly. Radio links used for commercial telecommunication circuits must therefore have signal/noise ratios well above the threshold.

would be a large difference between the signal/noise ratios of channels at the top and bottom of the band.

Because of its good signal/noise-ratio properties, FM is used for radio circuits whenever sufficient bandwidth can be provided. This is not possible for telephony in the more congested parts of the radio-frequency spectrum (VHF and lower frequencies), but it is standard practice at UHF and SHF. For telegraph signals, because of their narrow baseband, it is possible to use FM in the HF band [2]. This is known as *frequency-shift keying* (FSK). This form of transmission is also used for data transmission [17, 18] over telephone circuits in the switched telephone network [19, 20]. A typical data modem for use on telephone circuits [21] uses frequencies of 1·3 kHz and 1·7 kHz to transmit at 600 bauds or 1·3 kHz and 2·1 kHz to transmit at 1200 bauds.

In phase modulation (PM), the frequency deviation (βf_m) is proportional to the frequency of the modulating signal as well as its amplitude. Consequently, for signals (such as speech) that have the major proportion of their energy at the lower end of the baseband, PM makes inefficient use of transmission-path bandwidth in comparison with FM. Moreover, to demodulate a PM signal, the receiver must compare the phase of the incoming carrier with that of a locally-generated carrier which must be extremely stable. FM is therefore preferred to PM for the transmission of analogue signals.

Phase modulation is, however, used for the transmission of digital signals [17, 18]. This is called *phase-shift keying* (PSK). The carrier phase is switched to one of a number of possible values, e.g. 0° and 180° for a binary system,* or 0, 90°, 180° and 270° for a 4-level system. The receiver detects which level has been sent by comparing the phase of the incoming signal with that of a locally-generated reference carrier. The stability of this reference must be very high, so an alternative method, known as *differential phase-shift keying* (DPSK) is often used to avoid the need for a reference carrier. In DPSK the phase difference between successive intervals is used to convey the information, this difference being measured at the receiver by comparing the received waveform with the same waveform delayed by one interval [23].

2.7.4 Hybrid modulation

If a carrier is modulated by a digital signal, its amplitude and phase can only have a finite number of values. These can be represented in a signal–space diagram, as shown in Fig. 2.8. This shows an 8-level amplitude-shift keyed (ASK) signal and an 8-level PSK signal. The circles show the maximum permissible perturbation from the ideal signal before an error may occur in the transmission. It is obvious that, in this respect, PSK is superior to ASK.

It is possible to combine both ASK and PSK to obtain *hybrid modulation* [22], as shown in in Fig. 2.9. The 8-level system shown in Fig. 2.9*b* has a similar error threshold voltage to the PSK system of Fig. 2.8*b*. However, it corresponds to a lower mean transmitted power, because four of the 8 states

---

* This is sometimes called phase-reversal keying (PRK). It is equivalent to DSBSC with a modulating signal that is either +1 or −1.

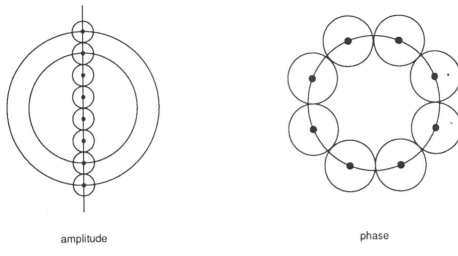

amplitude   phase

Fig. 2.8   Signal-space diagrams for 8-state amplitude and phase modulation

a Amplitude modulation
b Phase modulation

have a smaller carrier amplitude than those in Fig. 2.8*b*. These hybrid modulation schemes can be implemented using *Quadrature amplitude modulation* (QAM), i.e. DSBSC mdoulation of two carriers in quadrature (as discussed in Section 2.7.2). For example, if a 4-level digital signal is applied to each of the two modulators, the sum of their outputs has four values of its in-phase component and four values of its quadrature component; this produces the 16-state signal constellation shown in Fig. 2.9*c*. Two coherent demodulators reproduce the two 4-level signals at their outputs.

16-state QAM is used to provide 2400 bit/s data transmission over switched telephone circuits with a signalling rate of only 600 bauds (CCITT Recommendation V22 *bis*) and 9·6 kbit/s over private circuits with a signalling rate of only 2400 bauds (CCITT Recommendation V32). The output of the 4-state QAM system shown in Fig. 2.9*a* has a constant amplitude; it therefore provides phase-shift keying and is called quadrature phase-shift keying (QPSK). The input is coded so that phase changes of the transmitted signal are always 90° and never 180°. If 9-state QAM is used in conjunction with partial-response coding (which is sometimes called duobinary coding and is discussed in Chapter 7), the system is called a quadrature-partial-response system (QPRS) [23].

2.7.5 Pulse modulation

In the examples above, the carrier waveforms are sinusoidal. It is also possible to modulate carriers having other waveforms. For example, it is possible to modulate trains of pulses [14] to produce pulse-amplitude modulation (PAM), pulse-frequency modulation (PFM) or pulse-phase modulation (PPM). It is also possible to produce pulse-length modulation (PLM), which is sometimes known as pulse-width modulation (PWM).

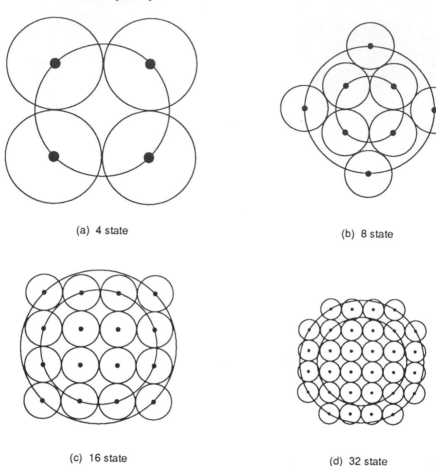

(a) 4 state

(b) 8 state

(c) 16 state

(d) 32 state

Fig. 2.9   Signal-space diagrams for m-state quadrature amplitude modulation

a 4-state
b 8-state
c 16-state
d 32-state

A basic PAM system is shown in Fig. 2.10a. If a train of pulses as shown in Fig. 2.10c is amplitude modulated by the baseband signal shown in Fig. 2.10b, the resulting PAM signal is shown in Fig. 2.10d. The baseband signal is represented by a sequence of samples of it, so the process is also called *sampling*.

A train of pulses having a pulse-repetition frequency (PRF) f_r may be represented by a Fourier series

$$v_c = \tfrac{1}{2}a_0 + \sum_{n=1}^{\infty} a_n \cos n\omega_r t$$

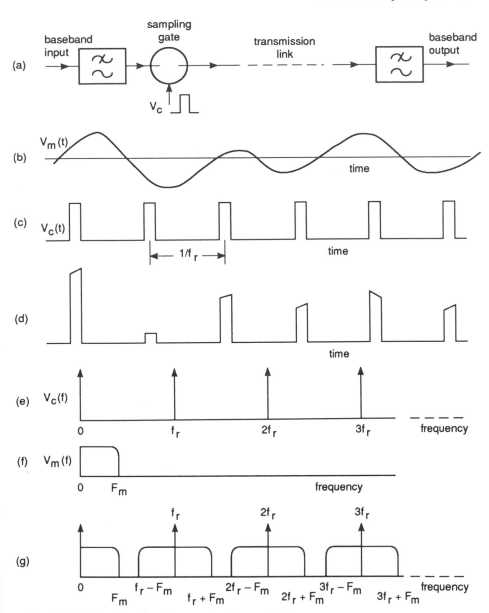

Fig. 2.10 Principle of pulse-amplitude modulation

a Basic system (one direction of transmission only)
b Baseband signal
c Unmodulated pulse train
d Modulated pulse train
e Spectrum of unmodulated pulse train
f Spectrum of baseband signal
g Spectrum of modulated pulse train

Its spectrum thus contains a DC component, the PRF and its harmonics, as shown in Fig. 2.10e. If the pulses are of short duration, the amplitudes a_n of the harmonics are approximately equal up to large values of n.

If the pulse train is amplitude modulated by a sinusoidal baseband signal, the resulting PAM signal is

$$v = (1 + m \cos \omega_m t)v_c$$

$$= \tfrac{1}{2}a_0 + \tfrac{1}{2}a_0 m \cos \omega_m t + \sum_{n=1}^{\infty} a_n \cos n\omega_r t$$

$$+ \tfrac{1}{2}m \sum_{n=1}^{\infty} a_n [\cos(n\omega_r + \omega_r)t + \cos(n(\omega_r - \omega_m)t]$$

Thus, the spectrum of the PAM signal contains all the components of the original pulse train, together with the baseband frequency f_m and upper and lower side frequencies $(nf_r \pm f_m)$ about the PRF and its harmonics. If the modulating signal consists of a band of frequencies, as shown in Fig. 2.10f, the spectrum of the PAM signal contains the original baseband, together with upper and lower sidebands about the PRF and its harmonics, as shown in Fig. 2.10g.

The PAM signal can be demodulated by means of a low-pass filter which passes the baseband and stops the lower sideband of the PRF and all higher frequencies. For this to be possible, we require

$$F_m \leqq f_0 \leqq f_r - F_m$$

where F_m is the maximum frequency of the baseband signal and f_0 is the cut-off frequency of the filter. Thus, we require:

$$f_r \geqq 2F_m \tag{2.16}$$

Eqn. 2.16 is a statement of the *sampling theorem*. This may be expressed as follows: if a signal is to be sampled and the original signal is to be recovered from the samples without error, the sampling frequency must be at least twice the highest frequency in the original signal. The sampling theorem is due to Nyquist and the lowest possible rate at which a signal may be sampled, $2F_m$, is often known as the *Nyquist rate*. If the sampling frequency is less than the Nyquist rate, the lower sideband of the PRF overlaps the baseband and it is impossible to separate them. The output from the low-pass filter then contains unwanted frequency components; this situation is known as *aliasing*.

To prevent aliasing, it is essential to limit the bandwidth of the signal before sampling. Thus, practical systems pass the input signal through an anti-aliasing low-pass filter of bandwidth $\tfrac{1}{2}f_r$ before sampling as shown in Fig. 2.10a. Practical filters are non-ideal; it is therefore necessary to have $f_r > 2F_m$ in order that the anti-aliasing filter and demodulating filter can have both very low attenuation at frequencies up to F_m and very high attenuation at frequencies down to $f_r - F_m$. For telephony, a baseband from 300 Hz to 3·4 kHz is provided and a sampling frequency of 8 kHz is used. Thus, $f_r - F_m = 4·6$ kHz and there is a *guardband* of 1·2 kHz to accommodate the transition of the filters between their pass band and their stop band.

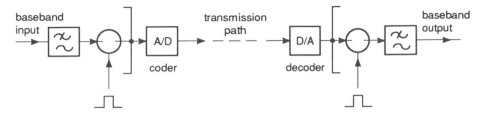

Fig. 2.11 PCM transmission system (one direction of transmission only)

When PFM, PPM and PLM are used, the sidebands include higher-order sideband components [14] at frequencies $nf_r \pm jf_m$ (where $j = 1, 2, 3, \ldots$), as in FM and PM. It is therefore necessary to use a sampling frequency greater than $2F_m$ to minimise aliasing.

These methods of pulse modulation all generate analogue signals. It is possible to produce a digital signal by applying a train of PAM samples to an analogue-to-digital (A/D) convertor, as shown in Fig. 2.11. Each analogue sample is thus converted to a group of on/off pulses which represents its voltage in a binary code. This process is called *pulse-code modulation* (PCM) [24]. It is discussed in more detail in Chapter 6. At the receiving terminal, a digital-to-analogue (D/A) convertor performs the decoding process. The combination of coder and decoder at a PCM terminal is often referred to as a *codec*. The group of bits (i.e. binary pulses) representing one sample is called a *word* or a *byte*. An 8–bit byte is sometimes called an *octet*.

For telephony, speech samples are usually encoded in an 8–bit code. Since sampling is at 8 kHz, a telephone channel requires binary digits to be sent at the rate of $8 \times 8 = 64$ bauds. Nyquist has shown [17, 25] that the minimum bandwidth required to transmit pulses is half the pulse rate. Thus, a bandwidth of at least 32 kHz is required to transmit a single telephone channel. The advantages of digital transmission are won at the expense of a much greater bandwidth requirement.

2.8 Multiplexing

2.8.1 Frequency-division multiplexing

In frequency-division multiplex (FDM) transmission, a number of baseband channels are sent over a common wideband transmission path by using each channel to modulate a different carrier frequency. Systems using this process are called *multichannel carrier systems.* They are commonly employed to transmit 24 telegraph channels over a single telephone channel, to transmit 24 telephone channels over a balanced pair or many hundreds of telephone channels over a coaxial pair or a microwave radio link.

A *channel translating equipment,* or *channelling equipment,* for multiplexing 12 telephone channels is shown in Fig. 2.12*a*. At the sending end, each incoming baseband signal $(0 \leqq f_m \leqq F_m)$ from an audio-frequency circuit is applied to a balanced modulator supplied with the appropriate carrier, f_c. The output of this modulator is a double-sideband suppressed-carrier signal,

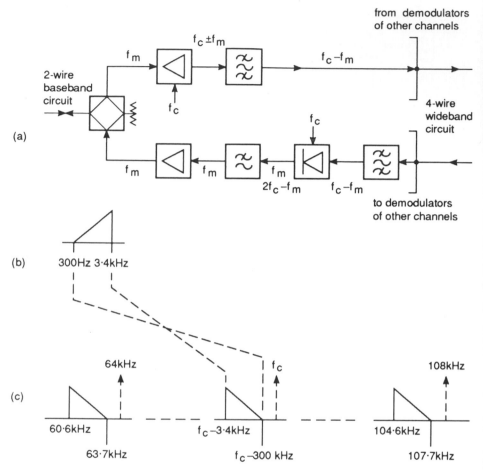

Fig. 2.12   Principle of frequency-division multiplexing

a Channel-translating equipment
b Frequency band of baseband signal
c Frequency band of wideband signal (CCITT basic group B)

as shown in Fig. 2.7*c*. This signal is applied to a bandpass filter which suppresses the upper sideband ($f_c + f_m$) and transmits the lower sideband ($f_c - f_m$). The outputs of these filters are commoned to give a composite output signal containing the signal of each telephone channel translated to a different portion of the frequency spectrum, as shown in Fig. 2.12*c*.

At the receiving end, the incoming signal is applied to a bank of band-pass filters, each of which selects the frequency band containing the signal of one channel. This signal is applied to a modulator supplied with the appropriate carrier f_c and the output of this modulator consists of the baseband signal and unwanted high-frequency components, as shown in Section 2.7.2. The unwanted components are suppressed by a low-pass filter and the baseband signal is transmitted to the audio-frequency circuit at the correct level by means of an amplifier.

Suppressed-carrier modulation is used to minimise the total power to be handled by amplifiers in the wide-band transmission system. The use of singe-sideband modulation (SSB) maximises the number of channels that can be transmitted in the bandwidth available. To avoid interchannel cross-talk the sidebands of adjacent channels obviously must not overlap, so the spacing between carrier frequencies is determined by the highest frequency F_m of the baseband signal. Practical bandpass filters cannot have a perfectly sharp cut off; so it is necessary to leave a small guardband between the frequency bands of adjacent channels. Fig. 2.12c shows the standard *basic group* of 12 channels (CCITT basic group B). The carrier spacing is 4 kHz; thus 12 channels occupy the band from 60 to 108 kHz. Each channel has a baseband from 300 Hz to 3·4 kHz. The frequency guardband between adjacent channels is only 900 Hz, so crystal filters are used to obtain the necessary sharp transitions between pass and stop bands.

In multichannel voice-frequency telegraph systems [26], 50 baud baseband signals are transmitted by carriers whose frequency separation is 120 Hz. In a 24-channel system, the carrier frequencies thus range from 420 Hz to 3180 Hz. Early systems used AM (on–off keying). Modern systems [27] use FM with a ±30 Hz frequency shift.

2.8.2 Time-division multiplexing

Another method of multiplexing the signals of a number of baseband channels to transmit them over a wideband transmission path is called *time-division multiplexing* (TDM), which is illustrated in Fig. 2.13. At the sending terminal, a baseband channel is connected to the common trans-mission path by means of a sampling gate which is opened for short intervals by means of a train of pulses. In this way, samples of the baseband signal are sent at regular intervals by means of amplitude-modulated pulses. Pulses with the same repetition frequency f_r but staggered in time, as shown in Fig. 2.13b, are applied to the sending gates of the other channels. Thus the common transmission path receives interleaved trains of pulses, modulated by the signals of different channels. At the receiving terminal, gates are opened by pulses coincident with those received from the transmission path so that the demodulator of each channel is connected to the transmission path for its allotted interval and disconnected throughout the remainder of the pulse-repetition period. The combination of a multiplexer and a demultiplexer at a TDM terminal is sometimes referred to as a *muldex*.

The pulse generator at the receiving terminal must be synchronised with that at the sending terminal to ensure that each incoming pulse train is gated to the correct outgoing baseband channel. A distinctive synchronising pulse signal is therefore transmitted in every repetition period in addition to the channel pulses. The complete waveform transmitted during each repetition period contains a number of time slots: one is allocated to the synchronising signal and the others to the channel samples. The complete waveform is called a *frame*, by analogy with a television-signal waveform, and the synchronising signal is called the *frame-alignment signal*.

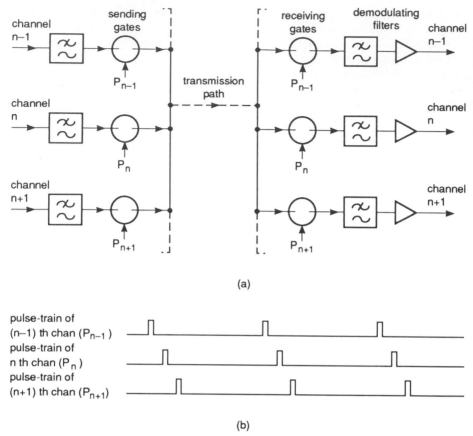

(a)

(b)

Fig. 2.13   Principle of time-division multiplex transmission

a Elementary TDM system (one direction of transmission only)
b Channel pulse trains

The elementary TDM system shown in Fig. 2.13 uses pulse-amplitude modulation (PAM). Other analogue methods of pulse modulation, including pulse-length modulation and pulse-position modulation can also be used [14]. However, these methods are not employed for line transmission. This is because attenuation and delay distortion cause the transmitted pulses to spread in time so that they interfere with the pulses of adjacent channels and cause crosstalk. To overcome this problem, pulse-code modulation (PCM) is used for TDM transmission of telephone signals.

In PCM, the coder used for A/D conversion and the decoder used for D/A conversion are required to perform their operations within the duration of the time slot of one channel. They can therefore be common to all the channels of a TDM system, as shown in Fig. 2.11.

2.8.3 Time-compression multiplexing

Time-compression multiplexing (TCM) uses time division [28]. However, instead of sending a sample from each baseband channel sequentially, as in conventional TDM, the signal of each channel is stored for a short time, this process being carried out concurrently for all channels. The signals are then read out from each store in turn and transmitted in a much shorter time over the wideband channel. At the receiving terminal, the received signals are written into another set of stores. They are then read out in parallel and fed concurrently to the corresponding channels.

If analogue pulse modulation is used, transmission of TCM requires less bandwidth than TDM. Because most adjacent samples are from the same channel, there is no need for guard intervals between them to prevent inter-channel crosstalk. The bandwidth required is thus only a few percent more than for FDM transmission. If digital transmission is used, there is no need for guard intervals between channels; thus, the bandwidths needed for TDM and TCM are the same. However, the frame structures are different. If a TCM system stores k samples of each channel before sending them, the transmitted frame contains k adjacent samples from each channel in turn. Consequently, the duration of the TCM frame is k times that of a TDM frame.

The TCM principle is used for *time-division multiple access* (TDMA) in satellite communication systems, as described in Section 13.2.3. Because the earth stations have different locations, the propagation times between them and the satellite differ. If TDM were used, it would be impracticable to ensure correct interleaving of samples arriving at the satellite. Using TDMA, if each earth station assembles k samples before sending them, the tolerances on timing are increased by a factor k. In a typical TDMA satellite system, the frame period is 2 ms, instead of the 125 μs used in TDM telephony.

A form of time-compression multiplexing is used in *packet switching*. Data from a slow-speed source are assembled to form packets; these are then transmitted over a data link at a higher speed, interleaved with packets from other sources. Since packets from different sources must be routed to different destinations, 'overheads' are incurred. Each packet contains a *header* which includes its address, together with other information required by the protocols which control the system.

A special form of packet switching is used in the *asynchronous transfer mode* (ATM) system. This is a switching system designed to handle traffic inputs with different bandwidths. It uses short fixed-length packets called *cells*. The cells, which consist of five header bytes and 48 data bytes, are short in order to minise delay (which is essential for telephony). The system operates at a fixed digit rate (154 Mbit/s) and different services are handled by changing the intervals between cells. For example, a 33 Mbit/s video codec will require many more cells in a given time than a 64 kbit/s speech codec.

2.8.4 *Code-division multiplexing*

In a multiplex system, the baseband signals, $s_1, s_2, \ldots, s_n$, of the channels multiply a set of carriers $c_1, c_2, \ldots, c_n$. The signal sent over the wideband channel is therefore

$$S(t) = \sum_{k=1}^{n} s_k(t) c_k(t)$$

At the receiving terminal, the output signal s_{kk} of the kth baseband channel, after demodulation with a coherent carrier, is:

$$s_{kk}(t) = s_k(t) \frac{1}{T} \int_0^T c_k^2(t) \, dt \qquad (2.17)$$

where $1/T$ is the cut-off frequency of the low-pass filter used. The output from the kth channel due to the input signal of the jth channel is

$$s_{jk}(t) = s_j(t) \frac{1}{T} \int_0^T c_j(t) c_k(t) \, dt \qquad (2.18)$$

Obviously, s_{kk} should be as large as possible and s_{jk} should be as small as possible (to minimise inter-channel crosstalk). The crosstalk is zero if

$$\frac{1}{T} \int_0^T c_j(t) c_k(t) \, dt = 0 \qquad (2.19)$$

This condition is satisfied if the n carriers form a set of orthogonal functions.

Eqn. 2.19 is satisfied in FDM, since sinewaves of different frequencies are orthogonal. It is also satisfied in quadrature modulation, since $\sin \omega_c t$ and $\cos \omega_c t$ are orthogonal. It is satisfied in TDM; since $c_j(t)$ and $c_k(t)$ are pulses occurring at different times, their product is zero.

In principle, any set of orthogonal functions can be used for the carriers of a multiplex system. If some crosstalk can be tolerated, then some departure from orthogonality can be permitted (i.e. quasi-orthogonal carriers). Eqn. 2.17 corresponds to the autocorrelation function (ACF), for $\tau = 0$, of the carrier of one channel and eqn. 2.18 to the cross-correlation function (CCF) for two different carriers. It is thus a requirement for multiplexing that the carriers shall have large ACFs and small CCFs.

Random noise voltages might appear ideal for use as carriers, since their ACFs would be impulses and their CCFs zero. However, it is impossible to generate the coherent carriers at the receiving terminal. Pseudo-random binary sequences (PRBS) have properties approximating to those of noise. Moreover, being periodic, they can be accurately generated at both sending and receiving terminal. A multiplex system can therefore be implemented by using a different sequence of binary pulses (instead of a single pulse) for the carrier of each channel. Moreover, these pulse trains do not need to be synchronised. This method is known as *code-divison multiplexing* (CDM).

Other forms of carrier coding can be used [28]. For example, each channel may use a single pulse which modulates several different carrier frequencies, or a sequence of pulses, each of which modulates a different combination of frequencies. These CDM systems are often called *random-access discrete-address* (RADA) systems. Since, in these CDM systems a single channel occupies a wider bandwidth than in traditional multiplex systems, they are also called *spread-spectrum systems* [29].

Fig. 2.14 shows how the channels occupy the wideband channel, in time and in frequency, for different forms of multiplexing. In FDM, each channel

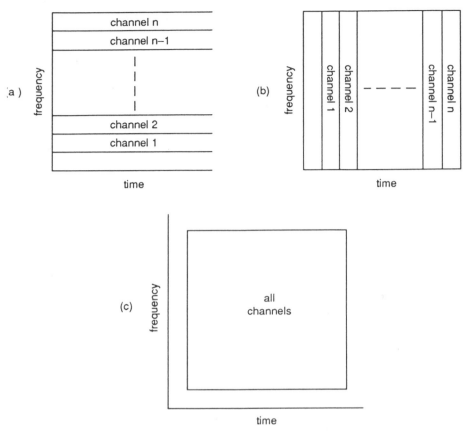

Fig. 2.14 Occupancy of the wideband channel by n-channel multiplex systems

a FDM system
b TDM system
c CDM system

uses only a small fraction of the available bandwidth, but occupies it all the time, as shown in Fig. 2.14*a*. In TDM (including TCM), each channel uses the whole of the bandwidth but only for a fraction of the time, as shown in Fig. 2.14*b*. In CDM, each channel occupies an appreciable fraction of both time and bandwidth, and the channels overlap in both time and frequency, as shown in Fig. 2.14*c*.

2.9 Bandwidth reduction techniques

Natural speech is a highly-redundant method of encoding information. Some redundancy is removed by restricting the band of a telephone channel to 300—3400 Hz. Nevertheless, the information capacity of a normal telephone channel is much greater than the information rates of the messages conveyed. Methods have therefore been invented to convey telephone speech through analogue channels having less than 3 kHz bandwidth and digital channels conveying less than 64 kbit/s. These methods make use of the *a priori* information that the signals to be transmitted have been generated by the human voice mechanism.

An early example was the vocoder [30]. At the sending end of a channel, the speech signal is analysed by a bank of filters into several frequency bands and the relatively-slowly-changing amplitudes of these components are transmitted over a narrow-band channel. At the receiving end, an output signal is synthesised by generating components in these frequency bands with the same relative amplitudes. This output is intelligible but unnatural. More recent methods, which reduce the bit rate for the digital transmission of speech, are described in Sections 6.8 and 6.9.

Television signals are also highly redundant and methods have been developed for processsing them in order to transmit much lower bit rates than those required for normal PCM. These methods are discussed in Section 6.11.

An obvious redundancy in telephone transmission can be removed more easily than the redundancy in speech signals themselves. Telephone connections normally provide a full-duplex circuit. Speech can be transmitted in both directions at all times, although the user at each end speaks for only about 40% of the time. The rest of the time is occupied by pauses in the speech and listening to the speaker at the other end of the connection. A system called *time-assignment speech interpolation* [31] [TASI] was developed to remove this redundancy and so reduce the number of channels required in each direction to less than half the number of the telephone connections provided.

At the sending end of a TASI system, a number of input channels (N) is connected to a smaller number of channels (M), provided over the transmission medium, by means of a high-speed switch. At the receiving end, these M channels are connected to N output channels by a similar switch. When speech activity is detected on an input channel, it is assigned to a channel of the transmission system and remains connected to it for the duration of the speaker's 'talkspurt'. A signal is transmitted to the switch at the receiving end, instructing it to connect the channel assigned in the transmission system to the output channel which corresponds to the active channel at the sending end. Since it takes a finite time to detect the presence of speech on an input channel, the first few milliseconds of the intial syllable of each talkspurt are lost. Also, which the number of active input channels exceeds the number of channels in the transmission system, then 'freeze out' occurs and complete talkspurts are lost. The TASI advantage, i.e. the ratio N/M, depends on the TASI activity factor, i.e. the percentage of time for which speech is present on an input circuit, and on the percentage of freeze out that is acceptable. A typical TASI system transmits 235 speech channels over a 96-channel transmission system [1].

It is economic to employ TASI if the cost of the necessary speech detection, switching and signalling is less than the cost of the additional telephone circuits that would otherwise be required. Considerable savings can be made if these circuits are very expensive, for example in a trans-oceanic cable or a satellite communications system.

When PCM transmission is employed, the switching and multiplexing can be combined, both using TDM. This is called *digital speech interpolation* [32] (DSI). It is used with satellite systems that employ time-division multiple access (TDMA), as described in Section 13.3. Using DSI, 240 speech channels at 64 kbit/s are transmitted over 127 satellite 64 kbit/s channels. When the latter are all busy, freeze out is avoided by changing from 8–bit to 7–bit coding in order to use fewer digits per time slot and thus provide up to 16 more channels.

Signal interpolation can also be applied in data transmission, when a number of low-speed data terminals are connected to a high-speed data link by time-division multiplexing. If the input data traffic is 'bursty', i.e. messages are short with intervals between them, the number of terminals served can be greater than the number of time slots provided by the data link. The multiplexer assigns time slots only to those terminals actively sending data. Such a multiplexer [22] is called a *statistical multiplexer*, or 'statmux' for short. No information is lost when 'freeze-out' occurs, because data can be stored at the input ports of the multiplexer and sent when time slots become free.

2.10 Digital transmission

2.10.1 Bandwidth requirements

The minimum bandwidth needed to transmit a digital signal at B bauds has been shown by Nyquist [25] to be $W_{min} = \frac{1}{2}B$ hertz. This can be demonstrated as follows. Consider a binary signal consisting of alternate '0's and '1's. This produces a square wave of frequency $\frac{1}{2}B$. Let this be applied to an ideal low-pass filter. If the cut-off frequency of the filter is $\frac{1}{2}B + \varepsilon$, where ε is small, the output is a sinewave of frequency $\frac{1}{2}B$ and the original square wave can be recovered by sampling it at its positive and negative peaks. If the cut-off frequency is reduced to $\frac{1}{2}B - \varepsilon$, the output consists only of the DC component and the signal is lost. Consequently, the minimum bandwidth required is:

$$\tfrac{1}{2}B - \varepsilon < W_{min} < \tfrac{1}{2}B + \varepsilon$$

In the limit when ε tends to zero,

$$W_{min} = \tfrac{1}{2}B \tag{2.20}$$

This result can also be demonstrated for the case when the transmitted symbols are very short pulses (which approximate to impulses), instead of the full-width pulses considered above. The impulse response, $h(t)$, of the ideal low-pass filter of bandwidth W is given by

$$h(t) = \frac{\sin 2\pi Wt}{2\pi Wt} \tag{2.21}$$

This response has its maximum at $t = 0$ (where $h(0) = 1$) and is zero for $t = \pm nT$, where $T = 1/2W$, as shown in Fig. 2.15. If pulses are transmitted at rate $B = 2W$ and each is detected by sampling at the time when it has its maximum output voltage, the outputs due to all preceding and following pulses are zero at that time, i.e. there is no *intersymbol interference* (ISI). Thus, it is possible to transmit pulses at rate B bauds through a channel of bandwidth $W = \frac{1}{2}B$ without any ISI.

In practice, it is not possible to obtain a channel with an ideal low-pass characteristic. (Eqn. 2.21 shows that, if it were possible, an output voltage would appear before the input pulse is applied!) However, Nyquist showed [25] that zero ISI can also be obtained if the gain of the channel changes from unity to zero over a band of frequencies with a gain/frequency response

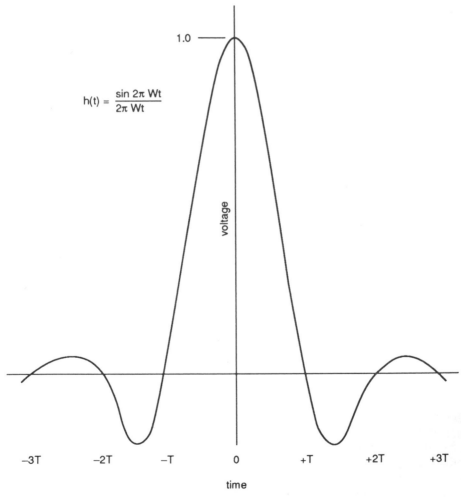

Fig. 2.15   Impulse response of ideal low-pass filter of bandwidth W (T = 1/2W).

that is skew-symmetrical about $f = \frac{1}{2}B$. It is also impossible to generate a perfect impulse (since it has zero duration and infinite amplitude). The transfer characteristic of the channel should therefore be equalised so that the output signal has the required spectrum. A commonly-used signal is that having the *raised-cosine spectrum* given by:

$$F(f) = 1 \qquad\qquad\qquad \text{for } 0 \leqq f \leqq (1-\alpha)/2T$$

$$= \frac{1}{2}\left(1 + \sin\frac{\pi}{2\alpha}(1-2fT)\right) \qquad \text{for } \frac{1-\alpha}{2T} \leqq f \leqq \frac{1+\alpha}{2T}$$

$$= 0 \qquad\qquad\qquad \text{for } f \geqq \frac{1+\alpha}{2T}$$

This gain/frequency response rolls off sinusoidally from unity to zero in the frequency band from $(1-\alpha)/2T$ to $(1+\alpha)/2T$. Thus, for 100% roll off (i.e. $\alpha = 1$), the spectrum occupies a bandwidth of $1/T$, which is twice the theoretical minimum requirement. Bandwidth is used most efficiently by using as small a roll off α as possible, but problems of timing and equalisation increase as is reduced.

2.10.2 Equalisation

Digital transmission systems can use gain and phase equalisation to obtain an output-signal spectrum corresponding to a pulse waveform with negligible inter-symbol interference, e.g. the raised-cosine spectrum described above. However, time-domain equalisers are often employed.

A common form of time-domain equaliser is the *transversal equaliser* (TVE) shown in Fig. 2.16. This consists of a delay line tapped at intervals equal to the inter-symbol interval T. Each tap is connected to an amplifier (which may be an inverter to obtain negative gain). The output of the equaliser is the sum of the outputs of these amplifiers. It is possible to adjust the gains of the amplifiers (in magnitude and sign) to cancel ISI by adding appropriately-weighted versions of preceding and following pulses at the time of each symbol, and thus cancel interference between them.

As a simple example, consider a pulse which has suffered dispersion during transmission. Its maximum height is 1 V, it is $\frac{1}{8}$ V one sampling interval earlier and $\frac{1}{4}$ V one sampling interval later, but it is zero elsewhere.

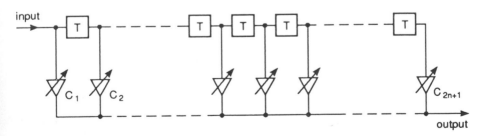

Fig. 2.16 Block diagram of transversal equaliser.

It is applied to a TVE with three taps. When the peak of the pulse is at the second tap of the equaliser, the output voltage is

$$v_0 = \tfrac{1}{4}c_1 + 1\,c_2 + \tfrac{1}{8}c_3$$

where the coefficients c_1, c_2 and c_3 are the gains of the amplifiers connected to the tapping points. One interval earlier, the output voltage is

$$v_{-1} = 1\,c_1 + \tfrac{1}{8}c_2 + 0c_3$$

and one interval later, the output voltage is

$$v_{+1} = 0c_1 + \tfrac{1}{4}c_2 + 1\,c_3$$

To cancel interference with the preceding and following pulses, we require $v_{+1} = v_{-1} = 0$

From the above equations, the required coefficient values are:

$$c_1 = -\tfrac{1}{8}c_2, \qquad c_3 = -\tfrac{1}{4}c_2$$

and the peak output voltage is $v_0 = \tfrac{15}{16}c_2$.

Two intervals earlier, the output voltage is

$$v_{-2} = \tfrac{1}{8}c_1 + 0c_2 + 0c_3 = -\tfrac{1}{64}c_2$$

and two intervals later, the output voltage is

$$v_{+2} = 0c_1 + 0c_2 + \tfrac{1}{4}c_3 = -\tfrac{1}{16}c_2$$

The output sequence of voltages is thus longer than the input sequence. Although ISI is cancelled at the adjacent sampling points, a small amount is introduced into the next-but-one intervals.

The residual ISI can be made arbitrarily small by increasing the number of taps in the TVE. In order to force $2N$ zeros at sampling points of the output voltage, the TVE requires $2N + 1$ tapping points. The design of the equaliser then requires the solution of $2N$ linear simultaneous equations to determine the relative values of the coefficients.

A TVE can be adjusted manually. The output voltage is observed on an oscilloscope and the amplifier gains are manipulated iteratively until the best output waveform is obtained. However, if the characteristics of the transmission path change with time, it is preferable for the equaliser to be adjusted automatically. A TVE which does this for itself during the course of normal operation is called an *adaptive equaliser*.

In principle, in order to calculate the coefficients of a TVE, one needs to know the impulse response of the transmission path. A well-known method of identifying the impulse response of a linear system is to cross-correlate its output voltage with a known input signal, such as a pseudo-random sequence of pulses. This suggests a method of implementing an adaptive equaliser. If the input signal to the transmission path is a random sequence of pulses, the input to the TVE can be cross-correlated with the output signal from it as if the latter were correct. The result can be used to adjust the coefficients of the TVE. As a result, the output signal becomes more nearly correct. If the process is repeated, the coefficients will converge to the correct values and the output signal to the correct waveform. This process has been called 'decision directed', since the equaliser goes through a learning process based on its own decisions.

A number of different algorithms have been implemented to perform this process [18]. Since the process depends on the input data being random, the equaliser settings will diverge from the correct values when the input signal becomes non-random, e.g. a long sequence of '0's or '1's. The equaliser settings will oscillate about their correct values until the repetitive sequence ends and will then converge again to the correct values.

2.10.3 Noise and jitter

The principal advantage of PCM and other forms of digital transmission is that it is possible to obtain satisfactory transmission in the presence of very severe crosstalk and noise. In the case of binary transmission, it is only necessary to detect the presence or absence of each pulse. Provided that the interference level is not so high as to cause frequent errors in making this decision, the output signal will be almost noise-free.

Consider an idealised train of unipolar binary pulses, as shown in Fig. 2.17a. If the symbols '0' and '1' are equiprobable, i.e. $P(0) = P(1) = \frac{1}{2}$, the mean signal power corresponds to $S = \frac{1}{2}V^2$. Thus, the signal/noise ratio is

$$\frac{S}{N} = \frac{V^2}{2\sigma^2}$$

where σ is the r.m.s. noise voltage.

The receiver compares the signal voltage v_s with a threshold voltage of $\frac{1}{2}V$, giving an output '0' when $v < \frac{1}{2}V$ and '1' when $v > \frac{1}{2}V$. If a noise voltage v_n is added, an error occurs if $v_n < -\frac{1}{2}V$ when $v_s = +V$, or if $v_n > +\frac{1}{2}V$ when $v_s = 0$. Thus, the probability of error, P_e, is given by:

$$P_e = P(0)P(v_n > +\tfrac{1}{2}V) + P(1)P(v_n < -\tfrac{1}{2}V)$$

But $P(0) + P(1) = 1$ and, for white noise, $P(v_n < -\frac{1}{2}V) = P(v_n > +\frac{1}{2}V)$. Therefore

$$P_e = P(v_n > \tfrac{1}{2}V)$$

$$= P[v_n/\sigma > (S/N)^{1/2}/\sqrt{2}] \tag{2.22}$$

Hence, the error probability can be calculated from eqn. 2.12 by using a normal probability table.

If a bipolar binary signal is used, as shown in Fig. 2.17b, the mean signal power is $S = V^2$ and the signal/noise ratio is $S/N = V^2/\sigma^2$. The receiver gives an output '0' when $v < 0$ and '1' when $v > 0$. An error occurs if $v_n < -V$ when $v_s = +V$, or if $v_n > +V$ when $v_s = -V$. Thus, the probability

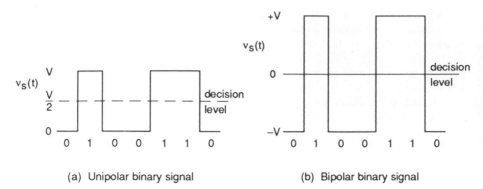

(a) Unipolar binary signal   (b) Bipolar binary signal

Fig. 2.17   Detection of digital signals
a Unipolar binary signal
b Bipolar binary signal

of error is given by:

$$P_e = P(0)P(v_n > +V) + P(1)P(v_n < -V)$$
$$= P(v_n > +V)$$
$$= P[v_n/\sigma > (S/N)^{1/2}] \tag{2.23}$$

Consequently, the same error rate is obtained with a 3 dB lower signal/noise ratio. Alternatively, a much lower error rate can be obtained for the same signal/noise ratio.

The above analysis can be extended to multilevel digital signals, as considered in Section 7.2, by replacing the pulse amplitude with the spacing between adjacent signal levels. If the number of levels is large, the error rate is nearly doubled. This is because all the intermediate levels can be misinterpreted in either direction, owing to noise voltages of either polarity.

For the case of a unipolar binary signal disturbed by white noise, the error rate, calculated from eqn. 2.22, varies with the signal/noise ratio as shown in Fig. 2.18. For example, if the signal/noise ratio is 20 dB, less than one digit per million is received in error. For telephone transmission, an error rate of 1 in 10^3 is intolerable, but an error rate of 1 in 10^5 is acceptable. Lower error rates are required for data transmission; if the error rate of the transmission link is inadequate, it is necessary to use an error-detecting or error-correcting code for the data [18, 22].

On a long transmission link, it is possible to use regenerative repeaters instead of analogue amplifiers. A regenerative repeater [1] samples the received waveform at intervals corresponding to the digit rate. If the received voltage at the sampling instant exceeds a threshold voltage, this triggers a pulse generator which transmits a pulse to the next section of the line. If the received voltage is below the threshold, no pulse is generated. If both positive and negative pulses are transmitted, the regenerator is required to compare the received voltage against both a positive and negative threshold and to retransmit pulses of either polarity.

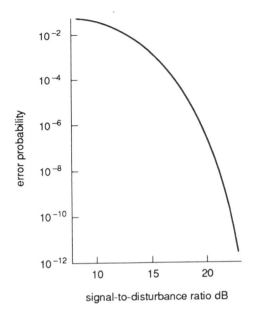

Fig. 2.18 Error rate for transmission of unipolar binary signal disturbed by white noise.

If the regenerators in a PCM link operate with a negligible error rate, the signal received at the far end of the link will be almost identical with that which was sent, regardless of the number of repeater sections. Thus, the requirements of each repeater section are no more severe than those of the overall system. This gives a considerable advantage over analogue transmission. For example, in an analogue system with 100 repeaters, the signal/noise ratio for each repeater section needs to be 20 dB better than that which would be needed if there was only one section. For a PCM system with 100 regenerators and an error rate of 1 in 10^6, the error rate per repeater section needs to be 1 in 10^8. Fig. 2.17 shows that the signal/noise ratio for each repeater section need only be approximately 1 dB better than that required if there was only a single section. In practice, since the error rate changes so rapidly with signal/noise ratio, it is likely that the error performance of one regenerator in the link will be much worse than the others. Its errors will then predominate and determine the overall perform-ance of the transmission link.

Other impairments can be caused by transmission over a link containing a number of regenerative repeaters. The instants at which pulses are retrans-mitted by a regenerative repeater are determined by a local oscillator synchronised to the digit rate, which must be extracted from the received waveform. Variations in the extracted frequency can cause a periodic vari-ation of the times of the regenerated pulses, which is known as *jitter* [33—35]. This is not a great problem as long as the tolerance to jitter of any subsequent equipment in a link exceeds the amount of jitter produced by preceding equipment. However, at the end of the link it causes phase modulation of the reconstructed sample, which produces distortion products in the analogue output signal. There can also be a long-term variation in the times of the regenerated pulses due to changes in propagation time. This is known as *wander*.

2.11 References

1 Bell Telephone Laboratories Staff: 'Transmission systems for communications' (Bell Telephone Laboratories, 1970) 4th edn
2 HILLS, M.T., and EVANS, B.G.' 'Transmission systems' (Allen and Unwin, 1973)
3 FREEMAN, R.L.; 'Telecommunication transmission handbook' (Wiley, 1981) 2nd end
4 SHANKS, P.H.: 'A new echo suppressor for long-distance communications', *Post Office Electr. Eng. J.*, 1968, **60**, 288
5 SONHI, M.M.: 'An adaptive echo canceller', *Bell Syst. Tech. J.*, 1967, **46**, pp. 497
6 JOHNSON, J.B.: 'Thermal agitation of electricity in conductors', *Phys. Rev.*, 1928, **32**, pp. 97
7 NYQUIST, H.: 'Thermal agitation of electric charge in conductors', *Phys. Rev.*, 1928, **32**, pp. 110
8 BENNETT, W.R.: 'Electrical noise' (McGraw-Hill, 1960)
9 BULL, C.S.: 'Fluctuations of stationary and non-stationary electron currents' (Butterworth, 1966)
10 CCITT Recommendation P53
11 COHRAN, W.T., and LAVINSKI, D.A.: 'A new measuring set for message circuit noise', *Bull Syst. Tech. J.*, 1960, **39**, pp. 911
12 CARTER, R.O.: 'Theory of syllabic compandors', *Proc. IEE*, 1964, **111**, pp. 503–513
13 WATT-CANTER, D.E., and WHEELER, L.K.: 'The Lincompex system for protection of HF radio telephone circuits', *Post Off. Electr. Eng. J.*, 1966, **59**, pp. 163
14 BLACK, H.S.: 'Modulation theory' (Van Nostrand, 1953)
15 TUCKER, D.G.: 'Modulators and frequency changers for amplitude-modulated line and radio systems' (Macdonald, 1953)
16 CCITT Recommendation G135
17 BENNETT, W.R. and DAVEY, J.R.: 'Data transmission' (McGraw-Hill, 1965)
18 LUCKY, R.W., SALZ, J., and WELDON, E.J.: 'Principles of data communication' (McGraw-Hill, 1968)
19 ALEXANDER, A.A., GRYB, R.M., and NAST, D.W.: 'Capabilities of the telephone network for data transmission', *Bell Syst. Tech. J.*, 1960, **39**, pp. 431
20 WILLIAMS, M.B.: 'The characteristics of telephone circuits in relation to data transmission', *Post Off. Electr. Eng. J.*, 1966, **59**, pp. 151
21 CCITT Recommendation V23
22 BREWSTER, R.L. (Ed.): 'Data communication and networks' (Peter Peregrinus, 1986)
23 TAUB, H., and SCHILLING, D.L.: 'Principles of communication systems' (McGraw-Hill, 1986) 2nd edn.

24 CATTERMOLE, K.W.: 'Principles of pulse code modulation' (Iliffe, 1969)
25 NYQUIST, H.: 'Certain topics in telegraph transmission theory', *Trans. AIEE*, 1928, **47**, pp. 617
26 FREEBODY, J.W.: 'Telegraphy' (Pitman, 1958)
27 CHITTLEBURGH, W.F.S., GREEN, D. and HEYWOOD, A.W.A.: 'A. frequency-modulated voice-frequency telegraph system', *Post Off. Electr. Eng. J.*, 1957, **50**, pp. 69
28 FLOOD, J.E.: 'Principles of multiplex communication' *in* SKWIR-ZYNSKI, J.K. (Ed.): 'New directions in signal processing', (Noordhoff, 1975) pp. 271–287
29 SKAUG, R., and HJELMSTAD, J.F.: 'Spread spectrum communication' (Peter Peregrinus, 1985)
30 SCHROEDER, M.R.: 'Vocoders: Analysis and synthesis of speech—30 years of applied speech research', *Proc. IEEE*, 1966, **54**, pp. 720–734
31 BULLINGTON, K., and FRASER, J.M.: 'Engineering aspects of TASI', *Bell Syst. Tech. J.*, 1959, **38**, pp. 353–364
32 CAMPANELLA, S.J.: 'Digital speech interpolation', *Comsat Tech. Rev.*, 1976, **6**, pp. 127–159
33 BYLANSKI, P., and INGRAM, D.G.W.: 'Digital transmission systems' (Peter Peregrinus, 1987)
34 BYRNE, C.J., KARAFIN, B.J., and ROBINSON, D.B.: 'Systematic jitter in a chain of digital regenerators', *Bell Syst. Tech. J.*, 1963, **43**, pp. 2679–2714.
35 KEARSEY, B.N., and McLINTOCK, R. N.: 'Jitter in digital telecommunications networks', *Brit. Telecom. Eng. J.*, 1984, **3**, pp. 108–116

Chapter 3
Transmission media
J. E. Flood
Aston University

3.1. Introduction

Circuits may be provided over open-wire lines, aerial cables or underground cables using metallic conductors or optical fibres, or over radio links. All of these media convey signals by means of electromagnetic waves. However, in the cases of metallic conductors and optical fibres, the medium is bounded and can be closely controlled. For radio transmission the medium is unbounded; it is provided by nature, subject to variations and cannot be controlled by engineers.

Transmission systems must be designed to be compatible with the properties of these media, which are described in this chapter. Line transmission systems are considered in Chapter 10 and optical-fibre systems in Chapter 11. Terrestrial radio links are considered in Chapter 12 and satellite communication systems in Chapter 13.

3.2. Lines

3.2.1. Line plant
Open-wire lines consist of bare copper conductors supported by insulators on crossarms mounted on poles. Compared with cables, they have the advantage of substantially lower attenuation, but this is increased by wet weather. Open-wire lines are also susceptible to electrical interference from all sources of natural or man-made electrical noise. Because open-wire lines are liable to mechanical damage from storms, they incur heavy maintenance costs. Open-wire circuits are still used in some rural areas where the number of circuits required is small (e.g. less than 10). In some countries, carrier systems have been used on open-wire lines to provide larger numbers of long-distance circuits [1].

Aerial cables involve lower maintenance costs than open-wire lines and prove cheaper unless the growth of the route in circuits per year is extremely small. They are more liable to damage by storms and man-made accidents than are underground cables. They are also subject to greater temperature variations and this must be taken into account in the engineering of transmission systems. However, an aerial cable is considerably cheaper than underground cable when a new route can use existing poles or when rocky terrain makes excavation expensive.

Underground cables have a higher cost of initial provision than overhead cables, but maintenance costs are considerably lower. Underground cables may be directly buried or ploughed in, or drawn into ducts. Direct burial is satisfactory if only one or two cables are required. However, if additional cables will be needed in the future, it is more economical to install multiway ducts having capacity for several cables, to avoid the costs of subsequent excavation and reinstatement. When large numbers of cables are required in cities, it may even be economical to bore tunnels.

Multipair cables [2, 3] have traditionally used copper conductors with paper insulation sealed in a lead sheath. Although the conductors are separated by paper wrappings, the dielectric surrounding them mostly consists of air, thus minimising capacitance between wires. While paper insulation has good dielectric properties, its insulation resistance drops rapidly if water or water vapour penetrates the cable. The cable sheath protects the cable core against water and humidity and against mechanical damage during transport and installation. Many cables are now kept filled with dry air under pressure. If a leak develops in the sheath, the higher pressure inside the cable keeps water out and the escape of air enables the presence of the fault to be detected at the exchange [4].

Use has also been made of aluminium conductors [5] because of the more stable price of this metal. To obtain the same loop resistance, aluminium conductors must have a diameter about 25% larger than copper conductors. Polyethylene is now used as an alternative to paper insulation [6]. It has good resistance to moisture. However, it has a higher permittivity than paper insulation; the wall thicknesses of the insulation and the diameter of the cable core are thus greater than for air-spaced paper-insulated cables having the same mutual capacitances. Because of the high cost of lead, aluminium and polyethylene are now the principal materials for cable sheaths. Although polyethylene has advantages over lead in terms of cost and weight, it does permit the ingress of some water vapour. This disadvantage is overcome by using a polyethylene sheath which incorporates an aluminium-foil water barrier.

During cable manufacture, the wire is first insulated by being wrapped with a helix of paper tape or by polyethylene being extruded around it. The insulated wires are then twisted into pairs or star quads, the rate of twist being different for different pairs. A *star quad* is formed by twisting four wires together in such a manner that the two wires lying directly opposite each other form a pair. The complete cable core is then formed by stranding pairs or quads in concentric layers.* Twisting the wires together reduces the mutual coupling between circuits to avoid crosstalk. Mutual

---

* For local-exchange-area cables, however, the pairs are usually first stranded into *units* containing a number of pairs [7]. This facilities the identification of circuits when cables branch off in a local distribution network.

decoupling between pairs in the same quad is not so easy to achieve as between pairs in different quads or pairs in twisted-pair cables. However, star quads occupy much less space than multiple-twin cable cores.

Finally, the sheath is extruded over the complete core. Sometimes the cable is still further protected by lapping the sheath with an armouring of steel wires or tapes. An armoured cable has increased transverse rigidity to enable it to resist pressure (e.g. for direct-buried cables and for submarine cables) and increased tensile strength (e.g. for self-supported aerial cables).

The attenuation of balanced-pair lines increases with frequency, and so does the crosstalk owing to mutual coupling between pairs in a cable. By arranging for each quad to have a different twist length from others in the cable and using great care during manufacture, it is possible to achieve smaller capacitance unbalances between pairs. It is thus possible to obtain multipair carrier cables that are suitable for analogue signals up to 250 kHz. Balanced pairs can transmit digital signals containing much higher frequencies (e.g. 2 Mbit/s PCM) because of their greater crosstalk immunity. At such high digit rates, it is preferable to use separate cables for the two directions of transmission in order to minimise the crosstalk. If a single cable is used, the two directions are chosen at opposite sides of the cable, with unused pairs between them to act as a screen. Alternatively, a cable may be used which has a transverse metal screen to separate the two directions of transmission.

For higher frequencies, it is necessary to use a different form of construction. This consists of an inner conductor surrounded by a concentric screen. The arrangement is known as a *coaxial pair* or 'tube'. To minimise dielectric losses, the insulation in a coaxial pair consists mainly of air. The conductors are separated either by polyethylene discs spaced at regular intervals or by a continuous layer of expanded polyethylene (which contains a large number of non-intercommunicating air cells). At high frequencies, the skin effect causes the line currents to flow on the outer surface of the inner conductor and the inner surface of the outer conductor. Thus, there is no radiation of the signal. Any interference currents flow on the outer surface of the outer conductor and therefore do not interfere with the signal current. This screening effect is ineffective at low frequencies, so coaxial cables are not normally used for frequencies below 60 kHz. The upper limit of frequency is determined only by attenuation, which increases approximately in proportion to the square root of frequency, and by uniformity of the structure which governs the generation of reflections from discontinuities.

Two sizes of coaxial pair are in common use. One has an inner conductor of 1·2 mm diameter and an inner diameter of the outer conductor of 4·4 mm and is known as 1·2/4·4 mm coaxial pair. The other has corresponding diameters of 2·6 mm and 9·5 mm and is known as 2·6/9·5 mm coaxial pair. Cables are manufactured having 4, 6, 8, 12, 18 and 28 coaxial pairs. The interstices between the coaxial tubes are usually filled with ordinary pairs or quads to make up a core of circular cross-section. These interstitial pairs can be used for audio-frequency circuits, or for control circuits between the terminal stations and repeaters of a coaxial-cable system.

Submarine cables use only a single coaxial pair. These cables are subjected to a very high pressure when laid on the bed of the ocean. A dielectric of solid polythene is therefore used instead of air spacing. This results in higher attenutation per km than for land coaxial cables of the same size. Early submarine telephone cables used conventional external armouring wires to provide the high tensile strength needed to withstand the weight of the cable during laying. Later transoceanic cables are of the light-weight type. The tensile strength is provided by steel wires inside the copper tapes that provide the inner conductor. The outer sheath is of polythene, thus resulting in a much reduced weight. However, it is still necessary to use conventional armouring for the portions of each cable which are close to shore, to provide protention against damage by trawls and ships' anchors.

3.2.2 Line theory

The *primary coefficients* of a uniform transmission line are:

R = resistance in ohms per unit length
G = leakance in siemens per unit length
L = inductance in henries per unit length
C = capacitance in farads per unit length

A steady-state solution to the partial differential equations for an infinitely-long line [8, 9] is given by

$$V(x) = V e^{-Px} \qquad (3.1a)$$

$$I(x) = I e^{-Px} \qquad (3.1b)$$

$$I = \frac{V}{Z_0} \qquad (3.1c)$$

where $V(x)$ and $I(x)$ are the voltage and current at a distance x from the source feeding power into the line and

$$P = [(R + j\omega L)(G + j\omega C)]^{1/2} \qquad (3.2a)$$

$$Z_0 = \left[\frac{R + j\omega L}{G + j\omega C} \right]^{1/2} \qquad (3.2b)$$

The *secondary coefficients* P and Z_0 are called the *propagation coefficient* and the *characteristic impedance* of the line, respectively.

The propagation coefficient P is complex, so we can write

$$P = \alpha + j\beta$$

The real part α of the propagation coefficient thus gives the attenuation of the line in nepers per unit length; it is therefore called the *attenuation coefficient*. The imaginary part β gives the phase in radians per unit length; it is therefore called the *phase coefficient*. The phase of the transmitted wave changes by 2π in a distance $\lambda = 2\pi/\beta$; this is thus the wavelength. The velocity of propagation is given by

$$v = f\lambda = \omega/\beta \qquad (3.3)$$

In general, the attenuation and velocity vary with frequency. The attentuation/frequency characteristic of a typical audio cable is shown in Fig. 3.1a.

Since $I(x) = V(x)/Z_0$, if a line of finite length is terminated in impedance Z_0, the voltages and currents at any point on the line are identical with those in an infinite line, given by eqns. 3.1a–c. Such a line is said to be *correctly termined.*

In practice, it is not always possible to obtain a terminating impedance matching Z_0 over the complete band of frequencies to be transmittted. A reflected wave is then produced at the termination. The ratio of reflected voltage to incident voltage and reflected current to incident current is called the *reflection coefficients* ρ. For a terminating impedance Z_L, it can be shown [8, 9] that

$$\rho = \frac{Z_L - Z_0}{Z_L + Z_0} \qquad (3.4)$$

The reciprocal of this ratio expressed in decibels (i.e. $-20 \log_{10} |\rho|$) is called the *return loss.* Obviously, for $Z_L = Z_0$, $\rho = 0$ and the return loss is infinite. For a short circuit, $\rho = -1$; thus the reflected wave is in antiphase with the

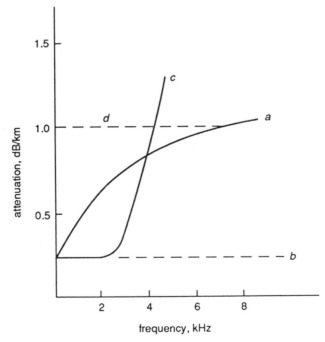

Fig. 3.1   Attenuation/frequency characteristics of an audio-frequency cable

a Unloaded line
b Continuous loading
c Lumped loading
d Equalised line

incident wave and no voltage appears at the end of the line. For an open circuit, $\rho = +1$; thus the reflected wave is in phase with the incident wave and doubles the voltage at the end of the line.

Reflections may be produced, not only by incorrect terminations, but also by impedance irregularities at intermediate points on the line. This causes attentuation and delay distortion, as shown in Section 2.3.3. Reflections also provide a convenient method of locating line faults. In pulse-echo testing, a short pulse is sent into the line and the impedance irregularity due to the fault causes a reflection. The time interval between the pulse being sent and the reflected pulse returning to the sending end corresponds to twice the distance to the fault.

From eqn. 3.2a, the attenuation and phase coefficients are

$$\alpha = [\tfrac{1}{2}(R^2 + \omega^2 L^2)^{1/2}(G^2 + \omega^2 C^2)^{1/2} + \tfrac{1}{2}(RG - \omega^2 LC)]^{1/2}$$

$$\beta = [\tfrac{1}{2}(R^2 + \omega^2 L^2)^{1/2}(G^2 + \omega^2 C^2)^{1/2} - \tfrac{1}{2}(RG - \omega^2 LC)]^{1/2}$$

These expressions are complicated, but the following approximations are valid in different frequency regions:

(*a*) For low frequencies (e.g. for telegraphy):

$$\omega L \ll R \quad \text{and} \quad \omega C \ll G, \quad \text{then}$$

$$\alpha = \sqrt{RG}, \qquad \beta = 0, \qquad Z_0 = \sqrt{R/G}$$

(*b*) For audio frequencies:

$$\omega L \ll R \quad \text{but} \quad \omega C \gg G, \text{ then}$$

$$P = \sqrt{j\omega CR}; \quad \text{therefore } \alpha = \beta = \sqrt{\tfrac{1}{2}\omega CR}$$

$$Z_0 = \sqrt{R/j\omega C} = \sqrt{R/\omega C}\ \underline{/-\pi/4}$$

Thus, the attenuation increases as the square root of frequency. The characteristic impedance is proportional to $1/\sqrt{f}$ and is capacitive.

(*c*) For radio frequencies:

$$\omega C \gg G \quad \text{and} \quad \omega L \gg R, \quad \text{then}$$

$$\alpha = [\tfrac{1}{2}\omega C]^{1/2}[(R^2 + \omega^2 L^2)^{1/2} - \omega L]^{1/2}$$

$$= [\tfrac{1}{2}\omega^2 LC]^{1/2}[1 + R^2/2\omega^2 L^2 - 1]^{1/2}$$

$$= \tfrac{1}{2}R\sqrt{C/L}$$

$$\beta = [\tfrac{1}{2}\omega^2 LC]^{1/2}[(1 + R^2/\omega^2 L^2)^{1/2} + 1]^{1/2}$$

$$= \omega\sqrt{LC}$$

$$Z_0 = \sqrt{L/C}$$

Thus, the characteristic impedance is purely resistive and independent of frequency. The velocity of propagation (ω/β) is also independent of frequency. The attenuation is not independent of frequency because skin effect reduces the depth of penetration of current and so causes R to increase with frequency. It can be shown [10, 11] that $R \propto \sqrt{\omega}$, so α is also proportional to $\sqrt{\omega}$.

Attenuation increases as the square root of frequency over a very wide range of frequencies (i.e. regions *b* and *c*). The increase of attenuation with frequency can be offset, over the required band, by using an equaliser. This results in attenuation that is independent of frequency but equal to the attenuation of the line at the highest frequency in the band, as shown in Fig. 3.1*d*. When it is necessary to preserve the waveshape of the signal (e.g. for television or digital transmission) the equaliser must correct phase distortion in addition to attenuation distortion.

A special case, which was first noted by Heaviside, arises when

$$LG = CR \tag{3.5}$$

If this condition is substituted in eqn. 3.2*a*, then

$$\alpha = \sqrt{RG} \qquad v = 1/\sqrt{LC}$$

Thus, both attenuation and velocity of propagation are independent of frequency. Eqn. 3.5 is therefore known as the *distortionless condition*.

For practical cables, the left-hand side of eqn. 3.5 is much smaller than the right-hand side. It is not possible to reduce *R* or *C*, since these will already have been made as small as is practicable. It is undesirable to increase *G*, since this increases the attenuation. It is possible to increase *L* to satisfy condition 3.5; this is called *loading*. As shown in Fig. 3.1*b*, it reduces the attenuation at all frequencies to that of the unloaded cable at low frequencies. (In contrast, equalisation increases the attenuation at all frequencies to that at the highest frequency in the required band, as shown in Fig. 3.1*d*.)

Loading can be done uniformly (e.g. by wrapping the conductors with tapes of magnetic material); this reduces the attenuation to a low value which is independent of frequency, as shown in Fig. 3.1*b*. This technique, called *continuous loading*, is expensive and rarely used. In practice, the inductance of audio cable pairs can be artificially increased by means of coils added at regular intervals. This is known as *lumped loading*. Since the added inductance is lumped instead of being distributed, this has the effect of inserting a low-pass filter in the line, as shown in Fig. 3.1*c*. It can be shown [8] that the cut-off frequency is $f_0 = 1/\pi \sqrt{L_S C_S}$, where L_S and C_S are, respectively, the total inductance and capacitance of a loading-coil section. In order to obtain an adequate cut-off frequency for telephony, it is only possible to increase the inductance by a factor of about 100, whereas an increase of about 1000 times would be required to satisfy eqn. 3.5.

Audio cables have extensively used 88 mH loading coils at a spacing of 1·83 km (2000 yards). This gives a cut-off frequency of 3·4 kHz for 0·9 mm conductors (20 lb/mile) and 3·9 kHz for 0·63 mm conductors (10 lb/mile). The loading reduces the attenuation at 800 Hz from 0·71 dB/km to 0·23 dB/km and from 1·05 dB/km to 0·45 dB/km, respectively. Many such circuits have now been converted to PCM transmission by replacing the loading coils with regenerative repeaters at the same locations. Thus, the traditional standard for loading has influenced the design of modern PCM systems.

Although the attenuation of a loaded cable is almost independent of frequency up to the cut-off frequency, the sharp frequency cut-off does introduce phase distortion. This has negligible effect on speech transmission, but can be harmful for data transmission [12]. The addition of inductance also reduces the velocity of propagation. Typical values are 220 000 km/s for unloaded cables and only 22 000 km/s for loaded cables. The resulting increase in propagation delay makes it undesirable to use loaded cables for very long circuits.

3.3. Radio

3.3.1 General

The properties of radio links are determined by the mechanisms affecting the propagation of the radio waves and these depend on the frequencies used for transmission. The different bands of radio frequencies and the standard nomenclature used for them are given in Table 3.1.

In radio transmission, the same basic principles apply as in line transmission, but in three dimensions [13, 14]. Propagation along a uniform line is replaced by radiation into free space. Wave fronts are spherical and reflections from discontinuities in the medium occur at other than right angles to the direction of propagation. The expressions corresponding to eqns. 3.2a and b for characteristic impedance and propagation coefficient are:

$$Z_0 = \left[\frac{j\omega\mu_0}{\delta + j\omega\varepsilon_r\varepsilon_0} \right]^{1/2} \tag{3.6a}$$

and

$$P = \alpha + j\beta = [j\omega\mu_0(\delta + j\omega\varepsilon_r\varepsilon_0)]^{1/2} \tag{3.6b}$$

Table 3.1 Nomenclature of radio frequencies

Classification	Abbreviation	Frequency	Wavelength
		kHz	km
Very-low frequencies	VLF	3—30	100—10
Low frequencies	LF	30—300	10—1
			m
Medium frequencies	MF	300—3000	1000—100
		MHz	
High frequencies	HF	3—30	100—10
Very-high frequencies	VHF	30—300	10—1
			cm
Ultra-high frequencies	UHF	300—3000	100—10
		GHz	
Super-high frequencies	SHF	3—30	10—1
Extra-high frequencies	EHF	30—300	1—0·1

The general term 'microwave' is used to describe radio waves of frequencies higher than 1 GHz.

where

$$\delta = \text{conductivity, siemens}/m$$

$\varepsilon_r = \text{relative permittivity}$

$\mu_0 = \text{permeability of free space, given by}$

$$\mu_0 = 4\pi \times 10^{-7} \text{ H/m}$$

$\varepsilon_0 = \text{permittivity of free space, given by}$

$$\varepsilon_0 = 8 \cdot 854 \times 10^{-12} \text{ F/m}$$

For free space, $\delta = 0$ and $\varepsilon_r = 1$. Thus

$$Z_0 = \sqrt{\mu_0/\varepsilon_0} = 377\Omega$$

$$P = j\omega\sqrt{\mu_0\varepsilon_0} = j\omega\beta$$

and

$$v = \omega/\beta = 1/\sqrt{\mu_0\varepsilon_0} = 3 \times 10^8 \text{ m/s}.$$

In an electromagnetic wave the electric and magnetic field vectors (E and H) are at right angles to each other and to the direction of propagation. The field stength is given by the RMS value of E in volts/m. (Eqns. 3.6a and b show that E is the significant component, since δ and ε are associated with it.) Thus, an EMF of $E\delta l$ is induced in an element of length δl in a receiving antenna which is in the direction of E.

If the electric field is vertical, the wave is said to be *vertically polarised.* (The magnetic field is then horizontal.) If the electric field is horizontal, the field is *horizontally polarised.* (The magnetic field is then vertical.) However, passage through the ionosphere can cause the polarisation to rotate. If the amplitude of the E and H vectors remain constant, their extremities trace a spiral whose projection on a plane normal to the direction of propagation is a circle. The wave is then said to be *circularly polarised.* If the amplitude of E and H decay with distance, their extremities trace a spiral whose projection is not a circle but an elllipse. The wave is then said to be *ellipitically polarised.* Vertically and horizontally polarised waves can be radiated by a vertical antenna and a horizontal antenna respectively. A vertically-polarised wave will induce an EMF in a vertical receiving antenna and a horizontally-polarised wave will induce an EMF in a horizontal antenna.*

The simplest form of radio propagation is direct propagation from a transmitting antenna to a receiving antenna through free space. In practice, however, perfect direct propagation is not obtained because of the presence of the earth and imperfections of its atmosphere. Propagation between a transmitting and receiving antenna can take place over a number of different

---

* A vertically-polarised wave induces zero EMF in a horizontal dipole and a horizontally-polarised wave induces zero EMF in a vertical dipole. It is therefore possible to transmit two signals using the same carrier frequency without mutual interference. This is known as *dual polarisation.*

paths, which are illustrated in Fig. 3.2. The received signal can therefore be the resultant of a number of the following:

(a)   a *direct wave* (i.e. free-space propagation)
(b)   a *reflected wave* (reflected from the ground)
(c)   a *sky wave* (reflected by ionised layers above the earth, known as the *ionosphere*)
(d)   a *surface wave* caused by diffraction round the earth associated with currents induced in the ground).
(e)   a wave reflected by variations in refractive index caused by local perturbations in the atmosphere (tropospheric scattering)

All these mechanisms can be present for propagation over a radio link. However, some of them are neglible in certain frequency ranges, so that very different propagation characteristics are obtained (see Sections 3.3.3 to 3.3.6).

At a boundary between two media, refraction and reflection can occur. The *refractive* index of a medium is

$$n = c/v$$

where

$$c = \text{velocity in free space } (3 \times 10^8 \text{ m/s})$$
$$v = \text{velocity in the medium}$$

For a pure dielectric, $v = c/\sqrt{\varepsilon_r}$, so $n = \sqrt{\varepsilon_r}$ and the refractive index is greater than unity. In the ionosphere, however, ionisation reduces n to be slightly less than unity and makes it frequency-dependent.

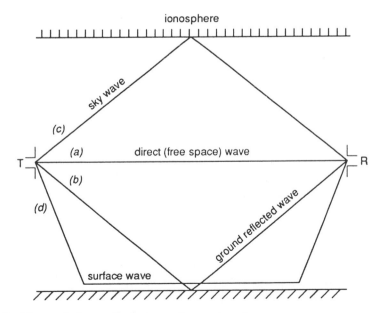

Fig. 3.2   Transmission paths between two antennas

If the direction of propagation of the incident wave in the first medium is at an angle θ_1 to the boundary between the media and the direction of propagation of the refracted wave in the second medium is at angle θ_2 to the boundary, then

$$\frac{\cos\theta_2}{\cos\theta_1} = \frac{v_2}{v_1} = \frac{n_1}{n_2} \qquad (3.7)$$

If $\cos\theta_1 = n_2/n_1$, then $\theta_2 = 0$. Thus, if the angle of incidence is less than this critical angle, total reflection occurs.

The concept of impedance is useful in considering reflection. The impedance seen by the incident wave is the component normal to its direction of propagation. At the boundary this is $Z_0 \sin\theta_1$ (where Z_0 is given by eqn. 3.6a). Thus, if the angle of incidence is small, this impedance is low. The situation is analogous to a short-circuited transmission line, for which the reflection coefficient (from eqn. 3.4) is -1. Thus, a flat surface gives a reflection factor approaching unity and a phase reversal.

3.3.2 Free-space propagation

If an antenna could radiate power uniformly into free space, the attenuation would be simply due to expansion of the wavefront as the distance from the transmitter increases. At distance r from the transmitter, the power is uniformly distributed over a sphere of radius r metres. Thus, the power density $P(r)$ at distance r is

$$P(r) = P_T/4\pi r^2 \text{ watts/m}^2$$
$$= E^2/377 \text{ watts/m}^2$$

where

$$P_T = \text{total radiated power, W}$$
$$E = \text{RMS electric field, V/m}$$

If a receiving antenna has an effective area A_R located on the surface of the sphere of radius r, the power received is

$$P_R = A_R P(r) = A_R P_T/4\pi r^2 \text{ watts} \qquad (3.8)$$

If the transmitting antenna is directional, it transmits more power towards the receiver than an isotropic antenna (and less in other directions). The gain g_T of the transmitting antenna is defined as

$$g_T = \frac{\text{Power density in direction of maximum radiation}}{\text{Power density when same power is radiated by isotropic antenna}}$$

It can be shown [15] that

$$g_T = 4\pi A_T/\lambda^2 \qquad (3.9)$$

where A_T is the effective area of the transmitting antenna.

Thus, the power received (from eqns. 3.8 and 3.9) is

$$P_R = g_T A_R P_T/4\pi r^2$$

Now, eqn. 3.9 can also be used to define the gain of the receiving antenna. Therefore,

$$P_R = g_T(\lambda^2 g_R/4\pi)P_T/4\pi r^2$$
$$= g_T g_R(\lambda/4\pi r)^2 P_T$$

The total path loss L is thus given by

$$L = 20 \log_{10}(4\pi r/\lambda) - G_T - G_R \text{ decibels} \qquad (3.10)$$

where G_T and G_R are the gains of the transmitting and receiving antennas in decibels.

3.3.3 Low-frequency propagation

In the VLF band (i.e. below 30 kHz), the wavelengths are so long that the simple optical ray theory is invalid. The wavelengths are so long that they are comparable to the height of the lowest ionospheric layer (approximately 50 km). The ionosphere and the surface of the earth thus act as conducting planes forming a waveguide [16]. Consequently, VLF signals can have worldwide coverage. This band is used for telegraph transmission, for navigational aids and for distributing standard frequencies.

In the LF band (i.e. 30 to 300 kHz), propagation by the surface wave provides the propagation mechanism. A vertically-polarised wave induces a current in the earth and a component of electric field parallel to the ground is thus produced. This causes the wavefront to tilt forward and enables it to pass round the curvature of the earth. The earth currents take power from the wave and cause attenuation. This diffraction also allows the wave to bend round obstacles. Stable transmission is obtained over distances up to about 1500 km. This band is used for sound broadcasting.

In the MF band, there is also surface-wave propagation. However, the amount of diffraction is proportional to the wavelength, λ, so the strength of the surface wave decreases with frequency. Consequently, MF transmission provides shorter ranges than LF transmission. The sky wave also has some effect and this increases the coverage, particularly at night when the absorption of the ionosphere is a minimum. This band is also used for sound broadcasting.

3.3.4 High-frequency propagation

Since the strength of the surface wave decreases with frequency, it has negligible effect in the HF band and propagation is mainly by means of the sky wave reflected by the ionosphere. The ionosphere is formed by ultra-violet and X-ray emission from the sun ionising the molecules of the upper atmosphere. There are several layers in which the ionisation density is a maximum; these are designated the D, E, F_1 and F_2 layers in order of height. The ionisation causes both absorption and refractive bending. The former causes attenuation which is proportional to $1/f^2$. The latter can cause the wave to return to earth, as shown in Fig. 3.2, at a distance from the transmitter known as the *skip distance*.

It can be shown [13] that the refractive index of ionised air is

$$n = (1 - 81N/f^2)^{1/2}$$

where N is the number of free electrons or ions per mm^3. From eqn 3.7, the critical angle for which total reflection occurs (and sky-wave transmission can be obtained), is

$$\cos \theta_i = (1 - 81N/f^2)^{1/2} \qquad (3.11)$$

Thus, there is a *maximum usable frequency* (MUF) above which sky-wave transmission is not possible. From eqn. 3.11 this is given by

$$MUF = 9\sqrt{N} \ \text{cosec} \ \theta_i$$

Since attenuation is proportional to $1/f^2$, the strongest signals at the receiver are obtained at the highest possible frequency. It is therefore desirable to operate as close to the MUF as possible. However, the properties of the ionosphere vary from hour to hour and from day to day throughout the year. Transmitting stations therefore need to be able to use several frequencies in order to provide reliable communication. Observations carried out at many stations throughout the world result in charts being prepared to enable the optimum frequency to be predicted for communication between any two points at a given time.

Fluctuations in the ionosphere cause the received signal to be made up of a number of components whose path lengths differ. When interference between these components results in a reduction of received signal strength, *fading* occurs. There are both slow fades caused by diurnal and seasonal variations in the ionosphere and rapid fades due to random fluctuations. These can cause reductions in received signal strength of up to 30 dB. Since different frequency components travel over different path lengths, modulated signals cans can have severe attenuation distortion, as explained in Section 2.3.3. This is known as *selective fading*. Fading is especially severe on paths exceeding 400 km, on which transmission takes place by multiple hops, as shown in Fig. 3.3

Fading can be combatted by *diversity reception* [18, 19]. In space-diversity systems, two or more antennas are used and the output is selected from the antenna receiving the strongest signal. The method relies on the theory that, since fading is due to small changes in the path lengths of multipath signals, it is unlikely to occur simultaneously at two separated receiving antennas. Since the path difference producing cancellation depends on frequency, protection against fading can also be obtained by using two or more carrier frequencies received on a single antenna. This is known as *frequency diversity* reception.

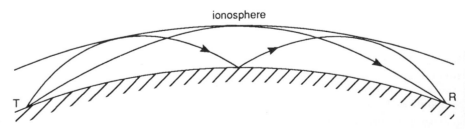

Fig. 3.3   Multiple paths involving reflections from the ionosphere

HF radio was the principal means of long-distance international communication, apart from submarine telegraph cables, until submarine telephone cables were introduced. It is still used for routes where there is insufficient traffic to justify a submarine cable or a satellite link. HF radio is still needed for long-distance communication to ships and aircraft, although satellites are now also used.

3.3.5 VHF and UHF propagation

Since the sky-wave mechanism does not operate at frequencies much above 30 MHz, radio communication at VHF and UHF depends on direct free-space wave propagation. If the wave travelled in a true straight line, the range of communication would be strictly limited to 'line of sight' distances. However, the path of the wave is affected by variation with height of the refractive index of the troposphere [13, 20]. The waves are refracted as they travel through the troposphere, as shown in Fig. 3.4a, and this increases the range beyond the optical horizon. In planning routes, this effect can be allowed for by assuming that the earth has an *effective radius* which is approximately 4/3 times its actual radius [13] as shown in Fig. 3.4b.

In addition to the direct wave, there is also a wave reflected from the surface of the earth, as shown in Fig. 3.5a. The received field strength is due to the resultant of these two waves. Reflection from the earth produces a phase reversal, as shown in Section 3.3.1. If the difference between the two path lengths is an odd number of half wavelengths the waves add, but for an even number they cancel. Thus, as shown in Fig. 3.5b, the signal passes through a series of maxima and minima as the distance from the transmitter increases, until the difference between the path lengths become less than half a wavelength. The field strength then decays steadily as cancellation of the two waves becomes more nearly complete.

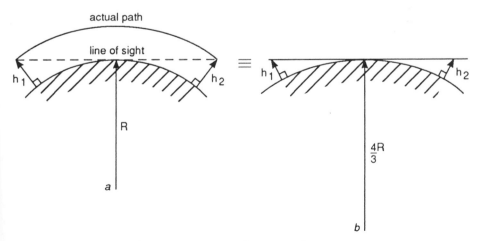

Fig. 3.4 Refraction of VHF wave in earth's atmosphere

a Actual conditions
b Equivalent line-of-sight path with modified earth's radius

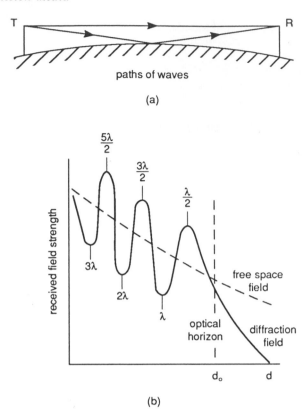

paths of waves

(a)

(b)

Fig. 3.5 VHF transmission by direct wave and wave reflected by ground

a Paths of waves
b Variation of field strength with distance (d) from transmitter

Transmission in the VHF band and the lower part of the UHF band is used mainly for television broadcasting and for short-range mobile communications.

3.3.6 Microwave propagation

Line-of-sight fixed radio links in telecommunication networks now use frequencies above 1 GHz, known as microwave frequencies. For these shorter wavelengths, highly-directional antennas can be used. If the antennas are mounted at a sufficient height, reflection from the ground is negligible and transmission approximating to free-space propagation can be obtained. The total path attenuation is then given by eqn. 3.10.

To obtain effective free-space propagation, the path profile must be planned carefully [21, 22, 23] to avoid intervening obstacles such as high ground, as shown in Fig. 3.6. The clearance between the direct ray and the highest obstacle should be sufficient to ensure that the difference between the path lengths of a reflected wave and the direct wave exceeds half a

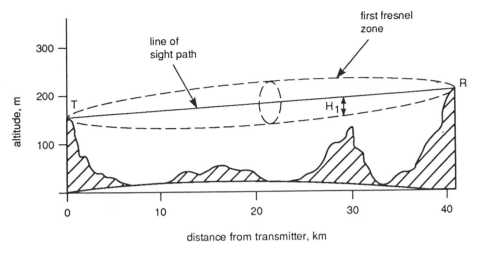

Fig. 3.6 Path profile for microwave link showing radius (F_1) of the first Fresnel zone

wavelength. This prevents interference between the waves causing cancellation. This minimum clearance distance, shown as H_1 in Fig. 3.6, is the radius of the first Fresnel zone. It can be shown [13] that

$$H_1 = [\lambda d_1(1 - d_1/d)]^{1/2} \qquad (3.12)$$

where d is the total path length and d_1 is the distance from the transmitter.

Fading occurs on microwave paths because of variations of refractive index due to temperature, humidity and pressure changes [20]. At frequencies above 10 GHz, the wavelength becomes comparable to the size of rain drops. Thus, at these frequencies, additional fading due to absorption is produced by rainfall [24]. Microwave transmission over distances beyond the optical horizon can be obtained by means of the mechanism known as *tropospheric scattering** [20]. It is believed to be caused by local perturbations in the atmosphere producing variations in refractive index. If a powerful transmitter sends a narrow beam to its horizon, it illuminates a region which is within line-of-sight of the receiving station, as shown in Fig. 3.7. A high-gain receiving aerial can obtain a signal refracted by the scatter region. The received signal is much less than would be obtained by free-space transmission and it is subject to considerable fading because the scattering is random. It is therefore necessary to use both space and frequency diversity reception [25, 26]. Bandwidths of several MHz can be obtained over distances up to 500 km.

For satellite comunication links [27, 28] it is necessary to transmit radio waves beyond the earth's atmosphere. Ideal free-space propagation conditions are then obtained. To escape through the ionosphere, frequencies

---

* The band 30 to 50 MHz is also used, employing scattering in the E layer. Narrow-band signals such as telegraphy are transmitted over distances of 2000 km.

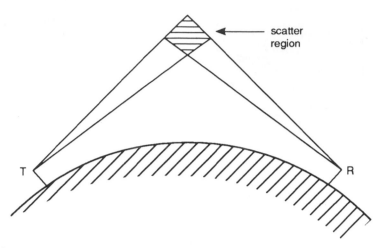

Fig. 3.7   Tropospheric scattering

well above 30 MHz must be used. However, absorption caused by water vapour occurs in the troposphere at frequencies near 26 GHz. Between these frequencies there is a 'window' which can be used for extra-terrestrial propagation. However, because of the high attenuation of the very long path length (36 000 km) of a satellite link, it is necessary to use antennas with very high gains. This restricts satellite systems to microwave frequencies.

Initially, satellite communication systems were used solely for intercontinental links and employed the 4 GHz and 6 GHz bands, which are also used for terrestrial microwave links. However, satellite systems are now also used for 'domestic' communications and direct broadcasting by satellite (DBS). Crowded conditions in the 4 and 6 GHz bands have led to the use of the 11 GHz and 14 GHz bands. Some use is also being made of frequencies below 4 GHz and above 14 GHz. At frequencies above 26 GHz there are some smaller windows, lying between various absorption bands due to water vapour and oxygen, which would also allow radio waves to pass through the troposphere.

3.4   Optical-fibre transmission

3.4.1   General

Electromagnetic waves can be propagated in a waveguide consisting of a dielectric bounded by another dielectric of lower permittivity instead of by a conductor. The use of such a waveguide, propagating waves of optical wavelengths, for telecommunication transmission was proposed by Kao and Hockham [29] in 1966. The practicability of such systems depends on the availability of materials for the guide having sufficiently low energy losses, and suitable optoelectronic devices for launching light waves into the guide and detecting them after transmission. A large research and development effort over the last twenty years has resulted in optical fibres with sufficiently low attenuation (typically <0·5 dB/km) and suitable solid-state transmitting devices (light-emitting diodes (LEDs) and lasers) and receiving devices (PIN and avalanche photodiodes). As a result, transmission over optical fibres has developed from a scientific curiosity to application in commercial systems. Today, most of the world's largest telephone operating organisations have systems in service and the major telecommunications equipment suppliers are manufacturing systems.

Transmission over optical fibres has several avantages over transmission by metallic conductors:

(a)   Since optical-fibre systems operate at much higher frequencies than coaxial-cable systems, they are potentially able to provide much wider bandwidth (e.g. 60 000 GHz).

(b)   The loss per km of optical fibres is now much lower than that of electrical cables operating at high frequencies. Thus, repeater sections can be much longer and even long-haul systems do not require any intermediate repeaters.

(c)   The fibres provide electrical isolation between transmitter and receiver.

(d)   Transmission is unaffected by electrical radiation. (Hence optical fibres can be used in noisy electrical environments.)

(e)   The light energy is contained within the fibre. Thus, no fibre-to-fibre crosstalk occurs.

(f)   Optical-fibre cables are smaller and lighter than electrical cables of the same channel capacity.

Disadvantages of optical fibres are:

(a)   Fibres must have very small diameters (about the size of a human hair). Care must therefore be taken to avoid subjecting them to mechanical stresses. In addition, making efficient connections to optical fibres and splices between them is more difficult than for metallic conductors.

(b)   Electric power cannot be sent over the fibres. However, this is no drawback when intermediate repeaters are not required.

(c)   Locating faults on cables is more difficult for optical fibres than for metallic conductors.

3.4.2. Fibres

An optical-fibre guide consists of a small-diameter cylindrical core of silica or glass surrounded by a cladding of lower refractive index [30, 31, 32, 33]. The usual type is silica-based fibre produced by deposition from the vapour phase (CVD fibre) [30, 31].

One form of fibre has a step change in refractive index between the core and cladding, as shown in Fig. 3.8a. A typical fibre has a silica core of 50 μm diameter surrounded by a glass cladding of 125 μm diameter.

Fig. 3.9 shows a light ray R_1 inside a cylindrical core which meets the surface at an angle θ_1. As shown by eqn. 3.7, there is total reflection of the ray if $\theta_1 < \theta_c$, where θ_c is the critical angle, given by

$$\cos \theta_c = n_2/n_1 \tag{3.13}$$

index
profile

typical ray paths

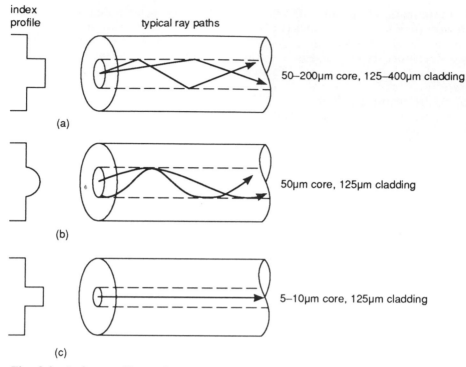

50–200µm core, 125–400µm cladding

(a)

50µm core, 125µm cladding

(b)

5–10µm core, 125µm cladding

(c)

Fig. 3.8   Index profiles and ray paths in optical fibres

a multimode step-index fibre
b multimode graded-index fibre
c single-mode step-index fibre

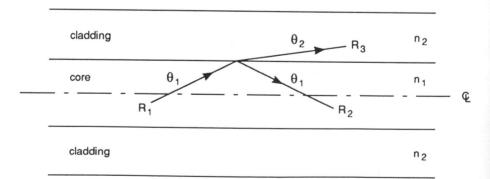

cladding

core

θ_2 R_3 n_2

θ_1 θ_1 n_1

R_1 R_2 ₵

cladding n_2

Fig. 3.9   Internal reflection of ray within fibre

where

n_1 = refractive index of the core

n_2 = refractive index of the surrounding cladding

(and $n_2 < n_1$).

All the power in ray R_1 is then reflected into ray R_2. However, if $\theta_1 > \theta_c$, some of the power in ray R_1 appears in an external refracted wave R_3. Thus, only rays having $\theta_1 < \theta_c$ will survive to travel for any considerable distance along the glass.

If the core has a plane end normal to its axis and is illuminated by a light source, then, as shown in Fig. 3.10, for a ray incident on the core at an angle θ_0 to its axis:

$$n_0 \sin \theta_0 = n_1 \sin \theta_1 = n_1 (1 - \cos^2 \theta_1)^{1/2}$$

If θ_1 is the critical angle, substituting from eqn. 3.13 gives

$$\sin \theta_0 = \frac{n_1}{n_0} [1 - n_2^2/n_1^2]^{1/2}$$

Thus, only waves that enter at angles less than the *acceptance angle* θ_0 will propagate within the fibre. When light enters the fibre from air, $n_0 = 1$ and $\sin \theta_0$ is called the *numerical aperture* (NA) of the fibre. Thus

$$NA = [n_1^2 - n_2^2]$$

$$\simeq n_1^{1/2}(2\Delta n)^{1/2}$$

where Δn is the index difference, $n_1 - n_2$.

If the diameter of the glass is large compared with the wavelength of the light, many different ray paths can be traversed, as shown in Fig. 3.8a. This is called *multimode propagation*. The propagation time for a ray at an angle θ_1 to the axis exceeds that of a ray parallel to the axis by a factor $1/\cos \theta_1$. Thus, if the end of the fibre is illuminated by a short pulse of light, the light will arrive at the far end as a pulse whose arrival time spreads between

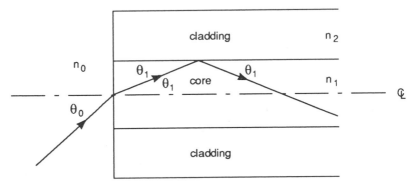

Fig. 3.10 Ray launched into fibre at angle less than acceptance angle

Ln_1/c and $Ln_1/c \cos \theta_c$, where L is the length of the fibre and c is the velocity of light in free space. The duration of the received pulse is thus:

$$\Delta t = \frac{Ln_1}{c} \left(\frac{n_1}{n_2} - 1 \right)$$

This dispersion of transmitted pulses increases with the length of the fibre; i.e. the bandwidth decreases. This is clearly disadvantageous for the design of wideband systems with long repeater sections. Nevertheless, multimode fibres are useful for short-distance systems of relatively low bandwidth. For example, if $n_1 = 1 \cdot 5$ and $\Delta n = 0 \cdot 01$, then $\Delta T/L = 34$ ns/km.

A method of preventing the large amount of dispersion that occurs in the simple multimode fibre is to use a *graded-index fibre*. The refractive index is made to vary continuously across the diameter as shown in Fig. 3.8b. Light in the outer regions of the core travels faster than light near its centre. Thus, the difference between the propagation times for steep-angle and shallow-angle rays is much smaller than for a fibre with uniform refractive index. It can be shown that the refractive index n should vary with radial distance r according to the law:

$$n = n_1 [1 - K(2r/d)^2]$$

A ray launched into the fibre at an angle to the axis oscillates sinusoidally with distance, as shown in Fig. 3.8b, and the propagation time is almost independent of the angle. In practice, it is difficult to match refractive index to the parabolic law exactly. Fortunately, large reductions in dispersion can be obtained with relatively poor approximations to this law.

Graded-index fibres provide a transmission path of almost uniform delay and have a core of reasonable diameter. Consequently, they were widely used in optical-fibre transmission systems until monomode-fibre cables became practicable.

If the core diameter of the fibre is so small (e.g. 5 or 10 μm) that it is comparable with the wavelength of light, only a single wave mode can propagate, as shown in Fig. 3.8c. This is called a *monomode fibre*. It is also desirable for the light to be provided by a coherent source (a laser) in order to obtain the high intensity needed to lauch sufficient power into such a small-diameter fibre.

For monomode propagation, the simple ray theory given above is inadequate; it is necessary to use field theory. It can be shown [31] that the necessary condition for monomode propagation is

$$r_1 < 2 \cdot 405 \lambda_0 / 2\pi (n_1^2 - n_2^2)^{1/2}$$

where λ_0 is the wavelength in free space, r_1 is the core radius and $(n_1^2 - n_2^2)^{1/2}$ is the numerical aperture of the fibre.

Although there is now no pulse dispersion due to multimode effects, small amounts of dispersion remain because of:

(a)   The dependence of n_1 and n_2 on wavelength λ_0, known as material dispersion or chromatic dispersion.

(b)   The dependence of the field distribution on λ_0, known as waveguide dispersion or modal dispersion.

Since practical sources emit light not at a single wavelength but over a range of wavelengths, these variations cause the components of a pulse to travel at different velocities and become dispersed in time. Thus, dispersion can be reduced by using a light source with a narrow spectrum. For a typical silica glass, the material dispersion at 0·85 μm may limit the signal bandwidth to 2·4 GHz for a kilometre length for a source having a spread in wavelength of 1 nm. Thus, for a laser having a spectral spread of 2 nm the bandwidth for 1 km is 1·2 GHz. For a light-emitting diode (LED) with a spectral spread of 33 nm, the bandwidth it only 70 MHz.

Unless the required bandwidth (or bit rate) is so large that dispersion is the limiting factor, the permissible distance between repeaters is limited by the attenuation of the fibre. Fig. 3.11 shows the variation of attenuation with wavelength for a typical monomode fibre. It has been shown that the loss is due to two causes:

(*a*) Density fluctuations frozen in when the molten glass solidifies. These produce scattering, which results in a loss that falls rapidly with increasing wavelength.

(*b*) Water contamination, which causes a peak of absorption at a wavelength of about 1·4 μm. There is also absorption produced by metallic ion impurities, which decreases rapidly with increasing wavelength.

Early systems operated in the infra-red region at wavelengths of about 0·85 μm giving attenuation of 2 or 3 dB/km. This wavelength matched the response of available aluminium–gallium-arsenide devices. However, sources have now been developed to operate in the region of 1·3 μm where the attenuation of very-high-purity fibres is less than 0·5 dB/km. This enables systems to operate with repeater spacings of about 40 km. As a result, few

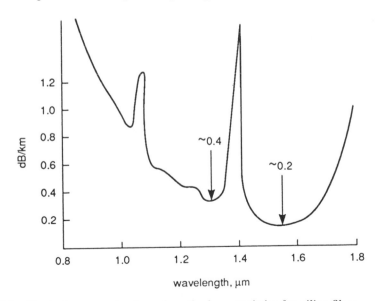

Fig. 3.11 Typical attenuation/wavelength characteristics for silica fibre

repeaters are needed and these can all be situated in buildings. Many inland routes do not need any repeaters and optical-fibre submarine cables are practicable [34].

The reduction of scattering attenuation with wavelength provides an incentive to develop systems operating at wavelengths of about 1·5 μm to achieve an attenuation of about only 0·2 dB per km. However, as shown in Fig. 3.12, the material dispersion per km increases at wavelengths beyond 1·3 μm [35]. Also, the increased repeater spacing made possible by the reduced attenuation will further increase the overall dispersion of a repeater section. Fortunately, Fig. 3.12 shows that the waveguide distortion is of the opposite sign (i.e. it makes delay increase with wavelength instead of decreasing). Therefore, appropriate choice of core diameter and index difference enables the waveguide and material dispersions to cancel at the wavelength of interest [35], as shown in Fig. 3.12. Such dispersion-shifted monomode fibres have already been made.

Since the scattering varies as λ^{-4}, materials with the fundamental absorption peak at much longer wavelengths may be possible [36] which in theory could offer attenuations of the order 0·001 dB/km. Other possibilities [36] include the use of wavelength-division multiplexing, optical amplifiers and coherent-detection systems in place of the intensity-modulated transmitters and receivers currently used. Thus, although the high performance of current monomode systems appears to indicate a mature technology, rapid development is likely to continue.

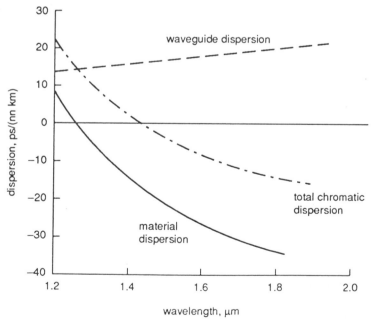

Fig. 3.12   Variation of material dispersion and waveguide dispersion, giving zero total dispersion near a wavelength of 1.5 μm.

3.4.3 Cables and splices

Optical fibres are coated in plastic to give them adequate mechanical strength. They can then be stranded into a cable in a similar manner to electrical conductors. However, metal wires or Kevlar cords are also included to enable the cable to have sufficient strength in tension. Typically, cabling may add 0·1 dB/km to the attenuation of the cable [31, 32, 33].

When completed, optical-fibre cables are as rugged as their copper-wire counterparts and they can withstand the normal abuses of installation. Cables are now available for use in ducts, for direct burial or for suspension from poles. Because they are smaller and lighter, optical-fibre cables can be reeled in longer lengths than conventional cables (e.g. a kilometre or more). This makes installation easier, since fewer splices are needed.

Permanent connections between fibres are made by splicing. Success depends critically on accurate alignment of the two fibres (because of their small diameter) and on square flat ends. Fortunately, good-quality ends can be obtained by cleaving the fibres under a mild bending stress [30]. In one method of jointing, the ends are accurately positioned by placing them in a V-shaped groove in a metal block and then joined with a quick-setting epoxy resin. Splicers are now available that fuse the ends together with a precisely-controlled gaseous arc [33]. Good splices can have insertion losses of less than 0·1 dB and a tensile strength of about half that of the fibre itself. Fusion splicing is thus rapidly displacing other techniques.

At the ends of the cable, where entry into a electronic equipment is required and the connection cannot be permanent, precision connectors are needed. Several organisations have developed components which connect two fibres with an insertion loss of less than 0·5 dB [33].

3.5. References

1 BOYCE, C.F.: 'Open-wire carrier telephone transmission (Macdonald & Evans, 1953)
2 LAWRENCE, A.J. and DUFOUR, I.G.: 'Audio junction cables', *PO Elect. Eng. J.*, 1977, **69**, p. 239
3 CCITT: Recommendation G541
4 MYNOTT, C.R., and HARCOURT, E.N.: 'A review of pressurization in the British Post Office network', *PO Elect. Eng. J.*, 1976, **69**, p. 169
5 PRITCHETT, J., and STENSON, D.W: 'Aluminium alloy as a conductor for local network cables', *PO Elect. Eng. J.*, 1975, **68**, p 131
6 WINDELER, A.S.: 'Design of polyethylene insulated multipair telephone cable', *Trans. AIEE*, Pt. 1 (Communcations and Electronics), 1960, (46), p 736
7 SPENCER, H.J.C.: 'Some principles of local telephone cable design', *PO Elect. Eng. J.*, 1970, **63**, p. 164
8 'Handbook of line communications: Vol. 1, (HM Stationery Office, 1958)
9 CHIPMAN, R.A.: 'Theory and problems of transmission lines' (McGraw-Hill, 1968)
10 GRIVET, P.: 'The physics of transmission lines at high and very high frequencies' (Academic Press, 1970)

11 JARVIS, R.F.J., and FOGG, G.H.: 'Formulae for the calculation of the theoretical characteristics and design of coaxial cables', *PO Elect. Eng. J.*, 1937, **30**, p. 138

12 JONES, I.O., and ADCOCK, R.C.: 'Group delay in the audio data network', *ibd.*, 1971, **64**, p. 9

13 GLAZIER, E.V.D., and LAMONT, H.R.L.: 'Transmission and propagation'. Services Textbook of Radio, Vol. **5**, (HMSO, 1958)

14 WAIT, J.R.: 'Introduction to antennas and propagation' (Peter Peregrinus, 1986)

15 KRAUS, J.D.: 'Antennas' (McGraw-Hill, 1950)

16 WATT, A.: 'VLF radio engineering' (Pergamon Press, 1967)

17 BULLINGTON, K.: 'Radio propagation fundamentals', *Bell Syst. Tech. J.*, 1957, **36**, p. 593

18 BETTS, J.A.: 'HF communications' (English Universities Press, 1967)

19 MASLIN, N.: 'HF communications: a systems approach' (Pitman, 1987)

20 HALL, M.P.M.: 'Effects of the troposphere on radio communication' (Peter Peregrinus, 1986)

21 CARL, H.: 'Radio relay systems' (Macdonald, 1966)

22 DUMAS, K., and SANDS, L.: 'Microwave system planning' (Hayden, 1967)

23 MARTIN-ROYLE, R.D., DUDLEY, L.W., and FEVIN, R.J.: 'A review of the British Post Office microwave radio-relay network', *PO Elect. Eng. J.*, 1977, **69**, p. 162

24 MEDHURST, R.G.: 'Rainfall attenuation of centimetre waves: comparison of theory and measurement', *IEEE Trans.* 1965, **AP–13**, p. 550

25 GUNTER, F.: 'Troposcatter scatter communication—past, present and future', *IEEE Spectrum*, 1966, **3**, p. 79

26 HILL, S.J.: 'British Telecom transhorizon radio services to offshore oil/gas production platforms', *Brit. Telecom. Eng.*, 1982, **1**, p. 42

27 EVANS, B.G. (Ed.): 'Satellite communication systems (Peter Peregrinus, 1987)

28 DALGLEISH, D.I.: An introduction to satellite communications, (Peter Peregrinus, 1989)

29 KAO, K.C., and HOCKHAM, G.A.: 'Dielectric fibre surface waveguides for optical frequencies', *Proc. IEE*, 1966, **113**, p. 1151

30 MIDWINTER, J.: 'Optical fibre communication systems' (Wiley, 1980)

31 CHERIN, A.H.: 'An introduction to optical fibres' (McGraw-Hill, 1983)

32 GOWAR, J.: 1984, 'Optical communication systems', (Prentice-Hall, 1984)

33 SENIOR, J., 'Optical fiber communications: principles and practice' (Prentice-Hall, 1985)

34 WORTHINGTON, P.: 'Design and manufacture of an optical fibre cable for submarine telecommunication systems', Proc. 6th European Conf. on Optical Communications, York, UK, 1980, p. 347

35 WHITE, K.I., and NELSON, B.P.: 'Zero total dispersion in step index monomode fibres at 1·3 and 1·5 μm', *Electron. Lett.*, 1979, **15**, p. 396

36 O'REILLY, J.J.: 'Approaching fundamental limits to digital optical-fibre communications', *Oxford Surveys in Information Technology*, 1986, **3**, p. 147 (Oxford University Press)

Chapter 4
Transmission performance

R. W. McLintock

BT, plc

BT, plc

4.1 Introduction

Network performance is an all embracing term which can encompass
many different characteristics of telecommunication networks. This
chapter is based primarily on how users of the network will perceive
performance.

Of all the impairments that can affect a particular switched service, they
can be split into three broad categories:

(*a*) Call-processing performance
(*b*) Transmission performance
(*c*) Availability performance

The first relates to the ability of the network to accept and interpret routing
information received from the customer and to establish a connection to
the required destination within certain response times. Equally, at the end
of the call, the connection should be correctly disconnected.

To a large extent, the call-processing performance is determined by the
performance of telephone exchanges and signalling systems, together with
the dimensioning of the network in terms of the number of transmission
circuits provided between exchange nodes to cater for the level of traffic
expected. Apart from complete failure of transmission circuits, the performance of transmission equipment has little impact on the call-processing
performance of a network. Thus, it is not within the scope of this chapter
to consider call-processing performance any further.

On the other hand, with the exception of switching noise in older
exchanges, the transmission performance experienced by a customer, once
the connection has been established, is mainly dependent on the performance of transmission equipment rather than exchanges. However, with the
emergence of integrated digitial networks, the traditional boundaries
between exchanges and transmission equipment are becoming rather
blurred.

Availability may be defined as the proportion of time for which satisfactory
service is given. It therefore encompasses the effects of both call-processing
and transmission performance, together with equipment reliability. A connection or service is deemed to be unavailable if any primary or directly-
observable performance parameter exceeds its permissible value for a

defined period of time. The parameters that describe availability performance (for example frequency and duration of outages) are known as secondary or derived parameters, since they are not directly measured but determined from primary parameters.

This chapter will mention some of the transmission performance impairments affecting the telephone service, and how they relate to existing analogue, mixed analogue/digital and completely-digital networks. With the trend towards integrated digital networks, the possibility arises to use them for a multitude of existing and new services, many of them relying on the transparent digital path existing virtually from end to end. In the latter context, there are a range of transmission impairments which can affect the transmission of digital signals, and these will also be discussed.

4.2. Telephony performance assessment

In telephone connections, the complete path from mouth to ear consists of:

(a) An air path from the talker's mouth to the telephone transmitter
(b) The talker's telephone
(c) A connection through a public switched telephone network (PSTN)
(d) The listener's telephone
(e) An air path from the telephone receiver to the listener's ear

The most difficult parts to quantify and assess are those involving changes between acoustic and electrical quantities, i.e. the telephones and their users. The method used is to make a comparison between a telephone under test and a defined reference system. The reference system recommended by the CCITT is known as NOSFER [1, 2] (nouveau système fondamental pour la détermination des équivalents de référence).

For many years, the concept of *reference equivalent* was used and was basically defined by the amount of loss inserted in a reference system (i.e. NOSFER) which led to the same perceived loudness to that obtained over the speech path being evaluated. The talker would speak alternately into the microphones of the reference and test systems, while the listener would alternate between the two corresponding receivers and adjust an attenuator until the received speech was judged to be equally loud from each system; the change in attenuator setting from its reference setting being the *overall reference equivalent* (ORE). Thus, if attenuation had been added to NOSFER, the ORE was positive. If it were subtracted, the ORE would be negative.

The same basic method was used to characterise the sidetone performance of a telephone, i.e. the attenuation of the path from a talker's mouth to ear via his own telephone. This measurement was made using a single telephone and the result was called the *sidetone reference equivalent* (STRE).

The ORE depended on the performance of both the sending and receiving telephones and these could be measured separately, and were referred to as *sending and receiving reference equivalents* (SRE and RRE) respectively.

The reference equivalent method was found to have a number of short-comings: reference equivalents cannot be added algebraically, replication accuracy is not good and increments of real (distortionless) transmission loss are not reflected by equal increments of reference equivalent.

The use of loudness ratings [3], based upon rather similar fundamental principles, has now essentially replaced reference equivalents in CCITT G-series recommendations. Loudness ratings are expressed in terms of the amounts of loss, independent of frequency, that must be introduced into an intermediate reference speech path to secure the same loudness of received speech as that defined by a fixed setting of NOSFER. Fig. 4.1 illustrates the basic measurement methodology.

Fig. 4.1a shows a complete unknown speech path subdivided into local telephone systems (LTSs) and an interconnecting junction. To determine the loudness rating, a suitable arrangement needs to be provided for inserting transmission loss within the junction portion.

Fig. 4.1b shows the complete intermediate reference system (IRS) with its adjustable element between the send and receive ends. The IRS is fully specified in CCITT Recommendation P. 48 [4]. The attenuators X_1 and X_2 are both adjusted to provide the same level of perceived speech sounds as the NOSFER with its attenuator set at 25 dB. The *overall loudness rating* (OLR) is given by $(X_2 - X_1)$ dB.

The *send loudness rating* (SLR) and *receive loudness rating* (RLR) are determined in a similar fashion, as shown in Fig. 4.1c and d, by replacing the sending and receiving parts of the IRS by the unknown LTS, as appropriate. SLR is given by $(X_2 - X_3)$ dB and RLR by $(X_2 - X_4)$ dB.

Finally, as shown in Fig. 4.1e, the *junction loudness rating* (JLR) can be determined and is given by $(X_2 - X_5)$ dB.

Experience has shown that this methodology achieves a satisfactory degree of accuracy in the following equality: OLR = SLR + RLR + JLR.

In order to determine the performance of complete connections, subjective laboratory tests have been carried out on a variety of typical connections. Subjects were asked to rate their opinions of the quality of speech transmission on a 5-point scale ranging from 'bad' to 'excellent'. In this way, it was possible to plot curves of the 'percentage difficulty' (i.e. the percentage of connections considered to be unsatisfactory) against the loudness ratings of the connections. These tests were carried out in the presence of impairments such as attenuation/frequency distortion and noise in addition to attenuation. Some results [1] are shown in Fig. 4.2. It will be seen that there is a preferred range of loudness ratings from about +5 to +15 dB.

Since there is an almost infinite number of possible connections in a PSTN, transmission impairment studies make extensive use of *hypothetical reference connections* (HRX) [5]. Such a model may be used as follows:

(a) To study the allocation of overall impairments to particular parts of a network

(b) To examine the effects of possible changes in routings or the allocations of impairments

(c) To test national planning rules for compliance with criteria recommended by the CCITT

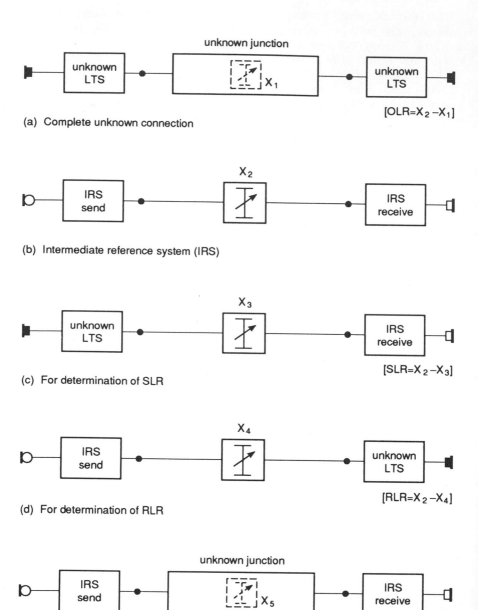

(a) Complete unknown connection

$[OLR = X_2 - X_1]$

(b) Intermediate reference system (IRS)

(c) For determination of SLR

$[SLR = X_2 - X_3]$

(d) For determination of RLR

$[RLR = X_2 - X_4]$

(e) For determination of JLR

$[JLR = X_2 - X_5]$

Fig. 4.1   Basic measurement methodology for determining loudness ratings

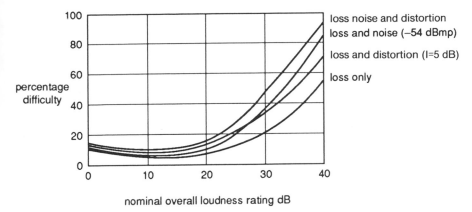

Fig. 4.2 Variation of percentage difficulty with nominal OLR for combinations of noise power level and attenuation distortion

The CCITT has also introduced the idea of a *hypothetical reference circuit* [6] for considering the allocation of impairments between the components making up one circuit within a multi-link connection. This is particularly applicable to studying the noise produced in FDM carrier systems, as described in Section 4.3.2. The CCITT has defined a long-distance hypothetical reference circuit of 2500 km and a very-long distance circuit of 5000 km in length.

4.3. Analogue switched networks

4.3.1 Attenuation

The starting point for establishing the transmission plan for a national network is a *national reference limiting connection*. This consists of two telephones and customers' lines of the limiting length permitted by their local exchanges with attenuation added between them representing the maximum loss to which trunk and junction networks will be designed.

As an example, consider the BT analogue switched telephone network, prior to digitalisation, shown in Fig. 1.2. This uses a transmission plan that was introduced in 1960 [7]. Allowing 1·5 dB for switching losses at group switching centres (GSCs), the limiting loss between local exchanges was nominally 19 dB for calls routed via the trunk transit network and 19·5 dB when routed over two GSC–GSC links in tandem. The limiting customers' lines have 10 dB loss, so the maximum attenuation could have been nearly 40 dB. The telephones with limiting customers' lines had a SRE of 12 dB and RRE of 1 dB, so the ORE for the national reference limiting connection was about $12 + 1 + 20 = 33$ dB. This corresponded to an OLR of approximately 29 dB.

If all customers' calls were to the limiting ORE of 33 dB, most customers would find the quality unacceptable. Nevertheless, it was considered to be the best limiting quality that could be financially justified within the economic constraints that applied to analogue networks. Fortunately, and this factor was very much taken into account, the limit was rarely encountered in practice because it required a combination of all adverse links which was statistically unlikely.

A feature of the plan was that the part of the trunk network based on the Group Switching Centres (GSC) used 2-wire switching. Because of the need to operate circuits between such exchanges at 3 dB loss for stability reasons, the number of cascaded GSC–GSC links was limited to two to avoid excessive losses. In order to provide for the many long thin traffic routings required, it was found to be economic to provide an overlay 4-wire switched transit network. Although the transit network carried only 6% of the total trunk traffic, it catered for approximately 60% of the total national routings.

4.3.2 Noise

In the case of analogue transmission techniques, noise tends to accumulate according to the length of the transmission circuits. The CCITT has identified noise objectives expressed in terms of fixed and per km allowances for multichannel FDM terminal equipment and line systems respectively [8] as follows:

line:	circuit	3 pWOp/km
	very long distance circuit	2 pWOp/km
modulators:	channel	200 pWOp
(per pair)	group	80 pWOp
	supergroup	60 pWOp

4.3.3 Signal power loading

For analogue systems it is important to operate at an appropriate input signal level so that the signal is neither so high that it overloads the equipment and introduces intermodulation products, or so low that the signal-to-noise ratio is excessively degraded [9]. For network design purposes, a transmission reference point is defined as being a hypothetical point used as the zero relative level point in the computation of nominal relative levels. By convention, the mean power level in each telephone channel at this point is taken to be -15 dBm. CCITT Rec. G.111 [9] provides details of how the concepts of loudness ratings are used to verify that international connections will provide an adequate loudness of received speech.

4.3.4 Delay

Some propagation delay is inevitable with all forms of transmission, although the actual values of delay can vary significantly depending on the transmission medium employed. The considerable distances involved in transmission to and from satellites operating in the geostationary orbit mean that satellites introduce significant signal delay (260 ms one-way). The CCITT recommends, except under exceptional circumstances, that the maximum one-way delay should not exceed 400 ms for international connections [10]. For national connections, one-way delays exceeding 15 ms are unlikely to be encountered in the UK. In larger countries, such as North America, delays in excess of 30 ms occur.

4.3.5 Echo and stability

Very much associated with the subject of delay is the effect of delay on echo performance. Chapter 2 has identified how echo can occur, together with the concepts of talker and listener echo. In terms of the subjective effect of a given level of echo, this is dependent on the delay associated with the echo signal. As the delay of the echo path increases, so the loss of the echo path must be higher if the use of echo-control devices (echo suppressors or echo cancellers) is to be avoided. Reference 11 recommends that echo-control devices should be incorporated if the one-way delay of the echo path exceeds 25 ms. An important objective in the planning of the UK network has been to avoid having to use echo-control devices, which are costly, for national connections. Echo-control devices, usually switched into the circuit at the international exchanges, are only employed for the longer international circuits and for telephone calls routed via satellite. However, within larger countries, echo-control devices are necessary on some long-distance national connections.

In order to minimise the effects of talker-echo on international connections, Reference 12 recommends that the mean value of echo loss presented by a national network to the international network should not be less than $(15 + n)$ dB when measured in a prescribed way, which includes a frequency weighting, where n is the number of 4-wire circuits in the national chain.

Listener echo causes hollowness on speech. However, higher values of echo loss are generally required for the satisfactory operation of voice-band data modems than for speech. Open-loop losses in excess of 22 to 27 dB, depending on the number of 4-wire loops in the connection, are required to obtain high-quality data transmission.

Reference 12 also includes a lower limit for the stability loss presented by a national network. Stability loss is defined as the lowest value of semi-loop loss (i.e. transmission loss of the so-called a–t–b path) [1] in the frequency band to be considered. The limit is $(6 + n)$ dB where n is, again, the number of 4-wire circuits in the national chain.

4.3.6 Sidetone

Telephone instruments effectively contain a 2-wire/4-wire hybrid. Any mismatch between the balance impedance of the hybrid and the impedance looking into the cable pair connected at the 2-wire port will cause some signal leakage between microphone and earpiece. Some deliberately-introduced leakage, known as sidetone, is essential to make the telephone feel 'live' to the user. However, an excessive level can have the undesirable effect of causing the talker to speak too softly. It can also impair performance at the receiving end due to ambient room noise transmitted as sidetone to the ear of the listener [2, 13].

4.3.7 Crosstalk

Both intelligible crosstalk and unintelligible crosstalk can occur owing to a number of mechanisms. Crosstalk between cable pairs and crosstalk due to the limited amount of channel filtering that is practicable in FDM multiplex equipment are two sources, although the levels are usually so low that they do not often cause problems. A further crosstalk mechanism is due to intermodulation in FDM systems, which can arise if the signal loading is excessive. Reference 14 recommends a limit of 65 dB for near- and far-end crosstalk ratio (intelligible crosstalk only) for 4-wire national and international circuits.

4.4 Digital transmission in the PSTN

The late 1960s saw the introduction of digital transmission techniques into the North American telephone network, largely as an alternative means of providing a number of circuits instead of using large multipair audio cables or FDM systems. Other countries soon followed. The initial emphasis was on providing digital transmission plant, which was justified on cost grounds, while retaining analogue exchanges. However, some countries initially concentrated on the deployment of digital exchanges while retaining analogue transmission. In both cases it was necessary to define an analogue-to-digital conversion scheme. The CCITT recommended 64 kbit/s pulse-code-modulation (PCM) speech coding using an 8 kHz sampling rate [15]. Amplitude quantisation is to 8-bit resolution, according to one of two logarithmic companding laws. Regrettably, the CCITT was not able to identify a single companding law that could be accepted by all administrations. The so-called μ-law is used in North America and Japan, and the A-law is used in Europe and elswhere. It is possible to convert digitally between the two laws, albeit with some slight additional impairment, by means of a look-up table specified by the CCITT.

In terms of transmission performance, the important impact of the new technology was that noise introduced into the connection was substantially determined by the performance of the PCM coding process. If the digital transmission facility between the encoder and decoder was error free, the performance of the circuit between analogue interfaces would be independent of length. In fact, it was found that the error performance was not particularly critical. Random errors at an error ratio better than 1 in 10^5 were generally recognised to be barely discernible.

The main form of noise associated with PCM coding is *quantising distortion* (QD), which is a multiplicative type of noise rather than additive. For a given signal level, a particular level of quantising distortion is less perceptible compared with the same level of added white noise [2]. For a defined test signal, the signal-to-total-distortion ratio (including QD) achieved by practical PCM codecs is typically 36 dB over a fairly wide range of input signal levels. This relatively high level of performance was considered necessary because, on a long international connection, it is essential to be able to connect a significant number of PCM coding stages in tandem (with interconnection at the analogue level) and still achieve an acceptable level of overall performance. A rule of thumb is that the signal-to-QD ratio is degraded by 3 dB for each doubling of the number of PCM coding stages in tandem.

As a planning guide, the CCITT has adopted the concept of *quantising distortion units* (qdu), where 1 qdu represents the QD introduced by a 64 kbit/s PCM coding stage. For planning purposes, the assumption is that the qdu from a number of coding stages in tandem can be simply added together. For an international connection, the maximum qdu recommended is 14, allocated on the basis of $5+4+5$ for the originating national, international and terminating national networks respectively. However, a short-term relaxation to 7 qdu is permitted for each national network contribution [16].

The CCITT has assigned qdu ratings to other processes which may be incorporated into connections. For example, a new 32 kbit/s adaptive differential PCM (ADPCM) coding law has been defined [17]. This coding law is defined in terms of a transcoding operation from and to 64 kbit/s PCM (either μ-law or A-law). 3·5 qdu has been assigned for the complete sequence of analogue-PCM-ADPCM–PCM-analogue. It can be seen that this advance in coding technology has made it possible to achieve a fairly good level of performance at half the bit rate. The main applications for 32 kbit/s ADPCM are where transmission costs are high, i.e. within geographically-large national networks, such as in North America, and on long international circuits, and also where digital bandwidth is limited, e.g. cordless or mobile telephones.

4.5 Integrated digital networks

The next development, which is now well under way in many countries, is the integration of digital transmission and digital exchanges to provide an *integrated digital network* (IDN). For example, in the British Telecom (BT) network, the trunk exchanges and trunk transmission facilities are already entirely digital, with digitalisation of the junction transmission network and local exchanges continuing at a very high rate [18].

This modernisation of the network using digital techniques can be justified in terms of reduced operating costs for the telephony service alone. However, there are other significant benefits in that the network will provide transparent 64 kbit/s switched paths between local exchanges. By extension of digital paths to customers' premises, a wide range of new services can be offered. This concept of providing digital customer access to a multipurpose digital network, with service integration, is known as the *integrated services digital network* (ISDN).

When an IDN is completed, the telephone transmission performance, in terms of overall quality, should be significantly enhanced, with much less variability between calls. The main objectives originally set by BT for its ultimate all-digital network can be summarised as follows [19]:

(*a*) All calls to have an overall loudness rating (OLR) not exceeding approximately 16 dB and a distribution which meets CCITT Rec. G.111

for the long term objective. These values are associated with ranges of loudness rating which yield minimum user dissatisfaction. Assuming no change in existing telephone designs, or line current feeding arrangements, this required that the 2-wire to 2-wire loss between local exchanges should be about 6 dB.

(b)   Adequate stability must be maintained on all classes of connection. Existing PCM systems, operated on the basis of a 2-wire to 2-wire loss of 3 dB, provide an open-loop loss of at least 6 dB, which has been found just adequate for voice-band data services in the UK. With the operation of the 2-wire to 2-wire part of the network at 6 dB loss, as described above, the open-loop loss of at least 12 dB is considered adequate for all likely requirements.

(c)   The use of echo suppressors or cancellers should be avoided on all national connections. As losses decrease in the digital era, care is necessary to ensure that echo performance will be adequate. Close attention has been given to the design of digital-local-exchange line-interface units, to ensure that the echo balance return loss is improved so that overall echo-path loss is not reduced. (Previously, 2-wire to 4-wire hybrid transformers used 600 Ω resistive balances almost exclusively, but this choice was not ideal for interfacing with local lines which are capacitive in nature.)

(d)   International recommendations on stability and echo should be met. The operation of 2-wire to 2-wire paths at 6 dB loss will ensure that the national network will contribute at least 6 dB stability loss between equi-level switching points (defined as a–t–b loss in CCITT Recommendation G.122) on international connections. With the use of a special hybrid balance network for the control of echo on national connections, this will also provide for acceptable performance internationally in terms of meeting the requirements of CCITT Recommendation G.122.

(e)   The traffic-weighted mean values of send loudness rating (SLR) and receive loudness rating (RLR) specified in CCITT Recommendation G.121 should be used as target values. The CCITT long-term preferred values of SLR and RLR up to the boundary between national and international networks, referred to a 0 dBr international switching point, are as shown below:

	Traffic weighted mean	Maximum value (average sized country)
SLR	7—9 dB	16·5 dB
RLR	1—3 dB	13 dB

(f)   The transmission plan should enable acceptable sidetone performance to be achieved. The opporunity is being taken to improve the sidetone performance of the network which has in the past been at a rather higher level than in networks of other countries. Too high a level of sidetone can cause the talker to lower the voice, which effectively reduces the send sensitivity of the telephone.

4.6 Digital transmission performance parameters

The interest in using integrated digital networks for a wide range of new services, many of which will rely on the availability of a customer-to-customer 64 kbit/s digital connection, has already been mentioned. With digital transmission facilities often being available somewhat in advance of digital exchanges, and because some customers (especially those in the business sector) have need of designated point-to-point transmission capabilities that do not have to be switched, a range of digital private-circuit services has emerged. Two such services operated by BT are Kilostream and Megastream. The former offers data rates at 2·4, 4·8, 9·6, 48 and 64 kbit/s, while the latter provides for transmission at higher bit rates, currently at 28 and 34 Mbit/s.

For both switched and private-circuit digital services, two questions arise from a transmission point of view:

(*a*) What types of impairment affect these services?
(*b*) What are the limits for these impairments?

Although standards are naturally still evolving, the following sections summarise the current position with regard to the standards applicable to ISDN as recommended by the CCITT.

4.6.1 Errors

An error is defined as a single bit inconsistency between the transmitted and received information. It is perhaps the most obvious impairment to affect a digital signal. However, in terms of specifying objectives and predicting the error performance of systems operating in a real network, there is much more than might be first imagined.

In the early days of digital facilities, error performance was expressed in terms of a long-term mean error ratio, i.e. an error probability. However, for some years now, it has been widely recognised that this is not very meaningful because, in practice, errors tend to occur in bursts. As well as the number of errors occuring in a given time, the distribution of errors with time has a major impact on how the errors affect the service being supported. Assuming some errors do occur, a stated long-term error ratio contains no information about the error distribution.

In order to take account of the distribution as well as the number of errors, three error-performance parameters are used to set error-performance objectives for the hypothetical reference connection shown in Fig. 4.3 [20, 21]. The parameters and objectives are shown in Table 4.1. All should be met concurrently. Parameters (*a*) and (*b*) are aimed at satisfying the needs of telephony-type services. Parameter (*c*) is more relevant for data-type services, especially those based on block transmission of data where one or more errors in a block results in the re-transmission of the entire block.

The overall permissible impairments have been subdivided into three grades of transmission plant, known as local, medium and high grades. Very approximately, the permissible impairments have been allocated on

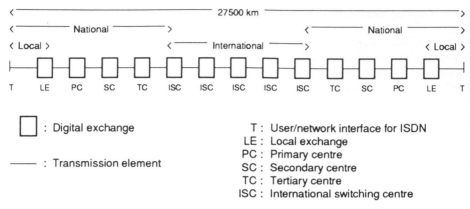

Fig. 4.3   Hypothetical reference connexion (HRX) used by the CCITT

the basis of 30% : 40% : 30% to the national : international : national portions of the international hypothetical reference connection. The local portion is assigned 15% out of the national allowance.

4.6.2   Slips

A slip is defined as the deletion or repetition of one bit, or a number of consecutive bits, of a digital signal. Slips may be classified into two types: controlled and uncontrolled. Digital exchanges by their very nature require that all the signals arriving at the exchange, all the signals leaving the exchange and the operating rate of the exchange should be synchronous or at least be very close in frequency (e.g. within 1 part in 10^{11}). In practice, this is usually achieved by a carefully-designed synchronisation network. This distributes timing information from a network reference clock, usually a replicated caesium-beam atomic standard accurate to better than 1 part in 10^{11}. However, in the case of a connection between different clock regions or in the event of a part of the network losing synchronisation, due for example to an equipment failure, the situation can arise where the digital

Table 4.1   Error performance objectives for an international HRX at 64 kbit/s

Performance Classification	Objective
(a) Degraded minutes	<10% of 1 min intervals to have a bit error ratio $>1 \times 10^{-6}$
(b) Severely errored seconds	<0·2% of 1 s intervals to have a bit error ratio $>1 \times 10^{-3}$
(c) Errored seconds	<8% of 1 s intervals to have any errors (i.e. >92% to be error free)

A total averaging period of any one month is suggested as a reference.

signals arriving at the exchange are no longer at the same rate as the exchange. The exchange can still correctly identify which bits belong to which channel, because of the frame-alignment signal associated with each digital stream. Under these conditions, faced with either too much or too little information arriving at the exchange, eventually the exchange implements a slip in a controlled manner, i.e. a controlled slip. As far as each 64 kbit/s channel is concerned, this is arranged to be the repetiton or deletion of one octet (8-bit byte). This unit of slip was chosen because it was considered to be subjectively acceptable for voice services based on PCM encoding. The slip corresponds to the deletion or repetition of one sample. Such an event is not normally perceptible unless the slip rate is very high. However, a slip can have a more significant effect on voice-band data signals, especially those associated with modems operating at the higher rates, e.g. 9·6 kbit/s.

Uncontrolled slips on the other hand are unplanned and unforeseen events where a bit (or a number of bits) is repeated or deleted. Such events can arise in the network as a result of, for example, protection switching of equipment or excessive jitter or wander. Whereas controlled slips are carefully implemented to avoid subsequent loss of frame alignment, uncontrolled slips are virtually certain to cause loss of alignment downstream. This will result in a relatively long recovery time, possibly lasting several milliseconds, while demultiplexers re-align. Thus, an uncontrolled slip usually has a more serious effect on the service being carried.

To date, the CCITT has only produced a recommendation which establishes objectives for controlled slips [22]. On the basis of the hypothetical reference connection shown in Fig. 4.3, Table 4.2 indicates the overall objective set. It is not yet clear whether specific recommendations will be developed in the future to embrace uncontrolled slips. Reference 22 should be consulted for more information about the subdivision of the total permissible impairments.

4.6.3 Short breaks

This impairment might be thought of as self-explanatory. However, it is not quite so straightforward as might be imagined. If a short break occurs somewhere in the network, it usually leads to a loss of signal being detected

Table 4.2. Slip rate objectives for an international HRX at 64 kbit/s

Performance category	Mean slip rate	Proportion of time
(a)	<5 slips in 24 h	>98.9%
(b)	>5 slips in 24 h and <30 slips in 1 h	<1·0%
(c)	>30 slips in 1 h	<0·1%

Total averaging time is at least 1 year

at the subsequent equipment, which begins to initiate an automatic fault-reporting procedure. This same equipment forwards downstream a special signal, on the signal path, known as an alarm indication signal (AIS), which usually corresponds to the equivalent of binary *all ones.* This procedure ensures that other subsequent equipment does not initiate alarms due to the original loss-of-signal event. The AIS also has the advantage of maintaining the presence of timing information downstream, which has the effect of reducing the recovery time once the traffic signal has been restored. Nevertheless, a short break of a few microseconds affecting a high-bit-rate system (e.g. at 140 Mbit/s) can result in a rather longer break of up to several tens of milliseconds as perceived at the end of the connection. The extension arises because of the time taken for the chain of subsequent demultiplexers to recover frame alignment, and for phase-locked loops, associated with demultiplex equipment and transmission systems, to stabilise. Much attention has been given to minimising recovery times; even so, this is an important characteristic of digital systems that was perhaps not quite so evident on analogue systems.

To date, little attention has been given to this impairment within CCITT, in terms of establishing objectives.

4.6.4 Jitter

Jitter is defined as the short-term non-cumulative variations in the significant instants (e.g. mid-points of pulses) of a digital signal from their ideal positions in time [24]. It can be viewed as affecting a digital signal in the same way as if the timing signal that was used initially to generate the digital signal had been phase modulated with a modulating signal which was the jitter waveform. Jitter signals occupying the frequency range from a few tens of Hz to several kHz are usually considered to be most significant [25, 26].

Jitter arises as a result of the way that bit timing, associated with a digital signal, is extracted from the signal itself [27]. It also arises as a consequence of the type of multiplexing, based on justification, used in higher-order digital multiplex equipment operating above the primary level in the digital hierarchy. Low-frequency jitter tends to accumulate as a signal passes along a chain of systems. The accumulation of higher-frequency jitter is much less because of the low-pass filtering action, on the jitter, of the timing-extraction filters present in each equipment. Jitter is a rather different impairment to errors and slips in that it can be reduced, relatively easily, to any desired level by retiming the digital signal in a jitter reducer. Typically, this would use a narrow-bandwidth phase-locked loop to attenuate most of the phase modulation affecting the timing signal.

Jitter itself is not necessarily significant or damaging. What is more important is that, at an interconnection interface, the input tolerance to jitter of the subsequent equipment should be compatible with the jitter generated by the preceding equipment. Otherwise, errors or uncontrolled slips could occur as a result of interconnecting the two equipments. However, jitter can be significant at the end of the digital connection in terms of causing phase modulation of the sample reconstruction process in any digital-to-analogue convertor. This mechanism can introduce distortion products into the final analogue output signal. PCM-encoded speech is fairly tolerant to this type of effect, but digitally-encoded TV is much more sensitive.

CCITT Recommendation G.823 [24] contains information about the jitter control philosophy adopted for digital networks based on the 2048 kbit/s hierarchy. Jitter limits have been derived for all the hierarchical bit rates from 64 kbit/s up to 140 Mbit/s. The basic approach is that a maximum network limit has been established which should never be exceeded anywhere in the network, even at the end of a long connection. All digital equipments have to be able to accommodate a test condition where sinusoidal jitter is applied at the input port at a level which ensures compatibility when connected to any equipment output port. Table 4.3 contains the network limits, jitter amplitudes being expressed in terms of unit intervals (UI), where unit interval is defined as the nominal difference in time between consecutive significant instants of the digital signal.

Jitter generated by individual systems, both multiplexers and transmission systems, is also limited by virtue of the requirements written into individual system Recommendations. These limits are significantly less than the maximum network limit to allow for the accumulation effects mentioned above. A jittter transfer characteristic, which limits the amount of jitter that can appear at the output of a system as a direct result of jitter at its input, is also included. This characteristic is highly relevant in terms of jitter accumulation through the network.

4.6.5 Wander

Wander is defined as the long-term non-cumulative variations of the significant instants of a digital signal from their ideal positions in time [24]. There is no definite frequency which can be considered as a boundary between jitter and wander. In practice, a distinction is often made on the basis of the mechanism producing the impairment. One of the main sources of wander is as a result of temperature variations of the transmission media

Table 4.3 Maximum permissible jitter at hierarchical interfaces

Digit rate kbit/s	Network limit (unit intervals)		Measurement filter bandwidth; bandpass filter having lower cut-off frequency f_1 or f_3 and upper cut-off f_4		
	Measured $f_1 - f_4$	Measured $f_3 - f_4$	f_1	f_3	f_4
64*	0·25	0·05	20 Hz	3 kHz	20 kHz
2048	1·5	0·2	20 Hz	18 kHz	100 kHz
8448	1·5	0·2	20 Hz	3 kHz	400 kHz
34368	1·5	0·15	100 Hz	10 kHz	800 kHz
139264	1·5	0·075	200 Hz	10 kHz	3500 kHz

* Limits for co-directional version

which, in turn, cause changes in propagation times. Because of the very-low-frequency nature of wander, it is not practicable to use phase-locked loops to remove it. Digital exchanges incorporate additional buffer storage capacity at their inputs to absorb wander; most other digital line and multiplex equipments are transparent to wander. CCITT Recommendation G.823 [25], as well as containing a requirement on input tolerance to jitter, also contains a requirement on input tolerance to wander. Unlike jitter, no maximum network limit has been established for wander. The input tolerance requirement is considered by the CCITT to be sufficient to ensure that controlled slips, occuring as a result of levels of wander exceeding this tolerance, will be insignificant.

Table 4.4 gives some estimated values of wander for various systems used by British Telecom. Wander amplitudes, like jitter, can be expressed in unit intervals, but it is more usual to use units of time. Note the relatively large value for geostationary satellites; this arises from cyclical variations in the position of a satellite when referenced to a fixed position on the ground. Special provision has to be made at earth stations to accommodate this wander.

4.6.6 Delay

With the introduction of various sophisticated digital signal-processing techniques into the network, there is a tendency for total end-to-end delays to increase. Table 4.5 contains some typical values for delays associated with different equipment. The values shown for digital exchanges are reproduced from CCITT Recommendations for digital exchanges [28].

4.6.7 Availability

At first sight, it may appear that availability is a fairly straightforward parameter to describe and to measure. Unfortunately, this is not the case and little progress has been made in CCITT, even for analogue systems, in establishing reliability or availability objectives. The most recent thinking in CCITT is based on the 3×3 matrix shown in Fig. 4.4, in which the available state is determined by observing the primary parameters. When a primary parameter, either singly or in combination with other primary

Table 4.4   Estimated values for wander

System	Typical wander	
	Monthly variation	Annual variation
2 Mbit/s; pair cable	30—42 ns/km	62—77 ns/km
140 Mbit/s; coax cable	0·5—0·7 ns/km	1·2—1·3 ns/km
140 Mbit/s; fibre optic	0·3—0·4 ns/km	0·6—0·8 ns/km
140 Mbit/s; radio	very small	very small
Digital mux/demux pair	negligible	negligible
Digital transit exchange	negligible	negligible
Satellite; geo. orbit	daily variation 0·15—1·5 ms	

Table 4.5   Typical values for propagation delays

System	Delay
2 Mbit/s; pair cable	4·3 μs/km
140 Mbit/s; coax cable	3·6 μs/km
140 Mbit/s; fibre optic	4·9 μs/km
140 Mbit/s; radio	3·3 μs/km
Digital mux/demux pair	1—12 μs (depending on hierarchical level)
Digital transit exchange (sum for both directions; digital to digital)	<900 μs (mean value) <1500 μs (95% percentile)
Digital local exchange (sum for both directions; analogue to analogue)	<3000 μs (mean value) <3900 μs (95% percentile)
Satellite; geo. orbit	260 ms

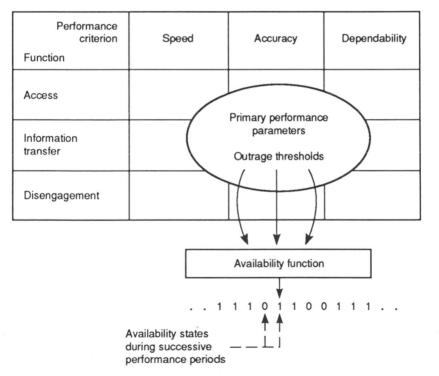

Fig. 4.4   Matrix to determine performance parameters and availability states

parameters, exceeds a threshold value for a defined period of time, the connection or service is deemed to be unavailable. It is then necessary to develop some parameters to describe the frequency and duration of outages.

An example of this concept is used by CCITT and CCIR [20, 29]. If the error ratio of a 64 kbit/s connection exceeds 1×10^{-3} in each second for more than ten consecutive seconds, a complete failure is assumed to have occured and the connection is deemed to be unavailable. The primary parameter in this case is the severely-errored second (see Table 4.1). Although in some respects the criterion in this example was an arbitrary one, it is particularly significant for the specification of certain radio systems, because it has an impact on whether impairments due to adverse propagation are covered by the error ratio or availability objectives. For digital equipment, the only availability objectives set are those established by CCIR for radio systems [29]. Those objectives amount to $>99·7\%$ (provisional value) availability for a 2500 km hypothetical reference digital path.

4.7 Conclusion

In this chapter it has only been possible to provide a brief insight into some aspects of network performance which is an all-embracing and rather specialised field. An attempt has been made to show how the original transmission plans for an analogue telephone network have evolved to cope with the provision of the telephony service using an integrated digital network. For those emerging services that will exploit the digital end-to-end transparency of either switched networks or private circuits, the range of transmission impairments arising has been given, together with some indications of numerical values. Performance standards will continue to evolve as new services and network capabilities emerge.

4.8 References

1 FRY, R.A.: 'Transmission standards and planning', *in* Telecommunication networks, (Peter Peregrinus, 1975) chap. 7, pp. 163–194
2 RICHARDS, D.L.: 'Telecommunications by speech: The transmission performance of telephone networks' (Butterworths, London, 1973)
3 CCITT Recommendation P.76: 'Determination of loudness ratings; fundamental principles'
4 CCITT Recommendation P.48: 'Specification for an intermediate reference system'
5 CCITT Recommendation G.103: 'Hypothetical reference connections'
6 CCITT Recommendation G.212: 'Hypothetical reference circuits'
7 TOBIN, W.J.E., and STRATTON, J.A.: 'A new switching and transmission plan for the inland trunk network', *POEEJ*, 1960, **53**, p. 57
8 CCITT Recommendation G.123: 'Circuit noise in national networks'
9 CCITT Recommendation G.111: 'Loudness ratings (LRs) in an international connection'
10 CCITT Recommendation G.114: 'Mean one-way propagation time'

11 CCITT Recommendation G.131: 'Stability and echo'
12 CCITT Recommendation G.122: 'Influence of national systems on stability, talker-echo, and listener-echo in international connections'
13 CCITT Recommendation G.121: 'Loudness ratings (LRs) of national systems'
14 CCITT Recommendation G.151: 'General performance objectives applicable to all modern international circuits and national extension circuits'
15 CCITT Recommendation G.711: 'Pulse code modulation of voice frequencies'
16 CCITT Recommendation G.113: 'Transmission impairments'
17 CCITT Recommendation G.721: '32 kbit/s adaptive differential pulse code modulation'
18 BOAG, J.F.: 'Learning to live with digital transmission' 3rd International Conference on Telecommunication Transmission, IEE, London, 1985
19 HARRISON, K.R.: 'Telephony transmission standards in the evolving digital network', *POEEJ*, 1980, **73**, pp. 74–81
20 CCITT Recommendation G.821: 'Error performance on an international digital connection forming part of an ISDN'
21 McLINTOCK, R.W., and KEARSEY, B.N.: 'Error performance objectives for digital networks', BTEJ, 1984, **3**, pp. 92–98
22 CCITT Recommendation G.822: 'Contolled slip rate objectives on an international digital connection'
23 SMITH, R., and MILLOT, L. J.: 'Synchronisation and slip performance in a digital network', BTEJ, 1984, **3**, pp. 99–107
24 CCITT Recommendation G.701: 'Vocabulary of digital transmission and multiplexing, and pulse-code-modulation (PCM) terms'
25 CCITT Recommendation G.823: 'The control of jitter and wander within digital networks based on the 2048 kbit/s heirarchy'
26 KEARSEY, B.N., and McLINTOCK, R.W.: 'Jitter in digital telecommunication networks', *BTEJ*, 1984, **3**, pp. 108–116
27 DUTTWEILER, D.L.; 'The jitter performance of phase-locked loops extracting timing from baseband data waveforms', *BSTJ*, 1976, **55**, pp. 37–58
28 CCITT Recommendation Q.551: 'Transmission characteristics of digital exchanges'
29 CCIR Recommendation 557, 'Availability objectives for a HRC and a HRDP', 1986

Chapter 5
Frequency-division multiplexing

D.J. Kingdom

Line and Radio Systems Division, BT

5.1 Introduction

It has long been established that the wideband analogue bearer circuits used to provide long-distance transmission facilities in telecommunications networks are most economically exploited when the telephone channels to be carried are assembled in frequency-division multiplex (FDM) using single-sideband suppressed-carrier (SSBSC) modulation. This is because the cost of these bearer circuits increases with bandwidth and power-handling capacity and in both respects SSBSC transmission minimises these demands. However, the generation and detection of SSBSC signals poses special problems for which solutions have to be found which are both economically and technically acceptable in the context of public telecommunication networks. An appropriate system architecture which minimises the technical demands is of prime importance.

The CCITT has standardised a range of channel assemblies containing from less than twelve to more than 10 000 channels and has outlined ways in which such assemblies should be built up. Most countries use systems conforming to these CCITT recommendations, although other assemblies are used, most notably in the USA. Although there are differences in detail, all these channel-multiplexing systems based on the hierarchical principle, in which the required channel capacity is built up from a number of basic channel assemblies using successive steps of modulation and combination. Each channel has a nominal passband of 300—3400 Hz and is allocated a 4 kHz slot in the transmission spectrum.

For short-haul bearer circuits, or where the number of channels to be multiplexed is small, the advantages of SSBSC transmission are less marked. As system length decreases below about 50 km or the number of channels below about twenty, the situation may be reached where the cost of SSBSC multiplexing equipment is several times that of the bearer circuit. This is especially so where low-cost bearers, such as overhead lines, are used. Application for short-haul low-capacity carrier systems have existed in the provision of junctions, trunks or toll circuits to rural exchanges, long lines to a group of rural subscribers, or even in urban situations when an increase in the number of lines is needed in a district but, for some reason, additional line plant cannot be provided. An economic system may then be realised by the use of conventional amplitude modulation with transmission of each channel's carrier together with both sidebands. The system length and size for which double-sideband becomes more economic than single-sideband transmission will depend very much upon local factors and the technology available. Consequently, short-haul low-capacity systems tend to be engineered for a specific field of application and there has not evolved any international standards for them.

This chapter then will deal principally with the architecture of SSBSC multiplex systems for which there is a large measure of international standardisation.

5.2 Elements of SSBSC multiplex systems

5.2.1 The SSBSC generator

There are a number of ways in which a SSBSC signal can be generated [1]. However, for telephony, some form of balanced, switching-type modulator together with filtration is almost always used. This is based on the concept of multiplying the baseband signal by a square-wave carrier and is the cheapest instrumentation, having an adequate technical performance, that has been available. The basic arrangement is shown in Fig. 5.1 where, with the baseband signal applied to the input port, the SSBSC signal appears at the output port.

Filtration is required to:

(*a*) Define the spectrum that is to be transmitted. Both pre- and post-modulation filters can contribute to this function.

(*b*) Select the sideband required for transmission and suppress all others. Only the post-modulation filter can perform this function.

(*c*) Reduce the level of any unwanted signals (see Table 5.1) that may be generated by the modulator. This function is performed by the post-modulation filter.

Essentially, the SSBSC generator performs the function of frequency translation, with or without frequency inversion depending upon the sideband selected for use. Where an increase in audio frequency gives rise to an increase in sideband frequency then that sideband is said to be 'erect'; where

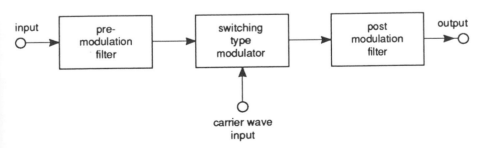

Fig. 5.1 SSBC generator using a switching modulator and filtration

the opposite occurs the sideband is said to be 'inverted'. Fig. 5.2 is a frequency plan showing the frequency translation performed by a SSBSC generator with a carrier-wave frequency of f_c kHz and a baseband spectrum extending from f_1 to f_2 kHz. Frequency bands on plans such as Fig. 5.2 are conventionally shown as triangles so that baseband frequencies can be related to corresponding sideband frequencies; no implication of power density is intended.

5.2.2   Demodulation of SSBSC signals

As shown in Section 2.7.2, a SSBSC signal can be demodulated by multiplying it by a carrier wave of the same frequency as was used in the generation process. The practical instrumentation used to achieve this is a balanced switching-type modulator together with filtration, as was the case with the generator. In the demodulator, the functions of the filters are to:

(*a*)   Select the required sideband from the multiplexed signal (see Section 5.2.3). This function is essentially performed by the pre-modulation filter, but the other can contribute.

(*b*)   Select one of the sidebands generated by the modulator that corresponds to the wanted baseband and suppress the others. Only the post-modulation filter can perform this function.

(*c*)   Reduce the level of any unwanted signals that may be generated by the modulator. This function is performed by the post-modulation filter.

At both generator and demodulator there is some degree of freedom as to the exact distribution of functions between post- and pre-modulation filters. Although the filter requirements of generator and detector are expressed somewhat differently, it is economically desirable to provide as much commonality between them as possible. To this end, it is usually technically

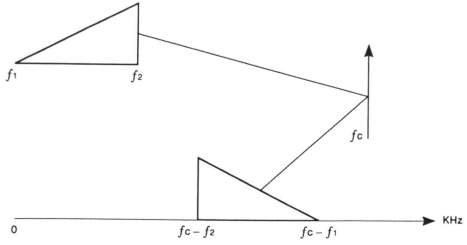

Fig. 5.2 Frequency plan showing translation between a baseband and an inverted sideband

possible to adopt a common design of bandpass filter for the post-modulation filter of the generator and the pre-modulation filter of the demodulator and to incorporate in it the most stringent performance demanded of the filtration. The other two filters may then be less-stringent lowpass networks. Because of this commonality, Fig. 5.1 can be taken as also representing a SSBSC demodulator. It performs a frequency translation on the SSBSC signal applied to the input port to produce the wanted baseband on the output port. Fig. 5.2 can be regarded as also indicating the operation of a SSBSC demodulator in translating from an inverted sideband to the required baseband f_1 to f_2.

5.2.3 Multiplexing

With suitable choice of carrier frequencies several SSBSC generators can be used, as shown in the 'send' end of Fig. 5.3, to combine a number of channels onto a common transmission medium in frequency-division multiplex. There must be a frequency separation between carriers which is at least equal to the passband defined for each baseband, plus a guard band in which the transition from passband to stopband response of the filters can take place. De-multiplexing equipment is shown at the 'receive' end of Fig. 5.3. A frequency plan for an N-channel multiplex using lower sidebands of the carriers is shown in Fig. 5.4.

5.2.4 Unwanted signals on output of SSBSC generator

Switching-type modulators can be regarded as multiplying the input baseband signal by a quasi square-wave carrier. The modulator output thus contains upper and lower sideband pairs about frequencies corresponding to the fundamental and each odd harmonic of the carrier wave. With ideal square-wave switching, the pair of sidebands about the Nth harmonic of

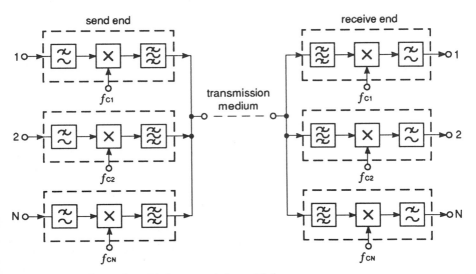

Fig. 5.3 *N-channel multiplexer and demultiplexer*

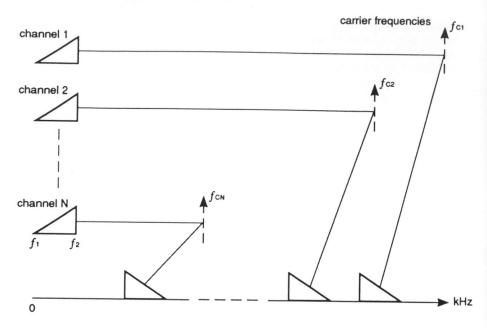

Fig. 5.4   Frequency plan for an N-channel multiplex

the carrier would be $20 \log_{10}N$ dB lower in level than the sidebands about the carrier fundamental frequency, and an infinite number of sideband pairs would be generated. In practice, with less than square-wave switching, there may still be thirty or more pairs of sidebands generated at levels which are of significance to the filter and system design. Generally, the faster the switching transitions, and fast switching makes for linear operation, the closer the levels of sidebands generated approach those that would be generated by ideal square-wave switching. In principle, any of the sidebands appearing at the modulator output could be selected for use; however, the lower sideband of the carrier fundamental frequency is almost always used, rendering the rest redundant.

Practical switching modulators will, depending upon the degree of balance achieved, present other unwanted signals at the ouptut port in addition to the unavoidable array of sidebands. These may include the following:

(*a*)   A component at the fundamental frequency of the carrier wave, known as 'carrier leak'
(*b*)   The input baseband signal, often referred to as 'through-signal leak'
(*c*)   Components at frequencies corrresponding to even and odd harmonics of the carrier wave
(*d*)   Sidebands about the even harmonics of the carrier wave

In a multiplex system, it must be ensured that the level of any of these unwanted signals appearing at the ouptut of each SSBSC generator is not such as to cause intolerable interference to other channels in the assembly.

Table 5.1 Unwanted signals that may be present at the output of a SSBSC generator using a switching-type modulator

Unwanted signals	Factors affecting level and spectrum
The unwanted sideband (about the carrier fundamental)	Spectrum shaped by pre-modulation filter; level dependent upon stopband attenuation of post-modulation filter
Through signal leak	Spectrum shaped by pre-modulation filter; level dependent upon modulator balance and post-modulation filter if this has a stopband at the relevant frequencies
Carrier leak (at the carrier fundamental)	Level dependent upon modulator balance and post-modulation filter
Carrier harmonics (odd and even)	Level dependent upon modulator balance and post-modulation filter
Sidebands about odd harmonics of the carrier	Level dependent upon switching speed of modulation and post-modulation filter; spectrum shaped by pre-modulation filter
Sidebands about even harmonics of the carrier	Level dependent upon modulator balance and post-modulation filter; spectrum shaped by pre-modulation filter

Table 5.1 summarizes these signals and indicates the important factors which determine their level and spectrum in practical generators.

The range of unwanted signals on the output of a translating equipment will depend upon the detail design of that equipment and the range of signals, both wanted and unwanted, that is applied to its input.

5.3 Hierarchical systems

In principle, systems of any required number of channels could be assembled by the method given in Section 5.2.3, i.e. by direct modulation of each channel to its position in the spectrum to be transmitted over the transmission medium. Such systems would, however, have four important problems associated with them, each one of which becomes greater with increase in system channel capacity. These problems are:

(a) Each channel would require a different carrier frequency
(b) Each channel would require a different design of filter
(c) Difficulties in realising sideband-selecting filters for channels having high carrier frequencies, the unwanted sideband being only twice the lowest baseband frequency away from the wanted sideband

(d) Lack of facilities for network planners to build up, from a limited range of standardised equipment, systems of differing channel capacities.

All these difficulties are minimised by building up channel assemblies using a hierarchical range of multiplexing equipments. Each equipment in the hierarchy multiplexes together a number of inputs obtained from equipments in the previous level of the hierarchy. Two procedures for building up hierarchical systems have been standardized by the CCITT [2] and the equipments used are classified as shown in Table 5.2.

Each equipment provides for both directions of transmission and, in suitable combinations, provides a range of basic channel assemblies. Some features of these assemblies are given in Table 5.3. Differences between the two procedures occur at the tertiary level and above.

Each of the basic channel assemblies shown in Table 5.3 can be used as blocks of speech channels for transmission. However, a further translating stage is usually necessary to place the assembly in the frequency band required by the transmission medium. This final stage of translation is dealt with in Section 5.5.

It is useful, when comparing assembly plans, to have a comparative measure of the filter performances that are required to realise the assemblies. One of the necessary filter functions, and one that can demand the most stringent performance in terms of the effective Qs required of the resonators in the filter, is that of sideband selection. The frequency band in which the transition from passband to stopband response must take place is centred on the carrier frequency and is no more than twice the lowest baseband frequency. For a given transition band, the higher the carrier frequency, the greater are the effective Qs required for the filter resonators. Thus, the ratio of the carrier frequency to transition band gives a good comparative measure of the required filter performance. Values range from 180 for channel-translating equipment to less than 4 in mastergroup translating equipment. A filter performing the function of sideband selection may also be used for other functions, such as reducing the effective

Table 5.2 Classification of equipment in CCITT recommended hierarchies

Assembly procedure	Hierarchical level	Equipment classification
Procedure 1 (mastergroup working)	Primary Secondary Tertiary Quarternary	Channel translating equipment (CTE) Group translating equipment (GTE) Supergroup translating equipment (STE) Mastergroup translating equipment (MTE)
Procedure 2 (15 supergroup working)	Primary Secondary Tertiary	Channel translating equipment (CTE) Group translating equipment (GTE) Supergroup translating equipment (STE)

Table 5.3 Basic channel assemblies provided by CCITT-recommended translating equipments

Equipment	Input to equipment	Multiplexed output		
		Number of speech channels	Frequency band (Bandwith) kHz	Assembly classification
CTE	12 speech channels	12	60—108 (48)	Basic group
GTE	5 basic groups	60	312—552 (240)	Basic supergroup
Procedure 1 STE	5 basic supergroups	300	812—2044 (1232)	Basic mastergroup
MTE	3 basic mastergroups	900	8516—12 388 (3872)	Basic super-mastergroup
Procedure 2 STE	15 basic supergroups	900	312—4028 (3716)	Basic 15 supergroup assembly (15 SGASS)

carrier leak from a SSBSC generator. However, this is at the discretion of the designer and will not be considered here when the filter performance of assemblies is being assessed.

An important feature of these CCITT hierarchies is that speech channels are allocated a nominal 4 kHz bandwidth and all carriers used have frequencies that are multiples of 4 kHz. This will be seen to have practical advantages when considering the effects of carrier leaks on a speech channel. The CCITT does not concern itself with the manner in which an equipment is realised and the carriers shown on frequency plans, such as those in Sections 5.4 and 5.5 are 'virtual' rather than real. A virtual carrier is a carrier of such a frequency as will perform the required translation in one stage. In a practical equipment, the translation may be carried out using more than one stage and using carrier frequencies other than the virtual.

Each of the basic channel assemblies recommended by the CCITT includes a reference pilot, shown by an open-headed arrow on frequency plans, and the functions of these are discussed in Section 5.4.6. These arrows and the triangles representing channel assemblies are identified by a stroke system according to the number of translating stages required to make up the assembly, i.e. one stroke for a group, two for a supergroup, etc. The various channel assemblies are described as 'basic' when they are in the frequency bands shown for them in Table 5.3. When translated to other frequency bands they are not then termed 'basic'. It will be seen on the various frequency plans for the assemblies that, as far as possible, common numbering schemes are used; for example, the supergroups in the basic mastergroup are numbered 5–8 corresponding with the supergroups in the same frequency band in the 15 SGASS.

A consequence of using a hierarchical system is that the overall performance is determined by a complex interaction between equipments in the various levels. For example, the significance of any unwanted signals at the output of a particular translating equipment can only be determined by considering the effects in detail on the equipment in the following level (higher or lower) of the hierarchy. On the other hand, the detailed filtering requirements of an equipment can only be determined from a knowledge of the range of signals applied to its input. Thus, great care is required in maintaining compatibility between equipments designed and manufactured at different times and from different sources.

5.4 Basic channel assemblies

5.4.1 The basic group

(i) *4 kHz channels*

Channel translating equipment [3] is used to assemble the basic group as shown in Fig. 5.5. Twelve speech channels, each having a defined passband of 3·1 kHz extending from 300 Hz to 3400 Hz, are assembled into the band 60—108 kHz. Virtual carriers of 64, 68, 72, 76, 80, 84, 88, 92, 96, 100, 104 and 108 kHz are used and the lower sidebands are selected for use. A spacing of 4 kHz between carriers, together with the nominal 3·1 kHz channel passband, gives a guard band between channels of 900 Hz. On Channel 1, which has a carrier frequency of 108 kHz, the carrier to transition-band ratio of the sideband-selecting filter is $108/2 \times 0·3 = 180$, which represents the most stringent filter performance required in the hierarchy of translating equipments.

Two unwanted signals on the output of each channel SSBSC generator are of particular importance: these are the unwanted upper sideband about the carrier fundamental and the carrier leak. At the transmit end, any residue of the upper sideband of a channel will fall into the passband of the adjacent higher-frequency channel and cannot be removed by any filtering at the receive end of that channel. It therefore appears as crosstalk but, because frequency inversion occurs, it will not be intelligible. However, it will have a speech-like rhythm and its level must be carefully controlled. A crosstalk power ratio of at least 60 dB has been set by the CCITT for this source of interference. This is with the disturbing channel loaded with a random noise signal having a power/frequency curve shaped to simulate speech and the crosstalk measured in the disturbed channel using a psophometer. An example of a channel filter characteristic is shown in Fig. 5.6.

Through-signal leak should be well out of the band of the basic group and does not normally represent a problem. However, if the bandwidths

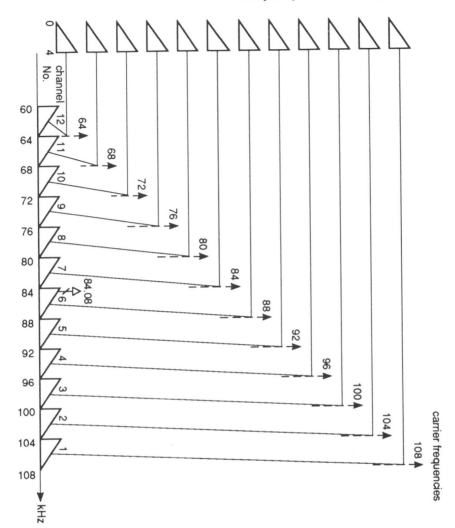

Fig. 5.5 Formation of the basic group

of the channels are mainly defined by the post-modulation filters, the possibility exists that significant through-signal leak can occur if signals higher than about 60 kHz are applied to the channel inputs. Impulsive noise coming in on the speech paths can give rise to this problem. For example, dial pulses can contain high-energy components over a wide frequency range. They should not be allowed to appear at the input to any channel.

Maximum carrier leak recommended by the CCITT from any one channel is −26 dB0 at the send end. This will be translated at the receive end to appear as zero frequency in the associated channel and as a 4 kHz tone in the adjacent higher-frequency channel. Zero frequency is of no concern.

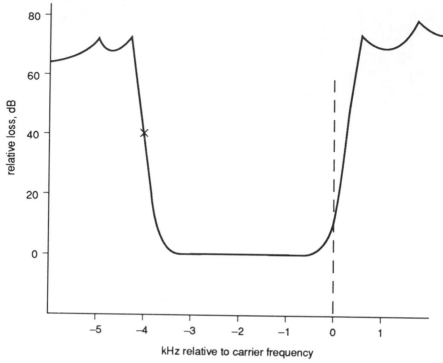

Fig. 5.6   Channel filter characteristic

However, the 4 kHz tone must be controlled to a low level, about −80 dBm0p, so as to be virtually inaudible. This can be achieved as follows:

(a)   About 40 dB loss at $f_c - 4$ kHz relative to the passband loss in the receive channel filter (see Fig. 5.6).
(b)   −26 dBm0 maximum carrier leak
(c)   15 dB weighting factor at 4 kHz

Because the basic group is contained well within an octave, the harmonics of carrier leaks, together with any sidebands about them, are at frequencies higher than 108 kHz and therefore outside the group band. As for through-signal leak, the possibility exists of impulsive noise generating a lower sideband on the second harmonic of Channel 12 carrier, which falls into the passband of Channel 1.

(ii) *Signalling*
A factor of importance to channel translating equipment is the type of signalling used between switching centres at each end of the transmission system. Where 'inband' signalling is used, this does not particularly affect the multiplexing equipment as all signalling information is carried, at a suitable level, within the speech band. If 'outband' signalling is used, provision for this must be made in the channel translating equipment. This

involves providing a signalling channel, associated with each speech channel, just outside the normal passband and centred at 3825 Hz. At the send end, a signal of 3825 Hz is applied to the channel under the control of a DC received over a signalling wire from the switching equipment. In the translating equipment at the receive end, the 3825 Hz is arranged to control a DC sent over a separate wire to the switching equipment. Signalling up to a rate of about 10 pulses per second can be achieved. Extra filtering is required to ensure that the 3825 Hz is not heard by the user and to prevent speech interfering with the signalling. In large installations using outband signalling, it must be ensured that the 3825 Hz signalling tones used on all channels are not in phase synchronism, as this may cause loading problems in the higher-order assemblies [4]. The use of 3825 Hz as a signalling tone effectively widens the passband required of each speech channel and this requires a rather more stringent design of through group filter (Section 5.8).

(iii) *3k Hz channels*
For long submarine cable systems, where the cost of the line equipment is very high, the basic group can be formed from 16 channels spaced at 3 kHz intervals across the basic group band [5]. A nominal passband from 250 to 3050 Hz is provided for each channel. The CCITT recommends [6] that the assembly be formed by taking lower sidebands of virtual carriers having frequencies of:

$$105 \cdot 15, 99 \cdot 15, 93 \cdot 15, 87 \cdot 15, 81 \cdot 15, 75 \cdot 15, 69 \cdot 15 \text{ and } 63 \cdot 15 \text{ kHz}$$

for odd-numbered channels and the upper sidebands of virtual carriers having frequencies of:

$$104 \cdot 85, 98 \cdot 85, 92 \cdot 85, 86 \cdot 85, 80 \cdot 85, 74 \cdot 85, 68 \cdot 85 \text{ and } 62 \cdot 85 \text{ kHz}$$

for even-numbered channels. Channels are numbered 1–16 in order of decreasing frequency. As the carriers are not multiples of 4 kHz, the carrier leaks from group and supergroup translating equipments have to be kept to a very low level to avoid unacceptable levels of tones being generated in some channels. Also, special group and supergroup reference pilots (Section 5.4.6) have to be used. These considerations and the lower bandwidths of the channels have precluded the widespread use of this 3 kHz channel assembly.

5.4.2 The basic supergroup
The basic supergroup is assembled by group-translating equipment [7] in accordance with the frequency plan shown in Fig. 5.7. Five basic groups are assembled into the band 312—552 kHz using the lower sidebands of virtual carriers having frequencies of 420, 468, 516, 564 and 612 kHz. The triangles on the frequency plan of Fig. 5.7 indicate that the sidebands are erect. In total, 60 channels are spaced at 4 kHz intervals over the entire 240 kHz band.

Each of the five basic groups applied to the input ports of the group-translating equipment have passband spectra that have been well defined by the channel filters in the channel translating equipment. There is no need then for filters at the send end of the group translating equipment to

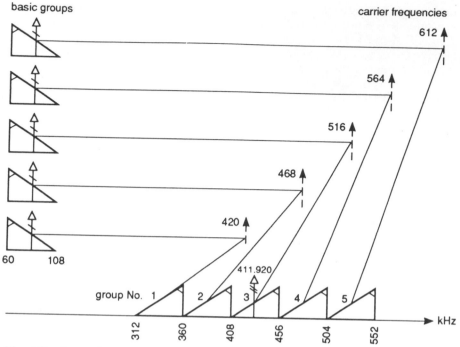

Fig. 5.7 Formation of the basic supergroup

perform this function. Sideband selection following each modulator is of course necessary. For each of the five carriers, the frequency separation between the wanted lower sideband and the unwanted upper sideband is $60 \times 2 = 120$ kHz. On the highest-frequency carrier this gives a carrier-frequency to transition-band ratio of about 4·6; this represents a very modest filter performance compared to that required for the channel translating equipment. Through-signal leak in the band 60—108 kHz does not cause interference in the passband of the supergroup. Maximum carrier leak for the send end of group-translating equipment has been set by the CCITT to be −47 dBm0. Referring to Fig. 5.7 and considering the send end of the equipment, it will be seen that the carrier leaks of groups nos. 1, 2 and 3 fall into the passbands of groups No. 3, 4 and 5. Because all carrier frequencies are multiples of 4 kHz, these carrier leaks are eventually translated by channel-translating equipment to fall at the same frequency as a channel carrier, in this case Channel 4 of each group. This does not add significantly to the carrier leak already allowed for channel-translating equipment, so it does not cause a problem for telephony channels. Group carrier leaks can, however, cause problems to some other types of signal and this is discussed in Section 5.10. Carrier leaks from groups nos. 4 and 5 fall outside the basic supergroup band, but they have to be taken into account when considering the formation of assemblies in the next level of the hierarchy.

5.4.3 The basic mastergroup

In CCITT Procedure 1, supergroup translating equipment [7] is used to assemble the basic mastergroup from five basic supergroups. Lower sidebands of virtual carriers having frequencies of 1364, 1612, 1860, 2108 and 2356 kHz are used to assemble 300 channels within the band 812— 2044 kHz. The translation is shown on the frequency plan of Fig. 5.8. There are guard bands of 8 kHz between supergroups and this is of importance for the through connections discussed in Section 5.8. The frequency spectrum occupied by the assembly is greater than an octave and some of the unwanted products listed in Table 5.1 fall inside the passband. At the send end, the upper sideband of the 1364 kHz carrier falls partly into the spectrum of supergroups nos. 7 and 8. Carrier leaks from supergroups nos. 4, 5 and 6 fall into the passband of supergroups nos. 6, 7 and 8, respectively, while the carrier leaks of supergroups nos. 7 and 8 fall out of band. These leaks should not be greater than −50 dBm0 each. The frequency of each carrier leak corresponds to a channel carrier frequency; the strategy for dealing with this is given in Section 5.4.1. The frequency spacing between upper and lower sidebands of each carrier is $312 \times 2 = 624$ kHz, giving a carrier-frequency to transition-band ratio of about 3·77 for the highest-frequency carrrier.

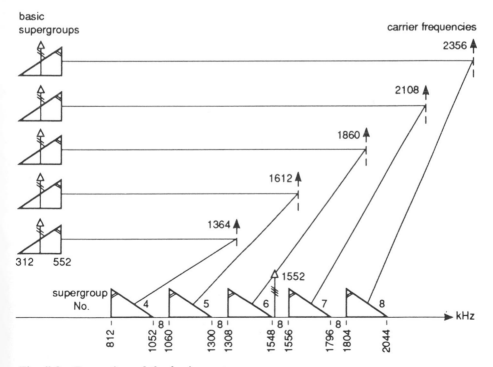

Fig. 5.8 Formation of the basic mastergroup

5.4.4 The basic supermastergroup

A basic supermastergroup is formed by mastergroup translating equipment [7] from three basic mastergroups using the lower sidebands of virtual carriers, having frequencies of 10 560, 11 880 and 13 200 kHz, as shown in Fig. 5.9. 900 channels are thus assembled within the band 8516— 12 388 kHz, with a gap between mastergroups of 88 kHz. At the send end, carrier leaks (which should not exceed −45 dBm0 each) from master-group 7 fall into mastergroup 8, and from mastergroup 8 fall into master-group 9. As in the formation of the mastergroup (Section 5.4.3), the unwanted upper sideband of the lowest-frequency carrier falls partly into the spectrum of the highest-numbered mastergroup in the assembly. Filtering require-ments are modest, with a carrier-frequency to transition-band ratio of approximately 8·13 for the highest carrier frequency.

5.4.5 The basic 15 supergroup assembly

In CCITT Procedure 2, supergroup translating equipment [7] assembles 15 basic supergroups into the band 312—4028 kHz to form a basic 15-supergroup assembly of 900 channels. The frequency plan is shown in Fig. 5.10. No carrier is required for supergroup no. 2, as it occupies the same frequency band as the basic supergroup. Consequently, while all the other supergroups in the assembly have inverted sidebands, that for supergroup no. 2 is erect. For the highest frequency supergroup the ratio of carrier

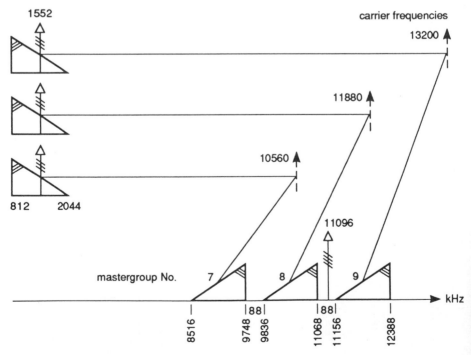

Fig. 5.9 Formation of the basic supermastergroup

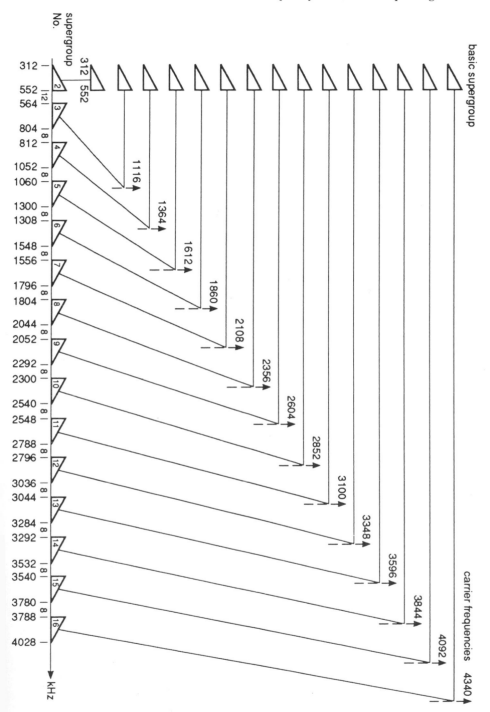

Fig. 5.10   Formation of the basic 15-supergroup assembly

frequency to transition bandwidth is only 6·95. However, filtration for this assembly requires careful design, as all the unwanted products listed in Table 5.1 play some part in generating interference within the passband of the assembly. Each carrier leak should be no higher in level than −50 dBm0.

5.4.6 Reference pilots

Each of the basic channel assemblies normally includes a reference pilot [8] having the frequency shown below:

Group reference pilot	84·08 kHz ± 1 Hz
Supergroup reference pilot	411·92 kHz ± 1 Hz
15 SGASS reference pilot	1552 kHz ± 2 Hz
Mastergroup reference pilot	1552 kHz ± 2 Hz
Supermastergroup reference pilot	11092 kHz ± 10 Hz

If the group or supergroup is used for wide-spectrum signals, as discussed in Section 5.10, then alternative pilot frequencies of 104·08 kHz and 547·92 kHz can be used. These pilots are level-stabilised to −20 dBm0 and added to the assembly at the send end of the appropriate translating equipment; i.e. the group reference pilot is added by the channel-translating equipment, the supergroup reference pilot by the group-translating equipment, etc.

These pilots are used in a number of ways. They provide a reference signal level for each assembly, which can be measured throughout a transmission system without the need to interrupt service, and so forms a useful aid to system maintenance and fault location. At the receive end of a link, they can be used as the control signal for automatic level regulators which compensate for the variation with time of the overall loss of the link. Failure of a pilot can be taken to indicate failure of the associated channel assembly; so circuits routed via it can be automatically busied out of traffic and an alarm given.

Both group and supergroup reference pilots have frequencies that are offset by 80 Hz from a value that is a multiple of 4 kHz. This avoids interference to the pilots by carrier leaks, but it results in 80 Hz or 3920 Hz tones in certain channels. Additional receive filtering is therefore required in those channels to reduce this interference to a level not greater than −73 dBm0p. At the send end of these channels, additional filtering is required to prevent speech signals interfering with the pilots. A further problem arises where channels are interconnected at audio frequencies. Here, it must be ensured that any pilot residue appearing at the audio output from one system is suppressed by an amount that will ensure that it is 40 dB lower in level than any pilot on the next system with which it might interfere.

Reference pilots for 15 SGASS, mastergroups and supermastergroups are located in the guard bands between channel assemblies. They can therefore be multiples of 4 kHz and do not fall into channel passbands.

As noted in Section 5.4.1, special pilots are required for the basic group and supergroup pilots when these assemblies carry 3 kHz channels. A frequency of 84 kHz is used for the group reference pilot and usually 544 kHz for the supergroup reference pilots. These fall in the guard bands between the 3 kHz channels.

5.5 Utilisation of the basic channel assemblies

The basic channel assemblies described in Section 5.4 can be used singly or in various combinations to provide a range of capacities between 12 and 10 800 channels, having bandwidths which match those of all the FDM metallic-line [9] and radio-relay systems recommended by the CCITT. Further frequency translation is usually [10] required to place all of the total channel assembly into the frequency band required either for transmission over metallic line links or for connection to the basebands of radio-relay systems. Table 5.4 lists examples of some of the channel assemblies that have been recommended by the CCITT. Included in the Table are channel assemblies for all recommended FDM line systems which use coaxial pairs and all the higher-capacity FDM radio-relay systems. As a result of there being two recommended procedures for building up basic channel assemblies, i.e. Procedure No. 1 based on the mastergroup and Procedure No. 2 based on the 15-supergroup assembly, there are generally at least two frequency plans for each transmission system. In some cases, mixtures of Procedure No. 1 and Procedure No. 2 have also been found useful.

60 MHz line systems carry the greatest number of standard telephone channels that the CCITT has been able to recommend for transmission

Table 5.4   Channel assemblies for use on coaxial-pair FDM line systems and FDM high-capacity radio-relay links

Total channel assembly				Recommended for		See Fig. no.
Channels	Basic assemblies used	Frequency Band (kHz)	Bandwidth (MHz)	Radio systems	Line systems (MHz)	
300	1 mastergroup	64—1 296	1·232	Yes	1·3	
300	5 supergroups	60—1 300	1·240	Yes	1·3	
900	1 supermastergroup	316—4 188	3·872	Yes	4·0	5·11
960	1 15 SGASSs plus 1 supergroup	60—4 028	3·968	Yes	4·0	5·12
1 200	4 mastergroups	312—5 564	5·248	Yes	6·0	
1 260	1 15 SGASSs plus 6 supergroups	60—5 636	5·570	Yes	6·0	
1 800	2 supermastergroups	316—8 204	7·888	Yes	No	5·18
1 800	2 15 SGASS's	312—8 120	7·808	Yes	No	5·19
2 700	3 supermastergroups	316—12 388	12·072	Yes	12·0	5·13
2 700	3 15 SGASSs	312—12 336	12·024	Yes	12·0	5·14
2 700	1 15 SGASSs plus 2 supermastergroups	312—12 388	12·076	Yes	12·0	5·15
3 600	4 supermastergroups	316—17 004	16·688	No	18·0	
3 600	4 15 SGASSs	312—16 612	16·612	No	18·0	
10 800	12 supermastergroups	4 332—59 684	55·352	No	60·0	5·16
10 800	12 15 SGASSs	4 404—59 580	55·17	No	60·0	5·17

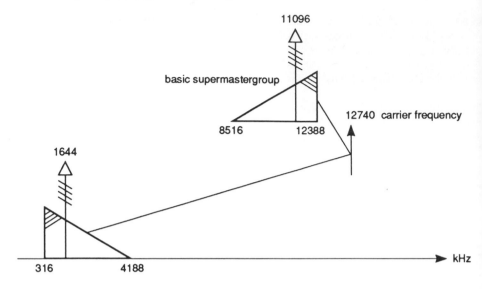

Fig. 5.11   4 MHz line system using mastergroups (*Procedure 1*)

over an analogue or digital transmission system using metallic pairs. Some 'stretched' non-standardised 60 MHz systems have capacities in excess of 10 800 channels.

CCITT recommendations for FDM line and radio transmission systems give frequency plans which show the further translation and combination procedures that need to be applied to the basic channel assemblies to form the channel assemblies required by these transmission systems. To illustrate the principles used in these recommendations, the procedures for some of the assemblies listed in Table 5.4 are shown in Figs. 5.11 to 5.19.

5.6 North American channel assemblies

Most channel assemblies used in North America differ from those recommended by the CCITT, although the important levels of basic group and

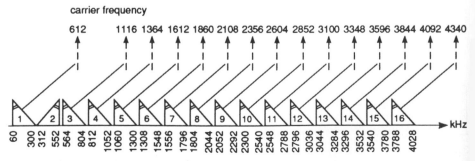

Fig. 5.12   4 MHz line system using 16 supergroups (*Procedure 2*)

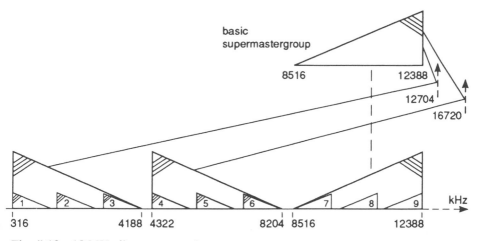

Fig. 5.13 12 MHz line system using supermastergroups

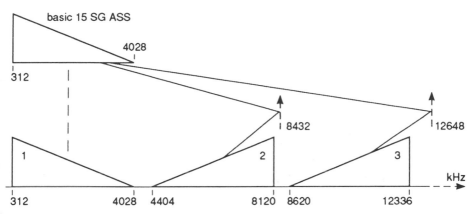

Fig. 5.14 12 MHz line system using 15-supergroup assembly

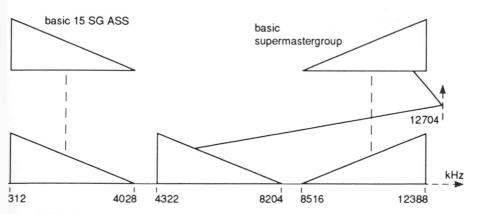

Fig. 5.15 12 MHz line system using hybrid assembly

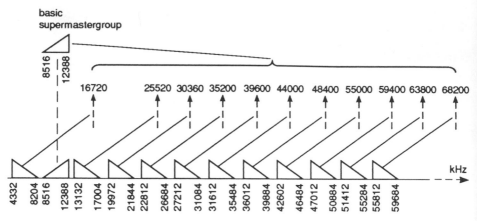

Fig. 5.16   60 MHz line system using supermaster groups

supergroup are identical. At the next level in the hierarchies, the master-group used in Bell systems differs significantly from that of the CCITT. The Bell System mastergroup contains 600 channels in the frequency band 564—3084 kHz and is formed from 10 basic supergroups, as shown in Fig. 5.20. Four important long-haul coaxial-cable systems used by Bell are shown in Table 5.5, together with an outline of the channel assemblies used.

The channels for the L1 system [11] are asembled from 10 supergroups in a manner similar to that shown in Fig. 5.12, except that supergroup 9 is the upper sideband of an 1860 kHz carrier and supergroup 10 is the lower

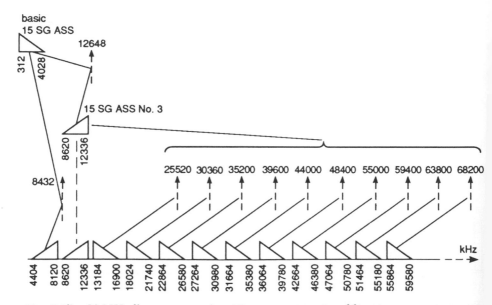

Fig. 5.17   60 MHz line system using 15-supergroup assembly

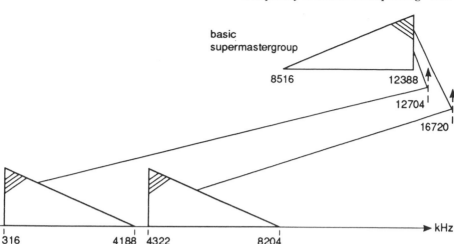

Fig. 5.18   1800-channel radio link using supermaster groups

sideband of 3100 kHz carrier. This assembly, occupying the band 60—2788 kHz, is known as the L600 multiplex.

For the L3 system [11], use is made of the Bell basic mastergroup, as shown in Fig. 5.21. One of these is transmitted to line without further translation together with a basic supergroup (in the band 312—552 kHz), where they form mastergroup 1. Two other basic mastergroups are separately modulated by a 13 000 kHz carrier and the lower sideband selected to place them in the band 9916—12 436 kHz. One of these is then further modulated using a 15 600 kHz carrier and the lower sideband is selected to place the assembly in the band 3164—5684 kHz, where it forms line mastergroup 2. The remaining mastergroup is further modulated by an

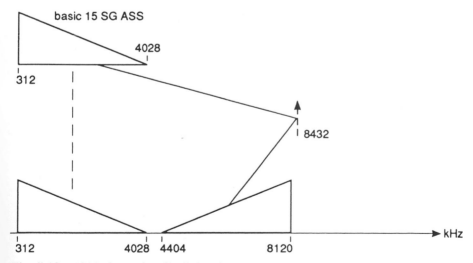

Fig. 5.19   1800-channel radio link using 15-supergroup assembly

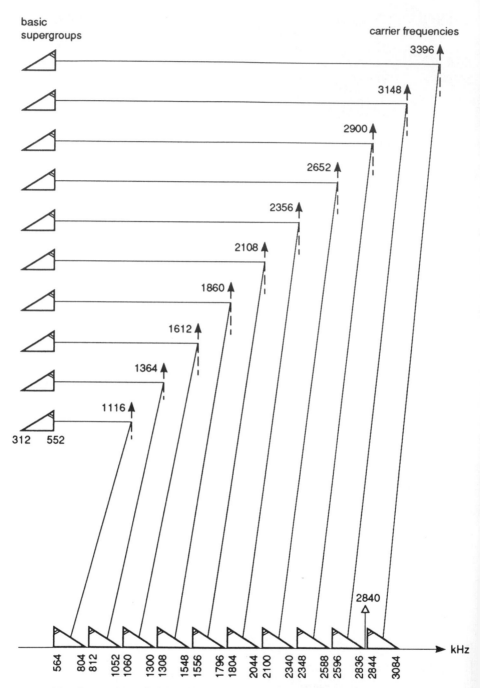

Fig. 5.20   Formation of the Bell System basic mastergroup

Table 5.5 Coaxial cable systems used by Bell

Coaxial cable system	Number of channels	Frequency band used (kHz)	Channel assemblies used
L1	600	60—2 788	10 supergroups
L3	1 860	312—8 284	3 mastergroups + 1 supergroup
L4	3 600	564—17 548	6 mastergroups
L5	10 800	3 124—51 532	18 mastergroups (3 jumbogroups)

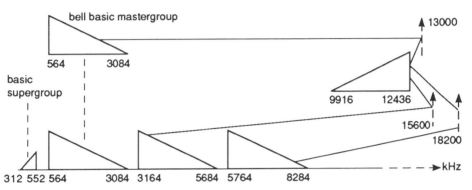

Fig. 5.21 Bell L3 line system

18 200 kHz carrier; the lower sideband 5764—8284 kHz is selected and transmitted to line as mastergroup 3. Thus, the whole assembly occupies the band 312—8284 kHz. Fig. 5.21 shows the L3 system frequency plan.

A frequency plan for the L4 system [12] is shown in Fig. 5.22. One basic mastergroup is transmitted direct to line without translation, while five others are translated as shown. Thus, the complete assembly occupies the band 564—17 548 kHz.

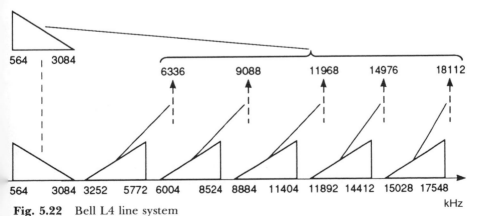

Fig. 5.22 Bell L4 line system

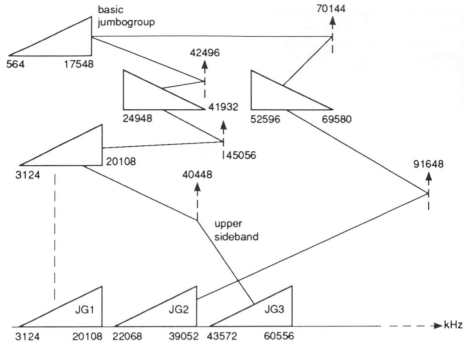

Fig. 5.23   Bell L5 line system

With the L5 system [13], the concept of a Basic Jumbogroup is introduced. This consists of six mastergroups in the same format as used for the line-transmission assembly of the L4 system (see Fig. 5.24). Three such jumbogroups are assembled (after multiple steps of modulation) into the band 3124—60 556 kHz, as shown in Fig. 5.23. This provides a 10 800-channel assembly.

5.7 Carrier supplies

5.7.1 Carrier generation
Distortionless waveform transmission over a SSBSC system requires the carriers supplied to the translating equipment at each end of the link to be in phase synchronism. For telephone speech transmission it is not necessary to preserve the waveform of the signal; indeed, acceptable quality is maintained with the speech changed in frequency by up to about 15 Hz. However, most administrations want speech channels to be also suitable for the tranmission of multi-channel voice-frequency telegraph signals and this requires the overall change in frequency to be not greater than about 2 Hz. The CCITT has accordingly adopted this value for the maximum frequency error allowed on any speech circuit [14]. To ensure a reasonable probability that this error will not be exceeded, the CCITT recommends [15] that the

frequencies of the various carrier supplies to translating equipment should
be within the following limits:

Channel carriers	± 1 part in 10^6
Group carriers	± 1 part in 10^7
Supergroup carriers	± 1 part in 10^7
Mastergroup carriers	± 5 parts in 10^8 for 12 MHz systems
Supermastergroup carriers	± 5 parts in 10^8 for 12 MHz systems
Mastergroup carriers	± 1 part in 10^8 for 60 MHz systems
Supermastergroup carriers	± 1 part in 10^8 for 60 MHz systems

Economic provision of large numbers of carriers having highly-stable and
closely-defined frequencies is a basic problem with SSBSC systems. In all
but the smallest installations it is solved by providing each repeater station
with a 'master oscillator' from which all carrier supplies are derived by
processes such as frequency multiplication, division and harmonic generation [16, 17]. Oscillators using quartz-crystal resonators operating in tem-
perature-controlled ovens are economically acceptable as master oscillators.
However, their frequency stability with time is such that they can only
provide the necessary accuracy for a limited period. This can range from
a few weeks to many months, depending upon the oscillator design and the
transmission system to be supplied. Long-term frequency accuracy in a
network has to be ensured by establishing, at a suitable location, a network
or national frequency standard against which all master oscillators in the
network can be corrected [15]. Suitable frequency standards having
accuracies better than 1 part in 10^{11} are rubidium or caesium clocks. Signals
derived from the frequency standard are distributed throughout the
network as 'frequency comparison pilots' (FCP) over the normal bearer
circuits without, of course, being allowed to pass through any frequency-
translation process using local carrier supplies. A number of frequency
allocations for FCPs have been made by the CCITT for various line systems,
with 60 300 and 4200 kHz being widely used. Any change in frequency
required to make an FCP conform to these allocations must be carried out
by means, such as harmonic generation, which do not detract from the
frequency stability.

With a frequency-comparison pilot available at a repeater station, the
master oscillator can be corrected, either manually at set intervals of time
or continuously, by being included in a frequency or phase lock loop. If
locking is used, it must be ensured that any noise (particularly of an impulsive
nature) picked up by the FCP in the distribution network does not cause
significant changes in the frequency of the controlled master oscillator. It
is also vital that the master oscillator continues to function with the required
frequency accuracy even if the FCP is disconnected by tranmission or other
difficulties. Frequency accuracy must be maintained for a period of time
long enough to allow restoration or replacement of the failed FCP. This
will probably require memory circuits to hold the correction signal that was
in force in the control loop immediately before the FCP failure.

All carriers have frequencies that are multiples of 4 kHz and, at various
levels in the hierarchy, they are often also multiples of 12, 124, 440 or

2200 kHz. The master oscillator itself is unlikely to be running at any of these frequencies. Its frequency will have been chosen on the basis of the stability required, cost and available technology. Frequencies from a few tens of kilohertz to several megahertz have been used. Often, the first step in deriving carriers from the master oscillator is the generation of these five base frequencies. From them all the others can be generated: e.g. channel carriers are harmonics of 4 kHz, group carriers are odd harmonics of 12 kHz and supergroup carriers are odd harmonics of 124 kHz. All the carriers shown in Fig. 5.17 for the 60 MHz system can be derived from the 124, 440 and 2200 kHz base frequencies. In any repeater station, the frequency stability of a master oscillator must conform to that of the most demanding equipment that is fed from it. Consequently, it is usual for the carrier supplies to equipment in the lower orders of the hierarchy to have a higher stability than the minimum recommended for them by the CCITT.

For each carrier frequency generated a number of outputs will be required, varying perhaps from several thousands for each channel carrier down to less than ten for each 60 MHz carrier frequency. Isolating amplifiers and networks are therefore used to reduce the probability that a fault on the carrier supply line to one modulator will affect any others. A centralised carrier-generating system inevitably contains many equipment units, the failure of any one of which could lead to the failure of many channels. This is guarded against by extensive use of duplication, together with automatic monitoring, alarm and change-over facilities. In larger stations, more than one master oscillator installation is usually provided and large-capacity systems, such as the 60 MHz, may be treated as special cases with their own master oscillators.

Carriers generated in a centralised system are subject to contamination by a variety of unwanted spectral components. Crosstalk between carrier supplies of different frequencies can occur via cables or equipment components; noise generated in the master oscillator can be increased, relative to the carrier, during subsequent processing; modulation of the carrier by power-supply frequencies can occur in active devices. These three potential sources of interference are considered below.

5.7.2 Crosstalk

Crosstalk between adjacent-frequency carrier supplies can give rise in turn to crosstalk between channels in an assembly. For example, if the carrier applied to the modulator of supergroup 4 (1364 kHz) in a supergroup translating equipment (see Figs. 5.8 and 5.10) is accompanied by a crosstalk component of the carrier for supergroup no. 5 (1612 kHz), then the signal applied to the supergroup no. 4 modulator will modulate the crosstalk component as well as the wanted carrier. A sideband will be generated in the 1060—1300 kHz position. Energy in the lower-frequency portion of this band will not be much attenuated by the post-modulation filter of supergroup no. 4 and will therefore appear as a crosstalk component in supergroup no. 5. To meet a crosstalk ratio of 80 dB in the channel assembly requires the crosstalk ratio between adjacent carriers to also be about 80 dB.

5.7.3 Noise

Master oscillators generate noise along with the wanted signal. This oscillator noise commonly has a power/frequency distribution that, over a limited range, increases at the rate of 6 dB/octave towards the carrier frequency as shown in Fig. 5.24.

Subsequent processing inevitably involves filtration which will modify the shape of the noise spectrum. Frequency multiplication will have the effect of increasing the level of the noise, relative to the carrier level, in the same ratio; i.e. if the carrier frequency is multiplied N times, then the level of the noise relative to the carrier level is increased by $20 \log_{10} N$ dB. Conversely, frequency division will reduce the relative level of the noise. Thus, the level of noise accompanying a carrier will depend upon the ratio of the carrier to master-oscillator frequency. The noise can be regarded as being a set of sidebands about the carrier, the result of modulating the carrier in both phase and amplitude. Any processing of the carrier involving amplitude limitation (and switching type modulators, among other devices, do just this) will reduce the sideband components representing amplitude modulation but will leave intact the phase-modulation sidebands. This will lead to phase modulation of any wanted signal modulated onto the carrier in a translating equipment and the spectrum of the phase-modulation sidebands about the signal will be similar to that of the carrier. A number of effects will arise. On speech, the noise in a channel will increase with speech loading; for MCVFT systems the noise in the channel will be above that of the idle-channel noise; data signals will have jitter impressed on them. In all cases, if the carrier noise spectrum is wide enough, noise will be generated in channels adjacent to the one directly affected. No specific limits have been set for noise from this source; therefore it must be controlled to a low level so as not to add significantly to noise from other sources on a link. Because of the noise-enhancement effects of frequency multipliers, carrier noise is likely to present the biggest problems on high-frequency carriers such as are used on 12, 18 and 60 MHz systems.

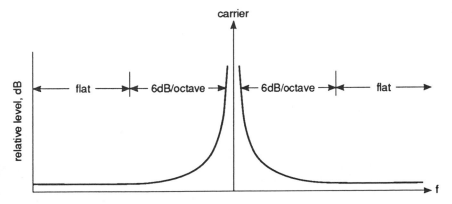

Fig. 5.24   Typical spectral distribution of noise about a carrier

5.7.4 Hum modulation

DC power supplies in a repeater station often have superimposed on them low levels of AC at the frequency of the mains power supply or its harmonics. When applied to active circuits in the carrier-generation equipment, this low-frequency AC or hum can modulate the carrier supplies. Any subsequent frequency multiplication or amplitude limiting will, as in the case of noise, increase the relative level of the hum sidebands and remove any amplitude-modulation components. Wanted signals modulated on to such a carrier in a translating equpment will themselves be hum modulated and give rise to noise and jitter in the channel. The CCITT has recommended [18] that, on a signal at the output of a translating equipment, the level of any low-frequency sidebands about that signal shall be at least 63 dB below it. If, in the translating equipment, the carrier supply is the only significant source of this interference, then the hum sidebands about the carrier must also be at least 63 dB lower in level than the carrier.

5.8 Through connections

Where several links from different routes terminate in a repeater station, there is likely to be a need to through connect blocks of circuits from one line link to another. Better technical performance and economy result if this is done without translating each channel down to audio frequency. For instance, if twelve channels need to be through connected, it is must better to through connect a basic group assembly rather than twelve individual audio channnels. Methods of through connecting all the basic channel assemblies have been recommended by the CCITT [19]. In each case, extra bandpass filters are required in the through-connection path to reduce the level of out-of-band components that accompany all basic channel assemblies at the receive end of a link. If these unwanted components were not reduced in level they would, after further translation, cause unacceptable crosstalk and interference to adjacent channel assemblies.

For the case of the basic group, Fig. 5.25 shows (for one direction of transmission) a 'through group filter' connected between two group-translating equipments. Each basic group at the output of the receive group translating equipment is accompanied by vestiges of adjacent groups in the receive basic supergroup as shown by the broken triangles in Fig. 5.25. These vestiges, if applied to the input of the send group translating equipment, would fall into the passband of the adjacent groups in the send basic-supergroup assembly. Similar vestiges are applied to the receive channel translating equipment, but they are here suppressed by the steep-sided response of the channel filters. From this it follows that the attenuation slope at each end of the passband of a through group filter has to be similar to that of a channel filter. In addition to vestiges of adjacent groups, the output spectrum of a group-translating equipment may contain other out-of-band components which, if not suppressed, could give rise to interference in a through-group connection. Particularly important are the upper sideband of the carrier fundamental and sidebands about harmonics of the

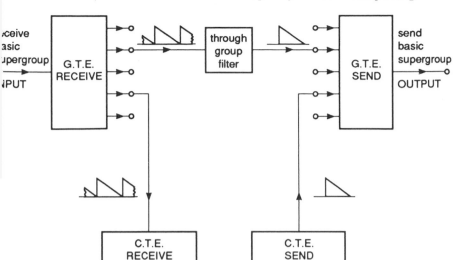

Fig. 5.25 Through group connection

carrier, all of which are also accompanied by vestiges of the adjacent groups. If applied to the input of the send group-translating equipment, they will be translated by the send group carrier and its harmonics to form possible interference products.

Similar considerations apply to through-connection filters for other basic channel assemblies [20]. These all have higher operating frequencies than the through-group filter and gaps of several kilohertz (see Figs. 5.8, 5.9 and 5.10) have to be left between the assemblies forming mastergroups, super-mastergroups etc., so that the through-connection filters can be economically realisable. All through-connection filters have to ensure that any crosstalk due to out-of-band components is at least 70 dB below the wanted signal.

Although through connection of channels is most often carried out using basic channel assemblies, the possibility exists of direct through connection at line frequencies without translation. In some circumstances, this may be the most economic solution. A wide range of possibilities exist, but no particular scheme has found widespread application. In some cases, sacrifice of part of the adjacent channel assemblies may be necessary if the filter design is to be economically realisable.

5.9 Noise and system loading

It is a CCITT objective that the total noise due to translating equipment on a hypothetical reference circuit should not exceed 2500 pW0p. To meet this, the maximum values of noise that should be produced by translating equipment have been recommended [21] as follows:

Channel translating equipment:	200 pW0p
Group translating equipment	80 pW0p
Supergroup translating equipment	60 pW0p
Mastergroup translating equipment	60 pW0p
Supermastergroup translating equipment	60 pW0p
15 SGASS translating equipment:	60 pW0p

This is the noise generated by the send and receive sides of the translating equipment taken together. It is meant to cover noise from all sources, including thermal, intermodulation, crosstalk and power supplies. To check this performance it is necessary, because of intermodulation noise, to load the equipment with a signal simulating the normal multiplex signal. The CCITT recommends [22] that this should be a white noise signal having a level of:

$$-15 + 10 \log_{10} N \, \mathrm{dBm0} \text{ for } N \geqslant 240$$

$$\text{or } -1 + 4 \log_{10} N \, \mathrm{dBm0} \text{ for } 12 \leqslant N < 240$$

where N is the total number of channels in the multiplex.

Considering the channel-translating equipment, the check is carried out by loading 11 channels at the send end with white-noise signals (from 11 uncorrelated sources) such that the level at the output (at the basic group point) is $+3 \cdot 3$ dBm0. The channels adjacent to the unloaded channel should not be loaded with noise at a level higher than -15 dBm0 on their inputs. At the receive end, the noise on the output of the unloaded channel should not exceed 200 pW0p. Other types of translating equipment are checked in a similar manner, except that 'quiet' channel slots have to be provided by means of bandstop filters.

The overload point of translating equipment should be such that the signal peaks of the multiplex signal can be carried. Because of the noise-like nature of a multiplex signal there is no unique peak value, so the 'peak' is usually taken to be a value which has a probability of being exceeded of 10^{-5}. A mean power of -15 dBm0 is assumed for each channel. This is the mean with time and the mean for a large number of channels. The peak/mean ratio of a multiplex signal depends upon the number of channels in the multiplex; it is large for small numbers of channels, becoming asymptotic to 13 dB as the number of channels increases. This is the same value as for a random noise signal having a Gaussian amplitude distribution. Based on the work of Holbrook and Dixon, [23] the CCITT has defined the power of a sinusoid whose amplitude is equivalent to the peak of a multiplex signal containing N channels. A range of values is given in Table 5.6.

Table 5.6   Equivalent RMS sine-wave power of the peak of a multiplex signal of N channels

Number of channels (N)	12	60	300	960	2700	10 800
Equivalent peak power (dBm0)	19	20·8	23	27	32	36

Thus, the transmission path through a translating equipment must have an overload point at least equal to the value shown in Table 5.6 plus some margin to allow for misalignment and variation of signal levels. Additionally, at the receive end of translating equipment, account should be taken of the actual number of channels carried at any point. This may be greater than the nominal number because of filter characteristics in a particular design of equipment.

5.10 Signals other than telephony

5.10.1 General

By far the greater part of the traffic carried by telecommunication networks is telephone speech and it is the characteristics and requirements of such signals that have been taken into account in the design of multiplexing equipment and systems. Nevertheless, the widespread availability of telephone transmission facilities has encouraged their use for the transmission of other types of signal. The transmission format adopted for such signals must be compatible with the requirements and performance of the telephone-transmission facilities, as only rarely has it been economically possible to change these to suit any special needs of signals other than speech. Some of the more important of these other services are discussed below.

5.10.2 MCVFT systems [24]

Multichannel voice-frequency telegraph (MCVFT) systems were early users of speech channels and, as pointed out in Section 5.7.1, limits on the overall frequency error of a channel have been set to meet their requirements. Historically, a loading greater than −15 dBm0, i.e. up to −8 dBm0 per channel, has been used for telegraph systems. Because the peak/mean ratio of speech in a single channel is much greater than that of a MCVFT signal, no problem has been caused to assemblies having a small number of channels. However, in larger assemblies, the number of channels used for this purpose must be carefully controlled to prevent overload. A loading of −12 dBm0 per channel has been accepted as a reasonable compromise between the noise requirements of the MCVFT system and the loading requirements of FDM systems. The MCVFT signal is, unlike speech, transmitted continuously and so has an activity factor of 1. Also unlike speech, the power of an MCVFT signal is spread almost uniformly across the channel passband. Hence, it may generate more crosstalk, and may itself be more susceptible to crosstalk than a speech signal. To ensure an adequate performance, crosstalk tests on circuits can be carried out using test signals at either end of the speech band.

5.10.3 Data

Data transmission has made much use of analogue telephone transmission facilities [25]. Usually it is sent over individual speech channels and a problem is that its activity factor is likely to be higher than that of speech. Operation at −13 dBm0 is possible if not more than 5% of channels carry data. Speech channels cannot transmit DC and may also introduce frequency error. Data signals are made compatible with these conditions by modulating a sub-carrier in the data modem, either in amplitude or phase, and transmitting this over the speech channel. Group-delay distortion has small effect upon speech signals, but it is of major concern to data. The CCITT has recommended limits on the group-delay distortion of channel translating equipment, but it is usually necessary for the data modem to provide some measure of overall correction. For higher-speed data, group links of 48 kHz bandwidth or supergroup links of 240 kHz bandwidth can be used. Again, it is necessary for the data to be modulated onto a subcarrier and overall correction of the group-delay distortion is required. An additional problem can be caused by carrier leaks from adjacent groups or supergroups. For example, referring to Section 5.4.2, in a group-translating equipment, the carrier leaks from groups No. 1, 2 and 3 fall into the passband of groups No. 3, 4 and 5 and will cause interference if these are in use for high-speed data transmission. Similarly, when the supergroup is combined into a mastergroup, the out-of-band carrier leaks of groups 4 and 5 fall into the passbands of groups 1 and 2 in the adjacent supergroup lower in frequency. A limit of −40 dBm0 for carrier leak into a group-band data circuit has been recommended by the CCITT. To ensure a high probability that this will be met on a circuit passing through a number of group links, the carrier leak allowed from group translating equipment is −47 dBmO.

5.10.4 Programme transmission

Sound programme signals can be sent over telephone transmission facilities using the bandwidth normally occupied by several telephone channels [26]. The CCITT has standardised the following:

> 50 Hz − 6·4 kHz using the bandwidth of 2 speech channels
> 50 Hz − 10 kHz using the bandwidth of 3 speech channels
> 30 Hz − 15 kHz using the bandwidth of 6 speech channels

These are for monophonic transmissions; for stereophonic signals a pair of 15 kHz circuits occupying the whole of a group band is used. To achieve a satisfactory performance, the signals are usually companded and emphasis/deemphasis networks are used. Single-sideband transmission is used and, in view of the low audio frequencies transmitted, a particularly-stringent performance is required of the sideband-selection filter. The 2 Hz frequency shift that can be encountered in a telephone network is just about tolerable for monophonic signals. However, for stereophonic signals, a pilot signal is transmitted to form a reference against which the signal can be demodulated. Certain crosstalk requirements, for instance for through-connection filters, are more stringent than for telephone speech. Also, on some 15 kHz circuits certain carrier leaks fall inband and have to be reduced to a tolerable level by means of notch filters.

5.10.5 Television

Analogue television signals can be transmitted over high-capacity FDM radio-relay links and 12, 18 and 60 MHz FDM line links [27]. In the case of radio-relay links, the video signal can be connected directly to the baseband input [28]. However, for transmission over line links, special television modulation/demodulation equipment is required that will ensure the television waveform is preserved. In the television modulating equipment at the send end of the link, the video signal is amplitude modulated onto a 'television' carrier. Vestigial-sideband (VSB) modulation is necessary to accommodate the very low frequencies found in video signals. Negative modulation is used, i.e. the carrier level is higher for black level than for peak white, disregarding any effects due to any special processing used such as video pre-emphasis, overmodulation or reference-carrier transmission. At the receive end, the television demodulating equipment must be capable of preserving the television waveform even though frequency shift may have occurred during transmission. This implies the use of synchronous or envelope detection.

For the 12 MHz line system [29, 30] the bandwidth of the television signal is restricted to 5·5 MHz. A television carrier of 6·799 MHz is used with a vestigial sideband width of 500 kHz. The modulated television signal therefore occupies the band 6·299 MHz to 12·299 MHz and no further frequency translation is required before this signal is sent to line. Up to 1200 telephone channels can be carried in the band below this, assembled in accordance with Procedure No. 1 or No. 2 or a mixture of the two. The television carrier is 'overmodulated' with a modulation ratio of 130% for a television luminance-bar signal. A video line clamp is generally required prior to the television modulator and this may have implications for a television signal which includes 'sound in synchs'.

Arrangements for television transmission on the 18 MHz [31, 32] and 60 MHz [32, 33] line systems are similar to each other. They are less restrictive and of better performance than on the 12 MHz system. This is achieved by providing a video bandwidth of 6·0 MHz and a vestigial sideband width of 1·0 MHz. Better loading of the line system and noise performance are achieved by the use of video pre-emphasis. Although VSB modulation is used, a high-level reference carrier is also transmitted to line and can be used in the demodulator to aid accurate waveform demodulation of the television signal. Low-frequency components in the television signal are reduced by up to 18 dB in a network at the send end and restored by a corresponding network at the receive end. This avoids interference to the reference carrier and makes the use of a video clamp unnecessary. Television transmission over these systems is therefore independent of the structure of the television signal. Because of the use of video pre-emphasis and a reference carrier, it is difficult to measure a modulation ratio; instead, the CCITT has defined a reference level of sideband that must be obtained with a defined video input. The 18 MHz system can carry two television channels or one television channel and up to 1800 telephone channels, using Procedure No. 1 or No. 2. The 60 MHz system can carry six television channels with no telephone channels, or one or two television channels each of which displaces two supermastergroups or two 15 SGASSs. Limitations in the performance of the line link prohibit a mixture of telephone channels with more than two television channels. For both the 18 MHz and 60 MHz systems, further frequency translation is required to place the television channels into the frequency bands required for transmission over the line link.

5.11 Transition equipment

5.11.1 General

Networks which are required to evolve from analogue to digital working may, depending on the evolution strategy adopted, pass through a transition period during which both methods of working co-exist and interconnection between them is required. Examples of such interconnections are shown in Fig. 5.26. The equipments which carry out the interconnections are known collectively as 'transition equipment' [34]. Three types have been standardised by the CCITT [35]:

(i)   Transmultiplexers (TMUX)
(ii)   FDM codecs,
(iii)   Modems.

Because they have been developed as solutions to certain problems which arise during the transition phase of a network, these equipments are often regarded as being useful for only a comparatively short period of time. However, in some circumstances, they may linger on — particularly trans-multiplexers. This can happen in a network which is evolving to digital working if it contains a major recently-installed analogue element that cannot be replaced until a reasonable return has been made on the investment. An example of this could be a long trans-oceanic analogue cable system. These systems are expensive and designed for long life. The use of transition equipments may enable them to live out their full economic lives in co-existence with a network that has otherwise fully evolved to digital working.

5.11.2 Transmultiplexers
A transmultiplexer (TMUX) is defined by the CCITT as:

> 'An equipment that transforms a frequency-division multiplexed signal (such as a group or supergroup) into a corresponding time-division multiplexed signal that has the same structure as if it had been derived from PCM multiplex equipment, and that also carries out the complementary function in the opposite direction of transmission'.

These functions could, of course, be carried out using conventional multiplexing and translating equipment with the crossconnections being

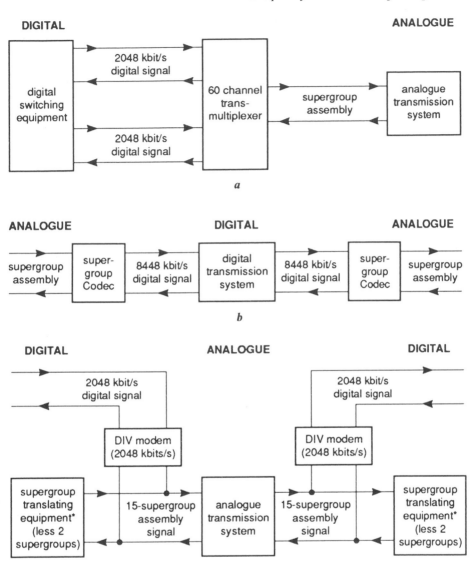

Fig. 5.26 Examples of the use of transition equipment

a Analogue/digital connection using a transmultiplexer: digital switching equipment interfaced to an analogue transmission system

b Analogue/digital interconnection using a pair of FDM codecs: a supergroup assembly carried over an 8448 kbit/s digital transmission path

c Interconnection using a pair of DIV modems: a 2048 kbit/s digital signal carried over an analogue transmission system. (The digital signal replaces two contiguous supergroups in the 15-supergroup assembly.)

made at audio frequency. However a TMUX, by using real-time digital signal processing, provides direct crossconnection without the need for an audio interface [36]. In doing so, it can achieve better transmission performance and stability with time, together with smaller size, power consumption and cost. Two types of TMUX have been recommended by the CCITT. The first one is a 60-channel equipment which transforms between an FDM supergroup and two 2048 kbit/s digital signals [37], while the second is a 24-channel equipment which transforms between two FDM groups and a 1544 kbit/s digital signal [38]. Transmultiplexers have potential for a wide range of applications. These include interfacing digital switching equipment to an analogue transmission system or interfacing analogue switching equipment to a digital transmission system. An example of the former is shown in Fig. 5.26a. Although a TMUX has no audio interface, it does process the signals on a channel-to-channel basis and this leads to some restrictions in utilisation. Services that cannot be carried include 3 kHz channels, sound-programme channels, group-band data and 64 kbit/s data. Special attention may be necessary for FDM assembly reference pilots and any outband signalling tones.

5.11.3 FDM codecs

An FDM codec encodes an FDM assembly, such as a supergroup, into a digital bit stream suitable for transmission over a digital link and it also carries out the complementary function in the opposite direction of transmission. A pair of such codecs enables an FDM assembly to be carried over a digital line system, as shown in Fig. 5.26b. The CCITT Recommendation [39] for these equipments covers the analogue interfaces and overall analogue-to-analogue performance. It does not make any recommendation concerning the encoding law or structure of the digital signal, except that it should be suitable for transmission over a standardised digital link. Consequently, these codecs are always used in pairs and the two codecs forming a pair must be of the same design. A link provided by these means has a performance which, for most characteristics, is similar to that which would be obtained if conventional translating equipment and analogue line systems had been used. An exception to this is the noise performance. On a codec link, the noise performance is substantially that of the codec pair (800 pW0p maximum is recommended by the CCITT) and it does not significantly increase with system length. This compares with the noise performance of a conventional FDM link, for which CCITT recommends 60 pW0p per translating equipment (see Section 5.9) plus 3 pW0p/km for the line system. A noise characteristic which is peculiar to FDM codecs is that of noise 'bunching'; under lightly-loaded conditions, the sampling process may result in the noise being concentrated around certain frequencies. The presence of reference pilots and carrier leaks is normally sufficient to obviate this potential problem. Practical FDM codecs [40] have met the CCITT recommended noise performance with some margin to spare. A pair of TMUX could provide almost the same facilities as a pair of FDM codecs and would make more efficient use of the digital transmission path. However, they may be more expensive than FDM codecs, may not be available with a capacity sufficient to carry the required FDM assembly and are subject to the utilisation restrictions outlined in Section 5.11.2.

5.11.4 Transition modems

A pair of transition modems enables a digital signal to be carried over an analogue line system as shown in Fig. 5.26c. The digital signal incoming to a modem modulates, after suitable processing, a carrier which is translated as necessary to occupy a frequency band that can be transmitted over the analogue line system. These modems are broadly comparable to the modems used for high-speed data transmission referred to in Section 5.10, although they may need to be of even higher speed. Two types of transition modem, a Data-In-Voice (DIV) and a Data-Over-Voice (DOV) modem, are covered by a CCITT Recommendation [41]. In the case of the DIV modem [42] the modulated signal is translated into a frequency band within the normal passband of the analogue transmission path. Therefore, the DIV signal displaces some or all of the FDM assemblies that may have been carried by the transmission system. In the case of the DOV modem, the modulated signal occupies a frequency band above that normally used for transmission of FDM assemblies and none of them need to be displaced. However, before a DOV system can be used, it may be necessary to verify that the required frequency spectrum is available and is of an acceptable quality. Both DIV and DOV systems can be affected by impairments normally found on analogue links. This is particularly so for impulsive noise, which may be at a level which is not noticed on analogue systems carrying speech signals but which may give rise to unacceptable interference to the digital signal. The CCITT Recommendation for these modems does not make any recommendation for the digital processing or modulation method to be used. These are left to the discretion of the designer; so it must be ensured that modems of identical design are used at each end of a link. As in the case of FDM codecs, a pair of TMUX could provide almost the same facilities as a pair of DIV modems, but similar utilisation restrictions apply. Modems provide bit transparency over the analogue path.

5.12 Final comment

Whilst it may be perceived that the trend in telecommnications is to go all digital, and some countries (like the UK) have transmission networks that are now almost 100% digital, analogue carrier systems still dominate globally and are likely to do so for perhaps another decade. However, major new analogue transmission systems are unlikely to be developed and CCITT no longer has any question under study specifically addressed to new analogue transmission matters. The CCITT Blue Book volumes referred to in this chapter are therefore likely to become definitive so far as analogue transmission systems are concerned and the 60 MHz line systems may well represent the apex of the art of the analogue line transmission engineer.

It is interesting to note that development of optical transmission systems is now moving towards wavelength-division multiplexing, a direct echo of the electronic FDM predecessors described in this chapter. The techniques, problems and limitations experienced with the FDM electronic systems now look set to be of direct relevance to their future wideband optical replacements!

5.13 References

1 KURTH, C.F.: 'Generation of single-sideband signals in multiplex communications systems' *IEEE Trans.* 1976, **CAS–23**, pp. 1–17
2 CCITT: Blue Book, Fascicle III.2 Recommendation G.211
3 *ibid.*: Recommendation G.232
4 GRONBERG, M., and JOHANNESSON, N.O.: 'Combination effects observed with out-band signalling', *Ericsson Rev.* 1967, **44**, pp. 28–33
5 LAW, H.B., REYNOLDS, J., and SIMPSON, W.G.: 'Channel equipment design for economy of bandwidth', *Post Office Elect. Eng. J.*, 1960, **53**, pp. 112–117
6 CCITT: Blue Book, Fascicle III.2 Recommendation G.235
7 *ibid.*: Recommendation G.233
8 *ibid.*: Recommendation G.241
9 *ibid.*: Recommendations G.311–G.371
10 *ibid.*: Recommendations G.411–G.473
11 'Transmission systems for communication' (Bell Telephone Laboratories, 1971) 4th Edn.
12 ALBERT, W.G., *et al.*; 'Terminal arrangements', *BSTJ*, 1969, **48**, pp. 993–1040
13 MAURER, R.E., *et al.*; 'L5 system: jumbogroup multiplex terminal' *BSTJ*, 1974, **53**, pp. 2065–2096
14 CCITT: Blue Book, Fascicle III.1 Recommendation G.135
15 *ibid.*: Fascicle III.2 Recommendation G.225
16 NJIO, W.F., and VERKOOIJEN, C.J.: 'Frequency generating equipment for FDM transmission systems', *Philips Telecommuncations Rev.* 1976, **34**, pp. 22–37
17 KUPKA, K., and PLUGGE, H.: 'Carrier supply for frequency multiplex equipment facilities', *Siemens Rev.* 1971, **38**, pp. 199–203
18 CCITT: Blue Book, Fascicle III.2 Recommendation G.229
19 *ibid.*: Recommendation G.242
20 DERAEMAEKER, R., PLUISTER, K., and VISSER, J.S.: 'Through-connection filters type 8TR314', *Philips Telecommunications Rev.*, 1973, **31**, pp. 23–30
21 CCITT: Blue Book, Fascicle III.2 Recommendation G.222
22 *ibid.*: Recommendation G.223
23 HOLBROOK, B.D., and DIXON, J.T.: 'Load rating theory for multi-channel amplifiers', *BSTJ*, 1939, **18**, pp. 624–644
24 CCITT: Red Book, Fascicle III.4, Part 1, Section 1.2
25 *ibid.*: Section 1.5
26 *ibid.*: Blue book, Fascicle III.6, Recommendations J.11–J.34
27 *ibid.*: Recommendations J.61–J.77
28 *ibid.*: Recommendation J.75
29 *ibid.*: Fascicle III.2, Recommendation G.332
30 *ibid.*: Fascicle III.6, Recommdneation J.73
31 *ibid.*: Fascicle III.2, Recommdendation G.334

32 *ibid.*: Fascicle III.6, Recommendation J.77
33 *ibid.*: Fascicle III.2, Recommendation G.333
34 KINGDOM, D.J.: 'Transition equipment—an overview', *British Telecommunications Enging.*, 1984, **3**, pp. 146–148
35 CCITT: Blue Book, Fascicle III.4, Supplement to Recommendations G.793–G.795
36 DICK, A.B.: 'Transmultiplexers'. *British Telecommunications Enging.*, 1984, **3**, pp. 149–153
37 CCITT: Blue Book, Fascicle III.4, Recommendation G.793
38 *ibid.*: Recommendation G.794
39 *ibid.*: Recommendation G.795
40 ANDREWS, M.J.: 'Supergroup and hypergroup codecs,' *British Telecommunications Enging.*, 1984, **3**, pp. 154–160
41 CCITT: Blue Book, Fascicle III.5, Recommendation G.941
42 HARRISON, N.: 'A 2048 kbit/s Data-in-Voice Modem', *British Telecommunications Enging.*, 1984, **3**, p. 3, pp. 161–168

Chapter 6

Analogue/digital conversion and the PCM primary multiplex group

A.D. Wallace, L.D. Humphrey and M.J. Sexton

BNR Europe, Ltd.

6.1 Introduction

Analogue systems operate on continuous signals. Digital systems, however, operate on signals which are defined only at specific time instants and signal voltages, both of which are usually uniformly spaced. The principal advantages offered by a digital signal are as follows:

(*a*) *Robustness*: Information can be carried from source to destination with degradation limited only by the quality of the analogue-to-digital convertor (ADC) and digital-to-analogue convertor (DAC). Analogue signals, by contrast, are subject to a wide variety of degradations, such as noise, hum, distortion, dispersion and crosstalk.

(*b*) *Repeatability*: Digital signal processing is completely predictable. Once an algorithm has been specified to bit level, all implementations conforming to that specification will have exactly the desired behaviour. Analogue signal processing, by contrast, suffers from such problems as distortion, drift, offsets, noise and the need to 'tweak' to achieve the desired performance.

(*c*) *Realisability*: Digital signal processing opens the door to new highly-complicated signal-processing algorithms, which would be impossible to achieve by analogue signal processing. For example, in telephony, complex speech-processing algorithms have been developed which halve, quarter, or reduce still further, the bandwidth required to transmit a single speech channel, without significant degradation in the quality.

These advantages have led to the development of integrated digital networks (IDN) in which all the telephone transmission and switching is carried out on 64 kbit/s channels employing pulse-code modulation (PCM) [1].

In order to obtain the above advantages, speech and other analogue signals must be converted into digital signals prior to transmission and subsequently converted back into analogue signals. They are also multi-plexed onto the transmission path using time-division multiplexing (TDM). The methods used to implement these processes are considered in this chapter.

6.2 Theory

6.2.1 Sampling

In order to produce a digital signal, an analogue signal must be sampled by a periodic train of pulses to convert it from continuous time to discrete time. It is shown in Section 2.7.5 that if a continuous signal has a bandwidth W, then the minimum sampling rate F_s which can be used is

$$F_s = 2W \tag{6.1}$$

This is called the *Nyquist rate*. The regular sampling results in a frequency spectrum which repeats with spacing $2W$. Subsequent removal of all spectral repetitions, bar that from the desired band, by a low-pass filter at the receiver reconstructs the original signal.

If the sampling rate is below $2W$, the spectral repetitions overlap, making it impossible to separate the original signal from the adjacent band. This problem is called *aliasing*. Aliasing is illustrated in Fig. 6.1, where two sinusoids are shown, one of frequency 1 Hz, and one of frequency 9 Hz. If a sampling frequency of 10 Hz is chosen, with sampling instants indicated by the marks on the time axis, it can be seen that both sinusoids result in the same set of discrete time samples, shown by the circles. In other words, with a sampling frequency of 10 Hz, the sinusoid at 1 Hz and the sinusoid at 9 Hz are completely indistinguishable. An anti-alias filter is needed, which will reject all signals of frequency 5 Hz or higher, in order to ensure that the desired 1 Hz signal cannot be mistaken for an irrelevant 9 Hz signal.

Aliasing can be avoided by passing the input signal through a suitable low-pass filter before sampling it, as discussed in Section 6.3.1. It is essential to include such an *anti-alias filter* at the input to any sampling system. Since the nominal band of telephone speech is 300 Hz to 3·4 kHz and the internationally-agreed sampling frequency for PCM systems is 8 kHz, there is a

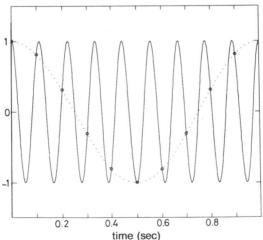

Fig. 6.1 Example of aliasing

guard band of 1·2 kHz between the top of the baseband and the bottom of the lower sideband of the sampling frequency. This must accommodate the transition between the pass-band and the stop band of the anti-alias filter at the sending end and of the demodulation filter at the receive end.

6.2.2 Quantisation

The analogue sampler performs the transformation from continuous-time to discrete-time representation, as described above. In order to provide digital transmission, the samples must also be converted from continuous voltages to discrete voltages. For example, if each sample is sent in a simple binary code, the transmitted signal has only two voltages, corresponding to '0' and '1', respectively.

The digital signal can only represent a finite number of input-signal voltages. For example, a binary code of k digits can only represent 2^k different input voltages. When the signal is reconstructed at the receiver, it is an approximation to the original signal, since it can only change in discrete voltage steps, as shown in Fig. 6.2a. This signal is said to be *quantised*, and a form of non-linear distortion, known as *quantisation distortion*, has been introduced.

The difference between the original signal and the reconstructed signal is the error voltage shown in Fig. 6.2b. Because the error voltage varies randomly with time, it is often referred to colloquially as 'quantisation noise'. Indeed, in PCM telephony, it sounds just like noise. If the RMS noise voltage considered acceptable in an analogue signal is σ, then a discrete representation with a voltage-level spacing of the order of σ will be adequate to represent the analogue signal.

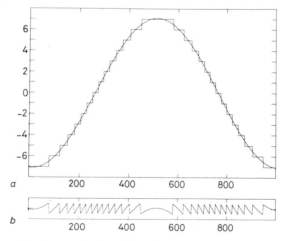

Fig. 6.2   Example of quantisation distortion

a Input and output waveforms
b error voltage

Obviously, the closer the spacing of the discrete levels, the smaller will be the quantisation distortion. However, the smaller the spacing, the greater is the number of levels. This increases the number of digits in the code words representing the levels (e.g. doubling the number of levels increases the length of a binary-code word by one bit). An increase in the number of digits representing every sample increases the bandwidth required to transmit them.

Consider a quantiser where the ith interval is centred around Q_i and possesses width Δ_i. Any input signal x in the range:

$$Q_i - \tfrac{1}{2}\Delta_i \leqq x < Q_i + \tfrac{1}{2}\Delta_i$$

will be represented by the quantised amplitude Q_i. The squared quantisation error will clearly be $(x - Q_i)^2$. Let us also assume that the probability density of the input signal is $p(x)$. Therefore, to obtain the mean square error over the ith range, the squared error value must be averaged over all possible values of the input signal for that quantisation interval, thus:

$$E_i^2 = \int_{Q_i - \frac{1}{2}\Delta_i}^{Q_i + \frac{1}{2}\Delta_i} p(x)(x - Q_i)^2 \, dx$$

If the signal varies widely in comparison to the size of a single quantisation interval, it can be assumed that $p(x)$ is constant across that quantisation interval, having value $1/\Delta_i$. Therefore

$$E_i^2 = \frac{1}{\Delta_i} \int_{Q_i - \frac{1}{2}\Delta_i}^{Q_i + \frac{1}{2}\Delta_i} (x - Q_i)^2 \, dx = \frac{1}{\Delta_i} \frac{\Delta_i^3}{12} = \frac{\Delta_i^2}{12}$$

We may now calculate the mean square noise over the entire quantiser dynamic range by averaging E_i over all values of i:

$$E^2 = \frac{\sum\limits_{i=1}^{N} P_i E_i^2}{\sum\limits_{i=1}^{N} P_i} = \frac{1}{12} \sum_{i=1}^{N} P_i \Delta_i^2$$

where P_i = probability of the signal voltage being in the ith interval, N = total number of quantisation intervals in the entire convertor dynamic range.

For the important case of a uniform quantiser, where all intervals have the same width Δ:

$$E^2 = \frac{\Delta^2}{12} \sum_{i=1}^{N} P_i = \tfrac{1}{12}\Delta^2 \tag{6.2}$$

For a sinusoidal input signal, the maximum amplitude which can be handled is

$$V_m = \tfrac{1}{2}N\Delta$$

so the signal power is

$$S = \tfrac{1}{2}V_m^2 = \tfrac{1}{4}N^2\Delta^2 \text{ volts}$$

Therefore, the signal/quantisation-noise ratio (SQNR) is

$$\text{SQNR} = \frac{S}{E^2} = 3N^2$$

Expressing this in decibels:

$$\text{SQNR} = 4\cdot8 + 20\log_{10} N \text{ dB} \qquad (6.3)$$

Now, for binary-encoded PCM, $N = 2^k$ where k is the number of bits in the codeword. Substituting in eqn. 6.3 gives

$$\text{SQNR} = 4\cdot8 + 20k\log_{10} 2 \text{ dB}$$

$$= 4\cdot8 + 6k \text{ dB} \qquad (6.4)$$

Eqn. 6.4 shows that adding one digit to the code improves the SQNR by 6 dB. Since the quantisation noise level is independent of signal level, a signal whose level is x dB lower than the maximum will have a SQNR which is x dB worse than that given by eqns. 6.3 and 6.4.

As explained in Section 6.7, the SQNR for low-level signals can be improved by non-uniform quantisation. This enables the wide range of levels encountered in telephone signals to be handled with low quantising distortion even when several PCM systems are connected in tandem (see Section 4.4). The international standard uses 8 bit coding, i.e. 256 quantising levels. Since the standard sampling frequency is 8 kHz, a PCM telephone channel therefore requires a bit rate of 64 kbit/s.

6.3 Sampling techniques

6.3.1 The anti-alias filter

The ideal anti-alias filter has a transfer function of unity for all frequencies in the range from DC to half sampling frequency, and zero thereafter, with no added noise, distortion or delay. In practice this is impossible, and the usefulness of an anti-alias filter can be characterised by its demerits, including:

(a) *Passband ripple*: How much the transfer characteristic varies across the region of interest

(b) *Stopband attenuation*: The amount by which signals outside the area of interest are reduced

(c) *Phase response and group delay*: The variation in group delay with frequency can set limits on how well equipment such as data modems can perform over a link including the anti-alias filter

(d) *Linearity and noise*: Active and passive filters alike introduce distortion, noise, crosstalk and other forms of pickup, all of which degrade the analogue signal prior to conversion

Anti-alias filters can be implemented in many ways. High-order passive *LC* filters have often been used, offering low distortion and low noise, but with the physical drawbacks of large size and weight, and inevitable insertion loss. A more up-to-date approach might be the use of a Sallen-and-Key active filter [2], which would be smaller, lighter and loss-free, but would add more noise, and can be more difficult to trim than the corresponding *LC* filter. Switched-capacitor filters [3] can be highly integrated, requiring only off-chip capacitors, and hence lead to a cheap package which does not require large amounts of PCB area. They do, however, add considerably more noise than the passive and active filters.

6.3.2 The analogue sampler

The analogue sampler, commonly known as the 'sample-and-hold', performs the task of sampling the input analogue voltage and maintaining that voltage until the next sampling instant. A simple sample-and-hold is illustrated in Fig. 6.3. When the SAMPLE command is given, the buffer is connected via the FET switch to the storage capacitor, the voltage on which tracks the input voltage thereafter. The capacitor is at all times buffered by a voltage follower. When the SAMPLE command is removed, the capacitor is isolated from the input buffer. The voltage across the capacitor, and hence the voltage follower, will then remain steady until the next SAMPLE command.

Such a simple design is marred by a number of adverse properties, limiting its usefulness. These properties, often used to characterise the quality of a commerical sample-and-hold, are briefly:

(*a*)   *Acquisition time*: A FET switch typically has an ON resistance of order 100 Ω. This, in combination with the slew rate and gain–bandwidth product limitations of the input buffer, limits the rate at which the capacitor voltage can follow the input voltage. The minimum ON time is called the acquisition time.

(*b*)   *Droop*: A FET switch has an OFF resistance, which may be of the order of 1000 MΩ, but this will still allow a leakage current to flow into the capacitor. Also, the output buffer will require a finite bias current, which

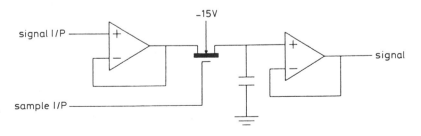

Fig. 6.3   Simple sample-and-hold circuit

will be drawn from the capacitor. These two factors cause the capacitor voltage to drift slowly with time, after the SAMPLE command has been removed.

(c)   *Glitches*: There will be capacitive coupling between the SAMPLE command input to the FET switch and the capacitor. Therefore, there can be a finite disturbance on the capacitor voltage when the SAMPLE command is removed. The size of the disturbance will also depend on the shape of the SAMPLE command edge, and may therefore change from sample to sample.

(d)   *Non-linearity*: FET switches do not behave as perfect resistors; they have an ON resistance which is dependent on the source-drain voltage. This will introduce distortion if the SAMPLE command is too short.

All of these problems can be reduced by careful design. For example, problems caused by capacitive coupling of the SAMPLE input to the storage capacitor can be reduced by using a pair of FET switches, driven by opposite senses of the SAMPLE command, hence balancing the charge coupling. Also, problems caused by leakage through the FET switch in its OFF state can be reduced by using a pair of FET switches in series, and bootstrapping the intermediate point via a high resistance to the output buffer.

6.4   The analogue quantiser, or ADC

6.4.1   General

The analogue sampler performs the transformation from continuous-time to discrete-time representation. The analogue quantiser performs the complementary operation of transformation from continuous-amplitude to discrete-amplitude. There are a variety of techniques which may be used in an analogue-to-digital convertor (ADC), such as dual-slope conversion, successive approximation and flash conversion. Each of these will be considered in more detail in this section, except for oversampling convertors, which will be considered separately in Section 6.6.

Just as the performance of an analogue sampler is characterised by how well the designers overcame the pitfalls mentioned above (i.e. acquisition time, droop etc.), the performance of an analogue quantiser may be characterised by its limitations, which are briefly:

(a)   *Conversion time*: The time taken between starting a conversion and producing the result

(b)   *Sampling rate*: The maximum rate at which samples can be presented at its input. This may be faster than that implied by the conversion time, if the design has been pipelined

(c)   *Linearity*: How accurately the quantiser decision values are placed

(d)   *Resolution*: The smallest change which can be detected

(e)   *Complexity/cost*: An important consideration for all equipment

(f)   *Power consumption*: An important consideration for battery-powered or heavily-populated equipment.

6.4.2 Dual-slope conversion

The essential building blocks within a dual-slope convertor are an integrator, analogue switch, voltage reference, comparator, clock source and a counter, configured as shown in Fig. 6.4. The circuit operates in two phases. In the first phase, the input voltage is fed via the analogue switch to the integrator. The voltage at the integrator output will rise from zero with a slope proportional to the input voltage. Phase 1 continues for a fixed number of clock cycles.

In the second phase, the analogue switch disconnects the input voltage, and instead connects the voltage reference to the integrator. The voltage reference is arranged to be of opposite sign to the input voltage; hence the integrator now starts to ramp down towards zero, with a constant slope fixed by the voltage reference. Phase 2 completes when the integrator output crosses zero. The duration of phase 2 is therefore proportional to the input voltage. All that remains to be done therefore is to measure that time. The counter is reset to zero at the start of phase 2, and counts steadily upwards at the clock rate thereafter. The counter output at the end of phase 2 is, therefore, proportional to the input voltage during phase 1. The process is illustrated in Fig. 6.5.

The dual-slope convertor has some very useful properties. The actual value of the integrator capacitor is not important, since the time taken to reach zero during phase two is completely independent of the capacitor value. Likewise, the value of the clock frequency is not critical, so long as the same clock is used to determine the duration of phase 1 and measure the duration of phase 2. The only block requiring high precision is the voltage reference. As a result, dual-slope convertors are easily implemented from low-cost components, and are widely used in precision digital Voltmeters (DVM), for example, offering the highest resolution for a given cost. Their principal drawback is long conversion time, and by implication, low sampling rate. One final point is that, in a DVM, the dual-slope convertor is often allied to a preamplifier, which has an (overridable) automatic selection of gain, typically in decades from 1 to 1000, to provide a signal which is always well placed within the dynamic range of the convertor, and extending the overall dynamic range accordingly.

6.4.3 Successive approximation conversion

The successive-approximation ADC is built from a digital-to-analogue convertor (see Section 6.5), a comparator, a storage register and controlling logic, is shown in Fig. 6.6. This type of convertor essentially operates by performing a binary search through the ADC dynamic range to find the input voltage.

Upon receiving the CONVERT command, the storage register is cleared. The output from the storage register is passed to the digital-to-analogue convertor (DAC), but with the most-significant bit set. This will cause the DAC to output its half-range voltage. This voltage is then compared with the input sample by the comparator. The comparator output is then stored in the most-significant bit (MSB) of the storage register. The convertor has

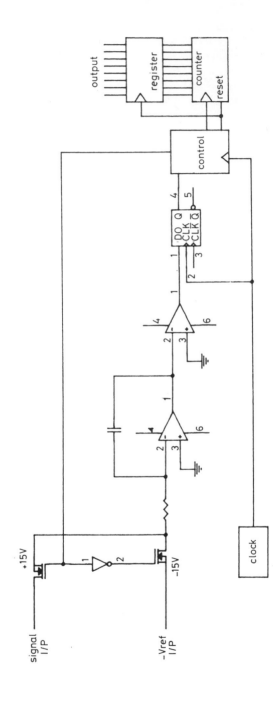

Fig. 6.4 The dual-slope converter

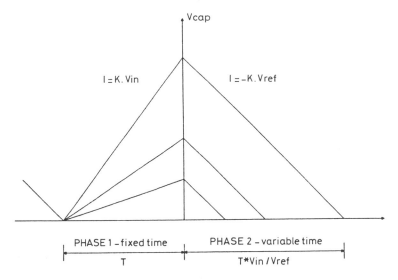

Fig. 6.5 The dual-slope conversion cycle

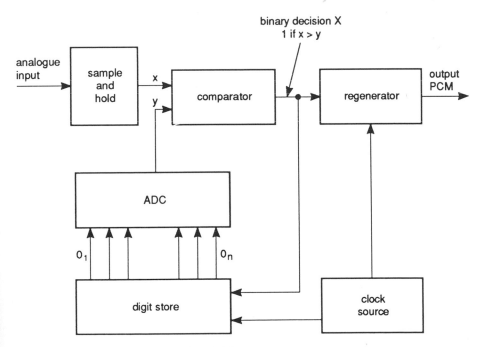

Fig. 6.6 Successive-approximation ADC

thus decided whether the input voltage is in the lower or upper half of the ADC's dynamic range.

On the next clock cycle, the storage register output is passed to the DAC with its 2nd-most-significant bit set. The comparator output then indicates whether the input signal lies in the upper or lower half of that portion of the dynamic range selected in the previous decision. The comparator output is stored in the 2nd-most-significant bit of the register.

This process continues, passing through as many cycles as there are bits in the storage register, with each pass determining whether the input voltage lies in the upper or lower half of the range selected in the previous cycle. The storage register then contains the ADC conversion value. The process is illustrated in Fig. 6.7. The upper trace shows the input signal value as a dashed line; it also shows the value stored in the convertor storage register, and hence the voltage developed at the output of the reference DAC. The lower trace shows the comparator output, and hence the ADC digital output, MSB first.

The successive-approximation convertor has a much faster conversion time than the dual-slope convertor, but its precision is limited by the reference DAC. Thus, it is more costly to manufacture a successive approximation convertor of a given precision, than a dual-slope convertor, because of the need to implement an N-bit precision DAC.

6.4.4 Flash conversion

The flash convertor consists of a chain of fast comparators, with one of each of the comparator inputs driven by the voltage to be converted, possibly buffered. The other input of each successive comparator is connected to the corresponding point of a resistor chain, containing as many resistors,

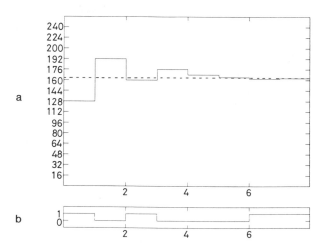

Fig. 6.7   An 8-bit successive-approximation cycle

a Input signal and value stored in register
b Comparator output

of equal value, as there are comparators, as shown in Fig. 6.8. The number of comparators, and hence resistors in the divider chain, is equal to the number of levels which must be coded. For example, for the case of a four-fit flash convertor, capable of coding voltages in the range 0 to 1 V, 16 comparators are needed, connected to a resistor chain containing 16 equal resistors, connected across a 1 V reference. The comparators thus compare the input voltage with $\frac{1}{16}$ V, $\frac{1}{8}$ V, $\frac{3}{16}$ V, and so on.

The comparator outputs are then passed to a priority encoder, which determines the position of the last comparator in the chain to indicate that its reference voltage is lower than the flash-encoder input voltage. This indicates directly which segment of the convertor's dynamic range contains the input signal and gives the corresponding binary-code output.

The main feature of the flash convertor is its extremely-rapid conversion time. They are often used for video ADCs, or any other application which requires a sample rate in excess of 1 MHz. The penalty for the speed is the rapid increase in complexity, and therefore power consumption and cost,

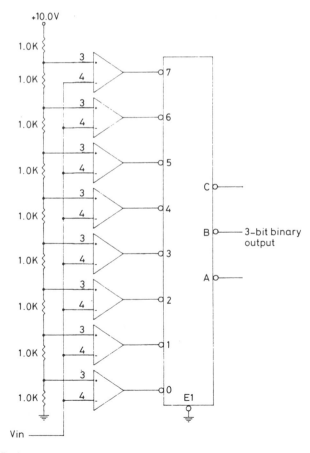

Fig. 6.8 A 3-bit flash converter

as the desired number of bits increases. Indeed, it is rare to find a flash convertor operating to more than 8 bit accuracy because of the cost of an array of more than 256 comparators. Variation of input offset voltage between adjacent comparators can set a further limit to ultimate resolution.

However, analogue and digital pipelining can be combined to implement two-stage flash conversion, giving accuracy up to twice the number of bits, with the same sample rate, and only double the conversion time.

6.4.5 Some general conversion issues

We have only given thought so far to convertors capable of coding voltages between ground and some upper limit. It is more usual to want to convert voltages of both polarities. This can be achieved either by

(a) the addition of a voltage equal to half the upper limit, with two's complement coding of the signal then easily achieved by inversion of the resulting MSB, or

(b) by the use of a precision full-wave detector, which results in an output signal of fixed polarity, together with information about whether the signal is positive or negative.

The latter scheme results naturally in the production of a sign-plus-magnitude code.

6.5 The analogue reconstructor, or DAC

After the digitally-encoded signal has been transmitted, it is often necessary to reconstruct an analogue signal at the receiver. Conversion from digital to analogue is generally easier to achieve than conversion from analogue to digital. There are again many techniques which may be used for this conversion, each possessing its own blend of merits and demerits. As with the analogue sampler and analogue-to-digital convertor, a digital-to-analogue converter (DAC) can be characterised by its demerits, i.e.:

(a) *Linearity/monotonicity*: A measure of how accurately the reconstructor outputs are positioned

(b) *Resolution*: The size of the smallest discernable change at the convertor output

(c) *Noise*: How much residual activity there is even when the digital codeword is not changing

(d) *Slew rate*: The analogue output will take a finite time to change from one value to the next, owing to charge-storage/capacitive effects

(e) *Glitches*: There will usually be impulsive signals added to the analogue output at the instant that the codeword is changed

(f) *Cost/complexity*: Important criteria for any application

(g) *Power consumption*: important for battery powered or heavily-populated equipment.

The simplest form of DAC is shown in Fig. 6.9. It consists of an array of resistors connected via an analogue switch either to a reference voltage, if

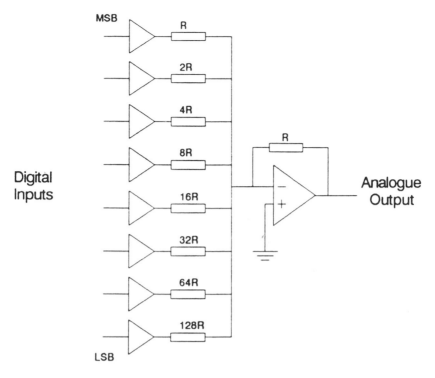

Fig. 6.9 Simple digital-to-analogue converter

the corresponding bit in the digital word is 'one', or to ground otherwise. The resistors are all connected to the virtual-earth input of a current-summing amplifier, and their values are chosen such that the currents flowing through each resistor when the corresponding bit is 'one' are binary weighted.

The output voltage of the summing amplifier is proportional to the sum of the input currents, and hence to the numerical value of the digital codeword.

The circuit illustrated in Fig. 6.9, is simple in nature, but it suffers some drawbacks. The linearity and monotonicity are both critically dependent on the precision of the resistors used. Any errors in resistor value will show particularly around the transition between the lower and upper halves of the convertor's dynamic range. The resistor values will also almost certainly drift slowly with time and temperature. Some more-elaborate designs based on the resistor ladder network overcome this drawback by making the current supplied by each resistor in the network slightly too small. The current for a particular bit can then be fine tuned by the optional addition of a second or third current source of much smaller value. The linearity and monotonicity of the convertor can be much improved by this self-calibration.

6.6 Oversampling ADCs: Squeezing a quart into a pint pot

6.6.1 Noise spreading

It is an established principle of physics that, if a measurement, subject to random error is repeated many times, then the error in the average value of the measurement is smaller than the error in each individual measurement. Indeed, if N measurements are taken, the error will be reduced by a factor $\sqrt{N}$.

In an oversampling convertor, the input signal is sampled at a multiple of the Nyquist rate. For a given resolution this process does not reduce the error in an individual sample (i.e. the mean square noise per sample remains unchanged), but it is now distributed over much greater bandwidth. Viewed in the frequency domain, oversampling by a factor N spreads the noise/error energy so as to reduce the average sampling error in the base band by $\sqrt{N}$. Recovery of samples of the required sampling frequency and resolution requires a digital low-pass filter, the output of which is subsequently digitally sampled to the required rate.

The process requires that the ADC be capable of sampling at a higher rate, but it allows convertor speed to be traded for resolution. In the extreme, it allows very-simple convertors to be used to provide high resolution. However, to work well, the quantisation noise in successive samples should be statistically independent. This means that the advantage can only be seen if the input signals are changing sufficiently rapidly, or if the real resolution of the ADC is controlled by noise effects which constitute dither greater than the theoretical step-size. Dither noise can, of course, be artificially injected to ensure that the latter condition is satisfied, and this forms the basis of a method of improving convertor linearity.

6.6.2 The delta modulator

Spreading the noise evenly across an enlarged frequency range is one method by which the baseband noise can be reduced. An even better method is to design an oversampling convertor which distributes the noise unevenly, pushing most of the noise power away from the baseband. Simple anti-alias filtering will then remove the unwanted noise components. The noise can be shaped by use of frequency-selective feedback to the ADC input via a digital-to-analogue convertor.

The simplest example of such a coder is the delta modulator [4] shown in Fig. 6.10. It is built around a one-bit quantiser. The corresponding decoder consists of one-bit DAC which drives an integrator. The encoder also contains a reference decoder. At each sample instant, the encoder compares the input signal against the reference decoder output. If the input voltage is less than the reference, then a '1' is transmitted and the integrator is 'charged' at fixed rate until the next sample time; otherwise a '0' is transmitted and the integrator is discharged at the same rate. The integrator therefore tracks changes in the input voltage, but never settles.

Delta modulators are cheap to build, and have a very useful property. The output of the comparator may be used as the signal for transmission down the digital link. If this is done, it should be apparent that all bits in

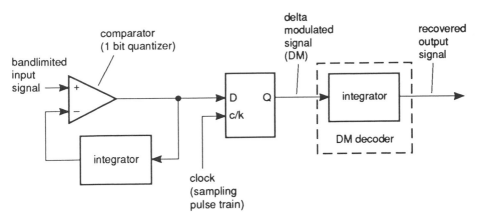

Fig. 6.10 Simple delta-modulation system

the digital signal carry equal weight, since every bit merely causes the integrator to increase or decrease by one. This makes delta modulation very robust in the presence of errors. Whereas frequent corruption of the most-significant bit on a conventionally-coded ADC sample could be devastating to channel intelligibility, a delta-modulated channel will still be usable at the same error rate.

It is interesting to consider the overload mechanism of the delta modulator, i.e. slope overload. This happens if the rate of change of the input signal exceeds the fixed rate of change of the integrator output, preventing the integrator from tracking the input voltage accurately. This results in a dynamic range which is very large for low frequencies, but which reduces as the input frequency is increased [4]. For speech, this is not much of a problem, because speech contains only small amounts of high-frequency energy.

To overcome the slope-limiting deficiency of the delta modulator, the continuously-variable-slope delta modulator (CVSD) was developed. In this coder, the rate of change of the integrator output is increased if a long sequence of ones or zeros is detected in the channel, indicating the onset of slope overload. This allows the CVSD coder to follow a rapidly-changing signal more accurately. The slope rate subsequently decays slowly to its normal lower limit.

6.6.3 The delta-sigma modulator

The delta-sigma converter [4] is another one-bit convertor, again using a charge-balancing technique. It is based around a comparator, a constant-current source and an integrator. However, in this design, the one-bit stream can be low-pass filtered to produce the digital codeword directly, whereas an integrator is needed in the delta-modulator application.

The input signal is buffered and applied through a resistor to the virtual-earth input of an integrator. A current proportional to the input voltage therefore flows into the integrator capacitor. The output of the integrator drives a low-hysteresis low-noise comparator, whose output is retimed to produce the one bit codeword. The retimed output from the comparator is also used to switch the constant-current source, which drives the virtual-earth input of the integrator in the opposite sense to the input current. A delta-sigma modulator is illustrated in Fig. 6.11, although the Norton equivalent of the input buffer amplifier has been shown.

Given a stable clock source at F_s, this will result in the removal of a fixed amount of charge from the integrator capacitor every time a 'one' is output. The overall effect of the circuit is to maintain an average of zero charge on the integrator capacitor, and the density of 'one's in the output bitstream will be proportional to the input voltage. Closer examination reveals that the noise in the one-bit stream is distributed unevenly, being suppressed at low frequencies, and rising thereafter at 6 dB per octave up to half the oversampling frequency.

If the number of 'one's occurring in the output stream is counted over a fixed period, by using the one-bit codeword as the enable on a counter, then the counter output at the end of that period will also be proportional to the input voltage, and hence can be used as an n-bit PCM sample. Obviously, for a given n-bit sample rate, the resolution becomes greater as the oversampling factor increases, as the counter output will have a wider range of potential values. This type of circuit, which converts a one-bit sequence into an n-bit codeword, is known as a conversion filter.

The conversion filter effectively has an impulse response equal to F_c/F_s for the duration of the summing period, $1/F_c$. Hence, it has a gain/frequency

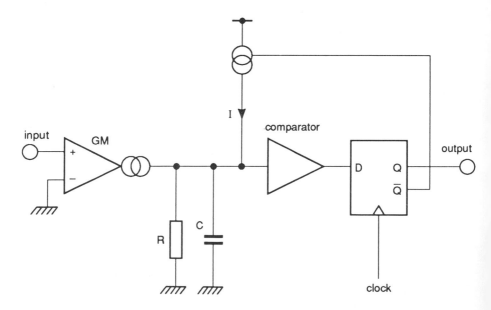

Fig. 6.11   Delta-sigma modulator

response given by

$$H(f) = \frac{\sin(\pi f/F_c)}{\pi f/F_c} \qquad (6.5)$$

This is approximately flat over the range $0 < f < F_c/8$, and has a falling characteristic for frequencies above $F_c/2$, falling at a rate of 6 dB per octave,

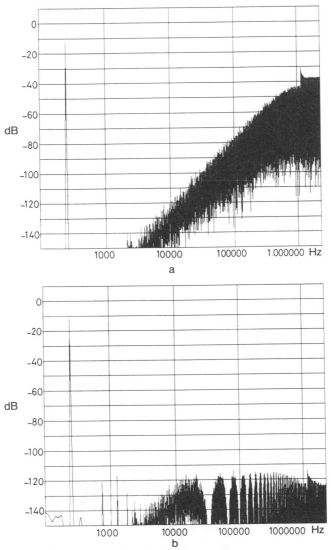

Fig. 6.12 Second-order one-bit ADC noise spectra

a Output spectrum
b Conversion-filter output spectrum

with zeros at all multiples of F_c. This means that the high levels of high-frequency noise present in the one-bit stream around multiples of the output sample rate F_c will be heavily attenuated. This is important, because the noise in these regions would otherwise fold down into the baseband when the sample rate is reduced to F_c.

The delta-sigma convertor offers several advantages over conventional convertors, such as the flash and successive-approximation convertors. Firstly, one-bit DACs and ADCs are very cheap to manufacture, do not require any trimming, and cannot suffer from non-linearity or non-monotonicity in the same manner as an n-bit convertor. Secondly, the input sampling rate is very high, which means that only low-order analogue anti-alias filtering is required. This could typically be achieved using a 3rd-order Bessel filter for good group-delay characteristics, with its cutoff frequency well removed from the final F_c. The bulk of the anti-alias filtering is performed by the conversion filter, but extra anti-alias filtering can be performed by digital signal processing (DSP) at the conversion-filter output if required. This leads to exceptional repeatability from one ADC to the next, in contrast to conventional analogue filters which often require trimming.

Modern BiCMOS processes can offer both the necessary analogue and digital processing sections on the same chip, giving very high levels of integration and hence low chip count, with few external components. Note that an oversampling DAC can be designed in exactly the same manner as an oversampling ADC, by simulating the analogue circuitry in the delta-sigma ADC in a purpose-built DSP block.

The convertors described above are classed as first-order convertors, because, in effect, the one-bit ADC is placed in a first-order feedback system. Even greater suppression of the baseband noise can be obtained by placing the one-bit ADC in the feedback loop of a higher-order low-pass filter, and using a conversion filter of equal order to suppress the high-frequency rising noise characteristic. Fig. 6.12 illustrates the performance which may be obtained from a second-order variant. Fig. 6.12(a) shows the output power spectrum of the delta-sigma modulator. Fig. 6.12(b) illustrates the action of a second-order conversion filter, showing how it effectively flattens the rising noise characteristic inherent in the delta-sigma modulator.

6.7 Speech statistics and companding

In typical telephone speech, the probability distribution of signal amplitudes has a maximum at zero amplitudes and becomes progressively smaller for high amplitudes, having a near Gaussian distribution. In addition, substantial level variations exist between one conversation and another, depending on such factors as the line length between the subscriber and the exchange, the sensitivity of the subscriber's handset and the loudness of the talker. These factors can lead to a variation of up to 30 dB between one speaker and another [5]. There is, therefore, a requirement for a wide dynamic range to be coded by the ADC for a speech signal.

As shown in Section 6.2.2, a uniform quantiser has a constant noise power, independent of the signal level. Eqn. 6.4 shows that to code a signal with a dynamic range of 50 dB, maintaining a signal-to-noise ratio of at least 30 dB over that dynamic range, would require a 13–bit uniform quantiser. The ear, however, is a very non-linear sensor and, in the presence of a speech signal, any broadband noise which is 30 dB quieter than the signal is masked by that speech. So, why use a uniform quantiser with its fixed noise power, unaffected by speech level? What is really needed is a quantiser which has a constant signal-to-noise ratio of 30 dB over a 50 dB dynamic range.

The signal/quantising-noise ratio (SQNR) for small signals could be improved by compressing the level range before coding and using expansion after decoding at the receiver (i.e. companding).

It is not necessary to use a 'syllabic' compander, because the process can be performed independently on each speech sample. Such an 'instantaneous' compressor could be implemented by a non-linear characteristic for which constant increments, δy, in output voltage correspond to input-voltage increments, δx, which are proportional to the input voltage x, as shown in Fig. 16.13b. Thus, $dy = (b/x) \, dx$, so $y = b \log cx$. However, $\log 0 = -\infty$ and a practical ADC must give zero output for zero input. A compression law is therefore needed which is logarithmic for large values of x but corresponds to a straight line through the origin for small values of x, as shown in Figs. 6.13c and d.

One compression law that is used is the *A law* shown in Fig. 6.13c. This is given by

$$y = \frac{1 + \log_e Ax}{1 + \log_e A} \quad \text{for } \frac{1}{A} \leqq x \leqq 1$$

$$= \frac{Ax}{1 + \log_e A} \quad \text{for } 0 \leqq x \leqq 1/A \tag{6.6}$$

The section below $x = 1/A$ is linear. The section between $1/A$ and 1 is logarithmic and, for $A = 87 \cdot 6$, this gives constant SQNR over a range of $20 \log_{10} 87 \cdot 6 = 38$ dB. For small signals, the A-law gives

$$\frac{dy}{dx} = \frac{A}{1 + \log_e A}$$

However, for uniform quantisation, $dy/dx = 1$. Thus, if the A-law companding advantage is defined as the improvement in SQNR compared with uniform quantisation, it is $20 \log_{10}(dy/dx)$. For $A = 87 \cdot 6$, this is 24 dB.

Another compression law that is used is the μ *law* shown in Fig. 16.13d. This is given by

$$y = \frac{\log_e(1 + \mu x)}{\log_e(1 + \mu)} \quad \text{for } 0 \leqq x \leqq 1 \tag{6.7}$$

This approximates to the logarithmic law for $x \geqq 1/\mu$ and to a linear law for $x \ll 1/\mu$ (since $\log_e(1 + \mu x) = x - \mu^2 x^2/2 + \mu^3 x^3/3 - \dots$).

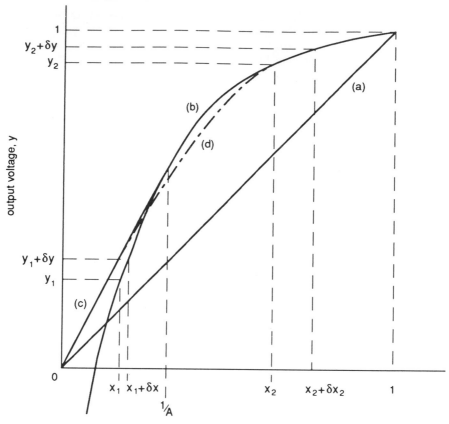

Fig. 6.13 Compression characteristics
a linear law
b logarithmic law
c A law
d μ law

From eqn. 6.7, the slope of the characteristic is

$$\frac{dy}{dx} = \frac{1}{\log(1+\mu)} \frac{\mu}{1+\mu x} \qquad (6.8)$$

Eqn. 6.8 shows that, when $\mu = 255$, the SQNR changes by less than 3 dB over an input-level change of 40 dB. For small signals

$$\frac{dy}{dx} = \frac{\mu}{\log_e(1+\mu)}$$

This corresponds to the companding advantage; thus, for $\mu = 255$, this is 33 dB. Early PCM systems [6] did use non-linear networks in conjunction with uniform encoding. Later systems [7] use piecewise linear approximations to the compression law, since these are easy to implement digitally.

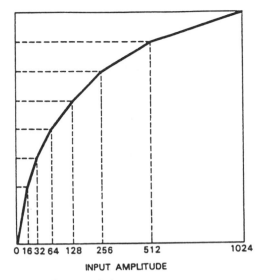

INPUT AMPLITUDE

Fig. 6.14 CCITT *A*-law characteristic

Two such logarithmic quantising schemes for 8–bit encoding have been defined by the CCITT [8]. They correspond, respectively, to the *A* law with $A = 86.7$ and the μ law with $\mu = 255$. The *A*-law is used across Europe, and the μ law is used across America and Japan. Both schemes use codewords which bear a remarkable similarity to 8 bit binary floating-point numbers. One bit is reserved for the sign, three bits for the exponent, and four bits for the mantissa, with the leading bit (which is always one) removed. The *A*-law characteristic is shown in Fig. 6.14.

In order to develop a PCM codec of good quality, performance guidelines must be laid down. The CCITT has studied the requirements for a standard speech-telephony channel in some detail, and the conclusions of the study groups are laid out in a CCITT recommendation [9]. This specifies acceptable limits for attenuation/frequency distortion, group delay, idle-channel noise, discrimination against out-of-band input signals, spurious out-of-band signals at the channel output, intermodulation, total distortion (including quantisation distortion), spurious in-band signals, variation of gain with input level, and more. These form an essential metric for the performance of any new codec.

6.8 Speech statistics and ADPCM

Logarithmic quantisers reduce the bandwidth required to transmit a toll-quality speech signal by reducing the number of bits per sample. A reduction can also be obtained by using differential PCM (DPCM). Instead of transmitting each sample, the difference between it and the previous sample is sent. Since signals very rarely change between their minimum and maximum possible voltages from one sample to the next, DPCM can use fewer digits per sample than simple PCM. Recent work has shown that the bandwidth required can be further reduced, with insignificant degradation in speech quality, using more sophisticated coding techniques. These are all based around a more detailed knowledge of the characteristics of the human voice and the vocal tract.

Speech utterances can be broken down into two classes: voiced and unvoiced sounds. Voiced sounds are produced by resonances within the vocal tract being stimulated by a regular train of pressure bursts from the vocal chords. Unvoiced sounds are mainly noise-like, albeit with some frequency shaping to allow 'f' and 's' to be differentiated. However, ruthless bit-rate reduction can usually be applied to any speech signal which is noise-like, without significant degradation.

Voiced sounds contain a large amount of redundancy, and this may be removed by a self-adaptive filter which attempts to predict the next speech sample based on previous speech samples. The speech estimate can then be subtracted from the input speech signal, leaving a small residual, which is coded by an adaptive quantiser prior to transmission. The residual signal is more noise-like and, again, we can reduce the number of bits below that which would have been used for the speech itself. A four-bit quantiser is typical for this type of system, having an inherent signal-to-noise ratio of approximately 20 dB for a sine wave.

The self-adaptive filter is very effective for nulling signals with large autocorrelation functions, such as a voiced sound. It is capable of providing a gain of around 20 dB to reconstruct the original redundant speech values from the much smaller residual signal. This figure adds directly to the signal-to-noise ratio of the quantiser, yielding an overall signal-to-noise ratio of about 40 dB for a sine wave.

This coding scheme is known as adaptive differential pulse-code modulation (ADPCM) and is viable at bit rates around 32 kbit/s. The building blocks contained within an ADPCM transcoder are illustrated in Fig. 6.15. ADPCM transcoders are used to double the capacity of existing telephony channels where bandwidth is at a premium, such as satellite links or submarine cables. For these applications, it is more cost effective to make better use of the available bandwidth than to provide more bandwidth by installing new links. The CCITT has been heavily involved in studies of various ADPCM algorithms, and has specified one such 32 kbit/s algorithm to bit level, with variants to allow operation at 24 and 40 kbit/s should channel capacity become scarce or plentiful [10].

Alternatively, ADPCM can be applied to make better use of the available 64 kbit/s bandwidth, by linking two ADPCM coders to a sub-band coder, with one coder operating over the range 0–4 kHz, with 6 bits per sample, and the other operating over the range 4–7 kHz, with 2 bits per sample, resulting in an overall bit rate of 64 kbit/s. The two frequency bands are subsequently recombined in the decoder. Allied to a 16 bit 16 kHz sampling ADC and DAC this system is capable of very-high-quality wideband speech transmission, and yet it still only requires 64 kbit/s transmission bandwidth. The CCITT has also been heavily involved with the development of the

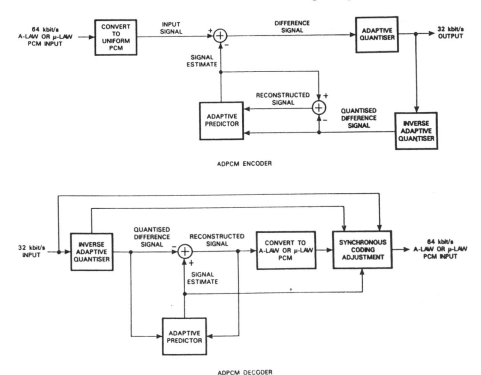

Fig. 6.15 ADPCM transcoder
a encoder
b decoder

ADPCM sub-band coder, and a bit-level specification of the ADPCM section of the algorithm is available [11]. The required performance of the band-split filters is also laid down, although these are not specified to bit level.

This coding scheme finds application in the Multi-point Interactive Audio-visual Conferencing (MIAC) system [12]. This is an audio-conferencing system with additional features for conference management, FAX trans-mission, electronic messaging, and still-picture transmission, over a single 64 kbit/s channel. This is currently undergoing evaluation at a number of engineering sites around Europe.

6.9 Speech statistics and LPC

Linear predictive coding (LPC), takes the idea of using a model of the vocal tract one stage further [13]. A short block of speech, typically 20 ms in duration, is processed to extract pitch period and formant information, which describe respectively the activity of the vocal chords and the reson-ances occurring in the vocal tract. Then, rather than transmitting the speech-sample information, an LPC coder transmits quantised versions of the pitch period and formant-model filter coefficients for that speech block. The decoder then uses this model of the vocal-tract activity to reconstruct the speech block.

Because of the small amount of information transferred between encoder and decoder, LPC coders are viable at bit rates from 16 kbit/s down to 4·8 kbit/s, although error protection for the formant coefficients is almost essential. The principal disadvantage of LPC coders is that they must obtain a block of speech samples before the speech can be processed. Thus, they possess severe processing delays, typically equal to two to three times the speech-block duration, so they must be supplemented by echo suppressors or cancellers.

The Groupe Speciale Mobile (GSM) a pan-European initiative with a specific interest in digital cellular radio, was formed by the governing body of the European PTTs, CEPT, and was given the task of specifying a standard algorithm for digital cellular radio across Europe. The chosen solution [14] uses a variant of linear prediction, known as regular-pulse-excitation long-term prediction (RPE–LTP). It will operate at 13 kbit/s with a further 3 kbit/s given to error protection, and it will use a speech-block length of 20 ms. As well as specifying the digital speech-coding algorithm, GSM has laid down extensive specifications for the network interface, data link and physical layers of the system in order to ensure compatibility between vendors.

6.10 Programme-quality sound

Broadcast programmes cover a wide range of audio material from speech through all sorts of natural, animal and musical sounds to the full orchestra. The dynamic range is much larger than for speech and the allowable degradations much smaller. As with telephony, the standards are decided on the basis of subjective evaluation. Sampling rate and initial quantising resolution have been set by CCIR at 32 kHz and 14 bits respectively for the transmission of 14 kHz channels.

The CCITT has recommended two alternative companding systems. 11-segment A-law gives a bit rate reduction from 14 to 11 bits per sample and 5-range near-instantaneous companding (NIC) gives a bit rate reduction from 14 to 10 bits per sample.

In the NIC system, blocks of 32 samples are stored and measured. One of five possible linear gain settings is selected, prior to encoding, according to the largest sample in the range. The selected range is signalled to the decoder for decoding purposes.

Parameters of the two systems are compared in Table 6.1. The NIC system has been adopted in the UK with a dedicated frame structure which takes six channels (three stereo pairs). A single channel can be justified to 384 kbit/s for incorporation in the telephony frame format displacing six telephony channels.

Table 6.1 Proposed coding methods for programme sound from draft report 647-1 (MOOI) CMTT

	NIC	A-Law	Units
Nominal bandwidth	0·04—15	0·04—15	kHz
PE/DE–CCITT J17	Yes	Yes	
(6·5 dB/0·8 kHz)			
Overload (at 2·1 kHz)	+13	+15	dBm0
Sampling rate	32	32	kHz
Companding law	5 ranges	11 segments	
Bit rate reduction	14—10	14—11	
Finest resolution	14	14	bits/sample
Idle noise	−65	−62	dBq0ps
Coarsest resolution	10	9	bits/sample
Total noise with +9 dBm0	−41	−32	dBq0ps
Total noise with	−55	−50	dBq0ps
+9 dBm0/60 Hz			
Source coding	323	352	kbit/s
Error protection	12	32	kbit/s
Framing/signalling	3	—	kbit/s
Service bit-rate/channel	338	384	kbit/s

6.11 Coding video signals

6.11.1 Video statistics and sub-Nyquist coding

Composite video signals, such as are used for broadcast television, e.g. PAL, SECAM and NTSC, contain a straightforward luminance component (Y) together with colour information (U&V) amplitude modulated onto two 4·43 MHz colour subcarriers held in quadrature. Experiments have shown that 8 bits are quite sufficient to allow broadcast quality to be maintained over several tandem encodings. The CCIR has laid down an internationally-agreed standard for the digital coding of studio-quality video signals [15]. This specifies a sampling frequency of 3 times colour sub-carrier frequency for the luminance component and 1·5 times colour sub-carrier frequency for each of the colour components, with 8 bits per sample for all components, which leads to a bit rate of 216 Mbit/s.

Since the composite video spectrum extends well beyond 5 MHz, it might be expected that the sampling rate must be at least 10 MHz, to satisfy Nyquist's sampling theorem. However, because of the line structure of a video signal, the power in the video spectrum around the colour sub-carrier tends to be concentrated in lines with a frequency spacing equal to half the line-repetition rate.

This fact can be exploited, and the sampling frequency F_s is chosen such that the aliases, which will inevitably be generated from those lines in the colour signal above $F_s/2$, fold down onto holes in the power spectrum below $F_s/2$. Comb filtering can then be applied to remove aliases caused by the sub-Nyquist sampling.

Assuming, therefore, a sampling rate of 8·86 MHz and the use of a flash convertor of 8 bit resolution, a channel capacity of approximately 70 Mbit/s is required to transmit a PCM video picture to broadcast standards. In practice, a 68 Mbit/s channel is used, replacing two 34 Mbit/s channels in a 140 Mbit/s digital transmission system. If error protection is applied, this pushes the bit rate still higher. The CCIR recommends 140 Mbit/s for television transmission [16] to ensure that the quality is better than that achieved by analogue systems.

6.11.2 Video compression techniques

If insufficient bandwidth is available to transmit the full 70 Mbit/s video signal, a number of techniques are available for reduction of the channel capacity required. Firstly, non-linear differential pulse-code modulation, or DPCM, can be used instead of linear PCM. The difference between successive video samples is coded by a non-linear quantiser and 4 to 6 bits can be sufficient for the luminance information. The colour information generally requires much lower bandwidth and resolution than the luminance. Thus, further rewards may be reaped by reducing the colour sample rate to one ninth of the luminance sample rate, and using 2 to 4 bits per colour-component sample. Rates of about 35 Mbit/s for PAL quality, or 8 Mbit/s for video-telephone quality have been used [17].

If this is still too high, more exotic video-processing algorithms abound [18]. One example is the application of conditional replenishment and motion prediction. An image is sub-divided into small blocks, with typically 400 such blocks per image, and a block is only retransmitted if significant changes have occurred within that block. Likewise, if the block is seen to be displaced slightly from one frame to the next, it is not retransmitted; instead, a vector describing the displacement will be transmitted, requiring far fewer bits. Such techniques allow the bit rate to be reduced to around 384 kbit/s, and give quite acceptable performance for applications such as full-colour video conferencing [19].

6.12 The PCM primary multiplex group

6.12.1 General description

PCM systems were first developed for telephone transmission over cables originally designed for audio-frequency transmission. It was found that these are satisfactory for digital transmission at rates up to about 2 Mbauds. Consequently, telephone channels are combined by time-division multiplexing to form an assembly of 24 or 30 channels. This is known as the *primary multiplex group*. The primary multiplex group is also used as a building block for assembling larger numbers of channels in higher-order multiplex systems, as described in Chapters 8 and 9.

The block diagram of a typical PCM primary multiplex equipment is shown in Fig. 6.16. Since the coder (ADC) and decoder (DAC) must each operate in the time-slot of one channel, they are made common to all the channels in the system. In order to use the system in switched telephone

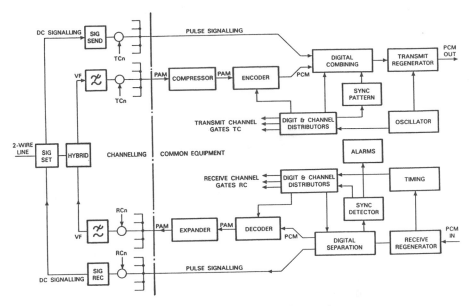

Fig. 6.16 Block diagram of PCM primary multiplex equipment

connections, provision must be made for transmitting the necessary signalling conditions. As described in Chapter 14, this may be done either by
channel-associated signalling or common-channel signalling. Fig. 6.9 shows
built-in channel-associated signalling.

In the sending direction, the signal from each incoming audio channel
passes through an anti-alias filter and a sampling gate before being encoded.
The sampling gate of each channel is operated by a pulse at a different time
TC_n in order to multiplex their pulse-amplitude modulated (PAM) pulses
onto a common highway to the encoder. The DC signalling condition on
the incoming line is sampled at the same time and the signalling samples
are subsequently inserted in the frame of speech samples, as described in
Sections 6.12.2 and 6.12.3. A distinctive *framing signal* is also inserted in
order to synchronise the receiving equipment at the far end of the transmission path.

In the receiving direction, the incoming digital signal is regenerated and
its timing is extracted to provide clock pulses for the decoder. The framing
signal is extracted and used to synchronise the channel-pulse generator to
ensure that the receive gate of each channel operates at the appropriate
time RC_n. Thus, each channel's receive gate selects the correct PAM pulse
train on the common highway from the decoder and this is demodulated
by a low-pass filter. Signalling pulses are similarly directed to the correct
channels.

The transmitted digit stream must contain, assembled by time interleaving, the following:

(*a*) Coded speech samples from each channel

(*b*)   Several identifiable signalling digits for each channel
(*c*)   Further digital patterns inserted for the purpose of identifying the message channels, and maintaining alignment between the receiver operations and the incoming signal (synchronising or framing digits).

The digit stream is organised into frames of duration 125 μs (i.e. the sampling interval). Each frame contains one coded speech sample from each channel, together with some supernumary digits which serve the needs of signalling and synchronizing. There are several frame structures in practical use: the current standard 30-channel system and the DS1 structure used in the USA are described in Sections 6.10.2 and 6.10.3. Both frame formats have been standardised in CCITT specifications [20].

6.12.2 The 30-channel system

As shown in Fig. 6.17, the 30-channel frame is divided into 32 time-slots each of 8 digits, so the total signalling rate is $8000 \times 8 \times 32 = 2 \cdot 048$ Mbit/s. Time-slots 1–15 and 17—30 are each allotted to a speech channel. Time-slot 16 is allotted to signalling. Since each contains only 8 digits, four such slots are needed to give one signalling digit to each speech channel. In fact, there are four signalling digits per speech channel so that up to 16 signalling conditions can be conveyed, either by suitable coding or by allotting each digit to an independent function. It follows that 15 slots are needed to accommodate all the signalling digits. These are distributed over a set of 16 frames occupying 2 ms; this set is called a *multiframe* (Fig. 6.17). Each signalling digit has a capacity of 500 bit/s, which is ample for any normal signalling function. The 16 signalling time-slots occurring in a multiframe include a spare capacity of one slot, which carries a pattern used for multiframe alignment and further signalling-related supervisory functions.

The process of achieving frame alignment is simple in principle, though complex in detail. The receiver has a clock which is recovered from the incoming digit stream and synchronised with it: counters driven from the clock define regular periods equal to a frame length. If these periods are aligned with the incoming frame, regular patterns (the frame alignment words) are observed at the beginning of the frame in time-slot 0. If they are not aligned, the counters can be slipped until alignment is detected.

Unfortunately, presence or absence of a pattern does not infallibly indicate alignment or misalignment. A pattern in the correct location may be mutilated by digital errors in transmission and fail to be recognised. Alternatively, the pattern (which is not unique and indeed cannot be made unique without excessive redundancy in the code) may be imitated elsewhere in the frame by fortuitous signal sequences. It is therefore usual for the framing mechanism to be given some inertia so that it searches for alignment only in response to several successive absences of the framing pattern. The subject of frame alignment in digital systems is dealt with in more detail in reference 21.

6.12.3 The 24-channel DS1 system used in the USA

The DS1 frame format is illustrated in Fig. 6.18. The basic frame consists of 193 bits; the first bit is used for framing purposes and is termed the *F*

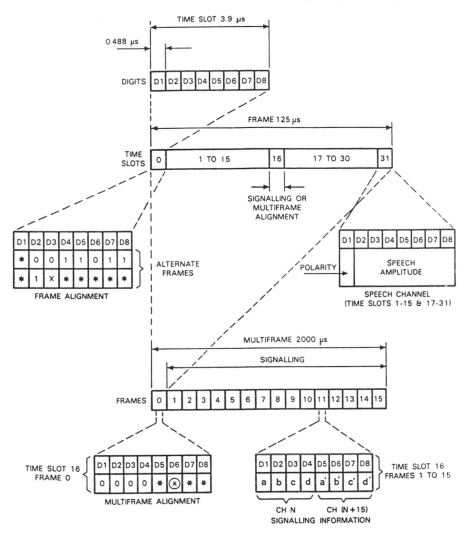

Fig. 6.17 30-channel frame format

bit. Bits 1 to 192 inclusive form twenty four 8–bit time-slots which are used to transmit basic PCM data. On all odd-numbered frames, the F bit takes on the alternating pattern 1, 0, 1, 0, . . . ; this is the pattern used for frame alignment. (It is a distributed frame-alignment signal as opposed to the block alignment signal used in the 30-channel system). The even-numbered

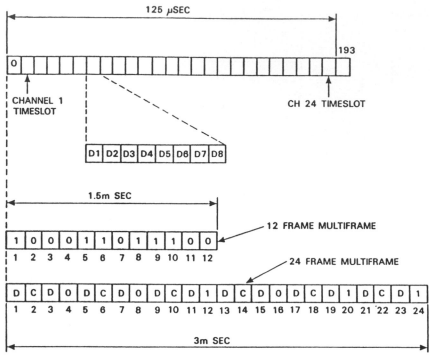

Fig. 6.18 24-channel frame format

frames carry the pattern 0, 0, 1, 1, 1, 0, which defines a 12-frame multiframe. On frame 6 of the multiframe, bit D8 in each of the channel time-slots is used for channel-associated signalling channel A, while on frame 12 it is used for signalling channel B.

The DS1 format has recently been extended to encompass a 24-frame multiframe of 3 ms duration. The 8 kbit/s F bit is now used to provide framing (and multiframing) error checking with 2 kbit/s for each, leaving 4 kbit/s for an auxiliary data channel which is used for various operations-support purposes. The 24-channel multi-frame will also support four signalling channels A, B, C and D using the same 'robbed bit' technique. Bit robbing causes a small degradation in quantising distortion, which is nonetheless considered acceptable.

6.13 References

1 CATTERMOLE, K.W.: 'Principles of pulse code modulation' (Iliffe, 1969)
2 SALLEN, R.P., and KEY, E.L.: 'A practical method of designing RC active filters', *IRE Trans.*, 1955, **CT-2**, pp. 74—85
3 SEDRA, A.S., and SMITH, K.C.: 'Microelectronic circuits', (Holt, Rinehart and Winston, 1982) pp. 642–648
4 STEELE, R.: 'Delta modulation systems' (Pentech Press, 1975)
5 PURTON, R.F.: 'Survey of telephone speech-signal statistics and their significance in the choice of a PCM companding law', *Proc. IEE*, 1962, **109B**, pp. 60–66
6 FULTZ, K.E., and PENICK, D.B.: 'T1 carrier system', *Bell Syst. Tech. J.*, 1965, **44**, pp. 1405–1451
7 VOGEL, E.C., and McLINTOCK, R.W.: '30-channel pulse-code modulation system', *PO Elec. Eng. J.*, 1978, **71**, pp. 5–11
8 CCITT Recommendation G711: 'Pulse-code modulation (PCM) of voice frequencies'
9 CCITT Recommendation G712: 'Performance characteristics of PCM channels between 4-wire interfaces at voice frequencies'
10 CCITT Recommendation G721: '32 kbit/s adaptive pulse code modulation'
11 CCITT Recommendation G722: '7 kHz audio coding within 64 kbit/s'
12 CLARK, W.J.: 'MIAC: a system for multi-point interactive audiovisual communications'. Int. Conf. on ISDN, vol. 1, London, 1986, pp. 87–98, Online publications, 1986
13 ATAL, B.S. and SCHROEDER, M.R.: 'Adaptive predictive coding of speech signals', *Bell Syst. Tech. J.*, 1970, **49**, pp. 1973–1986
14 ETSI: 'GSM full-rate speech transcoding', GSM 06.10 version 3.01.02, April 1989
15 CCIR Recommendation 601: 'Encoding parameters of digital television for studios'
16 CCIR Report 962: 'The filtering, sampling and multiplexing for digital encoding of colour television signals'
17 CCIR Report 629-2: 'Digital coding of colour television signals'
18 JAIN, A.K.: 'Image data compression: a review', *Proc. IEEE*, 1981, **69**, pp. 349–389
19 CCITT Recommendation H261: 'Code for audiovisual services at $n \times 384$ kbit/s'
20 CCITT Recommendation G704: 'Synchronous frame structures used at primary and secondary hierarchical levels'
21 BYLANSKI, P., and INGRAM, D.G.W.: 'Digital transmission systems' (Peter Peregrinus, 1987)

Chapter 7
Digital transmission principles
R.M. Dorward

GPT Limited, Coventry

7.1 Introduction

This chapter on digital transmission is intended to provide a basic introduction to the principles and practice of digital transmission in general. However, to avoid overlap with other chapters, it will be largely restricted to consideration of the transmission of digital information at rates above a few kilobits/second, in baseband form, over metallic conductors. Furthermore, it will concentrate on certain aspects specific to digital transmission, as the general features of line systems are dealt with in Chapter 10. Information at much greater breadth and depth is available in several fairly recent books devoted to digital transmission; for instance, Reference 1 which deals primarily with the theoretical and design aspects, and Reference 2 which emphasises the applications. Reference 3 on North American systems includes one chapter specifically on digital transmission; it provides a balance to References 1 and 2, which are largely orientated towards UK systems.

The primary purpose of a transmission system is to convey information from a source at one geographical location via some medium to a user at another geographical location with sufficient fidelity for the information to be acceptable to this user. Lack of fidelity of the signal is difficult to define precisely because of the variety of degradations in transmission paths. However, it may be considered as consisting of two parts: impairments arising from the presence of the intended signal (i.e. signal distortion) and spurious background signals (i.e. added 'noise').

The above considerations are quite general and presuppose no specific type or form of signal. However, the form of the signal transmitted can be broadly divided into analogue, where the information may be represented by a continuous variable, and digital, where it is a discrete variable, i.e. one which is confined to a finite number of states (and to changing between these states at defined time intervals). This discrete or quantised nature allows recognition of the information by the user in the presence of more noise and distortion than would be acceptable for analogue information. Furthermore, in the case of coarsely quantised digital information (i.e. with a small number of discrete states), the recognition may be carried out by relatively simple electronic circuits which may be incorporated in the repeaters distributed along the transmission link. Thus, these repeaters may be used not only to amplify and equalise the information signal but also to recognise the occurrence of the discrete states and re-generate the information, free from noise and distortion. This regeneration of the information prevents the accumulation of noise and distortion, thereby substantially eliminating the limit on the transmission path length that these normally impose on analogue line systems.

7.2 The regenerative repeater

The functions involved in a regenerative repeater for digital signals are shown in Fig. 7.1. It may be conveniently split into those parts which are similar to an analogue-signal repeater and those parts which are specific to a regenerative repeater. The functions of line build-out (to make all pairs look the same), equalisation and amplification in a regenerative repeater are the same as those carried out in, and comprise the whole of, an analogue repeater (see Chapter 10). The only difference is that, in a regenerative repeater, the equalisation need not be so accurate and the noise level may be higher than in an analogue repeater because any accumulation of distortion and noise is removed by regeneration. Thus, the greater bandwidth required for, and hence greater attenuation and distortion arising in, digital rather than analogue transmission are compensated by the much greater attenuation and distortion that can be tolerated. A system may include a mixture of analogue and regenerative repeaters, in which case it is called a 'hybrid'. A feature used in both types of repeaters is the means of minimising the signal distortion, i.e. a variable equaliser with an automatic-gain-control feedback loop. This is used to compensate for uncertainty in the line parameters, length or temperature. In its simplest form this consists of a signal monitor, e.g. a pilot-tone level detector in an analogue system or a peak-to-peak signal-level detector in a digital system, with feedback to control the characteristics of the variable equaliser. The control may simply provide for a variable 'flat' gain. However, more normally, it will provide for a variable gain shaped to correspond to the line, i.e. attenuation proportional to the square root of frequency.

The regeneration process is specific to regenerative repeaters and involves three stages:

(a) Timing extraction,
(b) Threshold detection and,
(c) Output-pulse generation.

The timing extraction circuit extracts from the equalised signal a continuous timing signal to define the intervals at which the signal may be expected to be in one of its discrete states. This is carried out by a narrowband filter tuned to the symbol rate of the signal and preceded by some non-linear processing to generate a spectral line at that frequency if one does not exist already in the equalised signal. The threshold detection is used, in conjunction with the extracted timing signal, to recognise in which state the equalised signal is at each time interval. In the simplest case of digital information

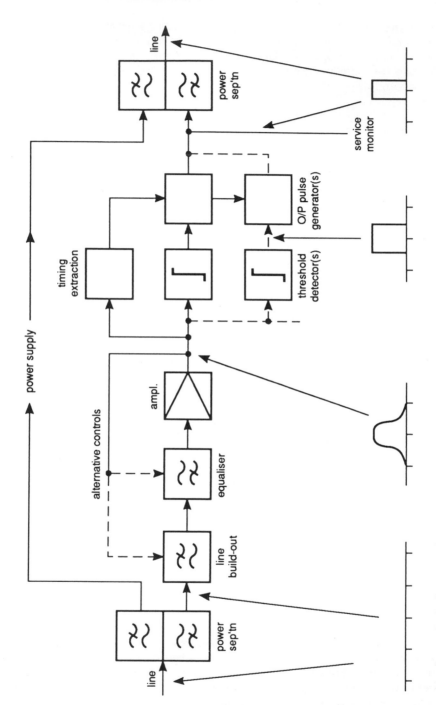

Fig. 7.1 Block diagram of regenerative repeater

with only two discrete states, the recognition process can be carried out by a single limiting amplifier whose output is sampled by the timing signal. The output-pulse generation produces unmutilated output states, as determined by the threshold detector, identical in form to those of the original transmitted signal.

In practice, the ideal situation described above does not exist and regenerative repeaters do not perfectly reproduce the transmitted information. They introduce two forms of degradation: digital errors and jitter. Although quantised information may be recognised in the presence of more noise and distortion than would be acceptable for analogue signals, there is always a finite probability that the combined effects of noise and distortion will occasionally cause the threshold detection to recognise the state of the signal erroneously and so produce an erroneous output pulse. This probability is shown in Fig. 7.2 for noise with a Gaussian (normal) probability distribution. It applies to the usual circumstances where there are a large number of independent noise contributions of similar magnitude. Fig. 7.2 is plotted in terms of level separation, i.e. voltage difference between adjacent signal states at a decision instant. It also assumes the thresholds are placed midway for good noise immunity and the effective level separation is that available upon which the threshold detector may operate after an allowance has been made for residual misequalisation and intersymbol interference (ISI). The lower and upper limits of the plot apply to the two extremes of binary and many-level signals respectively. In the latter case, the error probability is doubled because most signal states can be misinterpreted due to noise peaks of either polarity.

For most users, the required signal fidelity implies an error probability of 10^{-5} or even orders of magnitude lower, where the slope in Fig. 7.2 is very steep. This illustrates one of the most important features of a digital transmission system: the overall performance is critically dependent on worst-case conditions. Thus, in general, the system error rate will arise predominantly from one repeater. This leads naturally to great emphasis on attaining the maximum signal-to-noise ratio in the design of any repeater, and to the necessary bandwidth trade-off between noise and intersymbol interference. A typical result is shown in Figs. 7.3 and 7.4. The overall equalised gain/frequency response has significant attenuation even at half the symbol rate. Thus, although the transmitted signal may be rectangular pulses, the resultant waveform at the input to the threshold detector is quite 'rounded' and exhibits significant intersymbol interference as shown in Fig. 7.5.

This trade-off is to a large extent responsible for the generation of the second degradation: jitter (the deviation of the regenerated pulses from their ideally equally spaced intervals in time). From Fig. 7.3 it is obvious that there is no spectral component at the symbol rate, so some non-linear processing, such as full-wave rectification, will be required prior to the narrowband filter. This produces a series of spikes at the transitions between states, but with some (see Fig. 7.5) deviating from their nominal position due to the intersymbol interference. The narrowband filter extracts from this stream the symbol-rate component. However, because of its finite band-

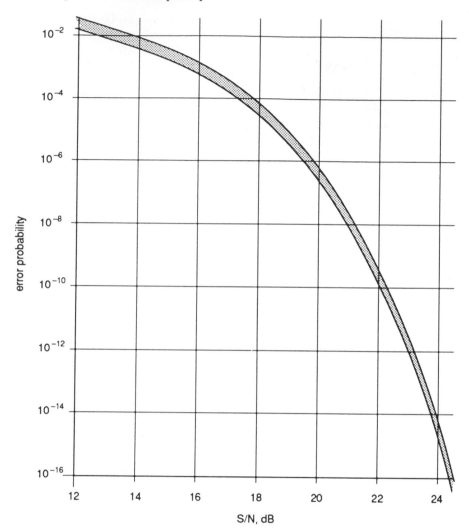

Fig. 7.2   Variation of error probability with signal-to-noise ratio for Gaussian noise. (Signal power corresponds to separation of adjacent signal levels)

width, it also lets through some of the sidebands, which produce phase modulation. This phase modulation, or jitter, of the continuous timing signal is then reproduced on the output pulses, and so it is cumulative along the transmission path [4, 5]. Unlike digital errors, however, jitter can be reduced subsequent to its occurrence, by the use of a filter with an extremely narrow bandwidth in a final timing operation, probably at the end of the transmission path. If the jitter has built up through a large number of repeaters, it may have an amplitude of several symbol periods. In this case, some buffer storage will also be necessary to prevent errors being introduced by the retiming operation.

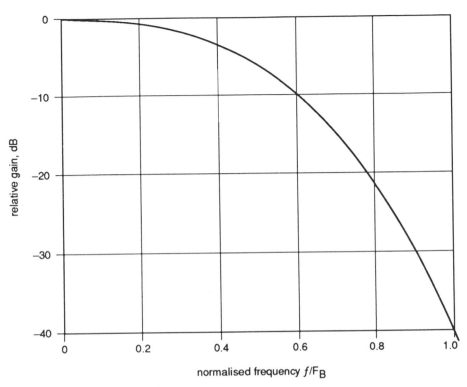

Fig. 7.3 Typical overall equalised gain/frequency response

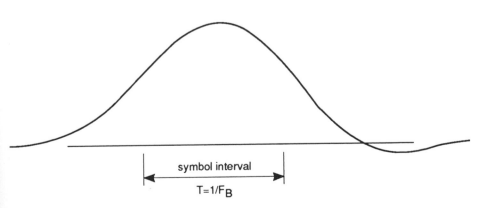

Fig. 7.4 Pulse response corresponding to equalised gain/frequency response of Fig. 7.3

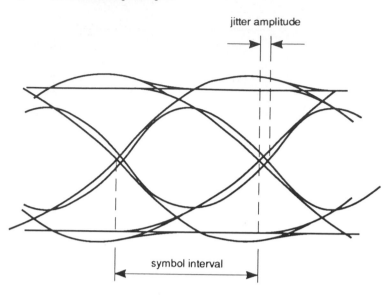

Fig. 7.5   Composite equalised waveform (eye diagram)

7.3 Pulse shape and equalisation

The maximisation of the signal-to-noise ratio at the threshold detection point, which is so central to the design of a regenerative system, is carried out in a metallic-pair line system in the presence of a cable attenuation in decibels which is predominantly proportional to the square root of frequency. Widely different cables have very much the same type of attenuation characteristic, apart from a scaling factor, as may be seen from Fig. 7.6. Within the constraints of this characteristic, and a given symbol rate and number of transmitted levels, the signal-to-noise ratio is primarily determined by the shape and amplitude of the transmitted pulse, and by the trade-off between noise and intersymbol interference (ISI) at the threshold-detection point in the repeater.

The form of the transmitted pulse is mainly relevant to a thermal-noise-limited system and so it will be considered first and on that basis alone. With the ideally 'flat' power spectrum of thermal noise, the transmitted pulse should contain as much energy as possible in the region of the frequency requiring the greatest amplification. This depends to some extent on the equalisation characteristic used; however, for systems with moderate intersymbol interference, it is generally somewhere between half the symbol rate and the symbol rate. Precise tailoring of the transmitted pulse shape is usually undesirable because of the extra power consumption which would be involved in a sophisticated repeater output-pulse generator or the losses involved in any shaping networks. Moreover, a simple 'rectangular' pulse may approach to within a few decibels of the optimum in terms of signal-to-

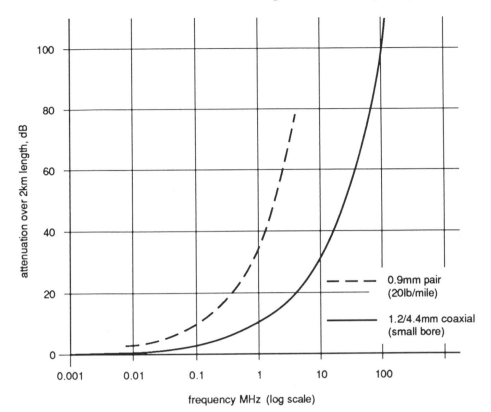

Fig. 7.6 Typical cable attenuations

noise ratio. With such a pulse shape, the two variables are pulse width (or duty cycle relative to the symbol interval) and rise-time. Fig. 7.7(*a*) shows how the required repeater output power consumption for a given signal-to-noise ratio varies with the first of these parameters for the two cases of:

(i)   Power consumption dependent on output-pulse power and
(ii)   dependent on output-pulse amplitude.

It is apparent that the usual choice of a 50% duty cycle pulse is close to the optimum in either case. In addition, the pulse rise and fall times have only a small effect, within reasonable limits, as shown in Fig. 7.7(*b*).

The equalisation characteristic in a repeater is required to compensate for the attenuation and distortion of the cable section preceding it and to provide a suitable trade-off between noise and intersymbol interference at the threshold detector or decision point. Despite fairly severe overlap of pulses, it is usually desirable to choose the overall transmission characteristic, i.e. from the output of one repeater to the decision point of the next, to have no intersymbol interference at the decision instants. Nyquist studied

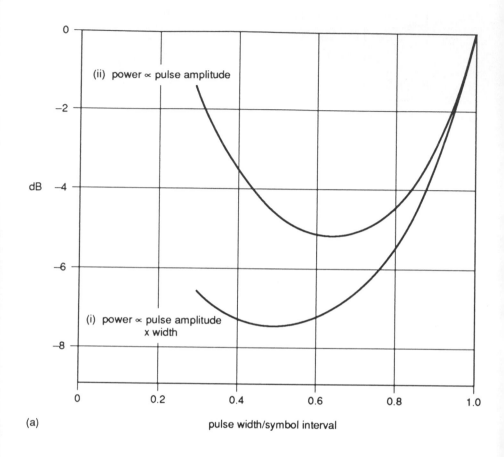

(a)

pulse width/symbol interval

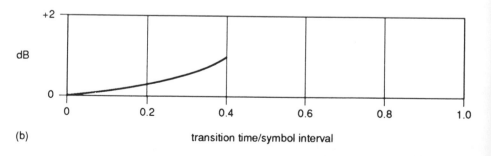

(b)

transition time/symbol interval

Fig. 7.7   Variation in power consumption of repeater output-pulse generator

a variation with ratio pulse-width/symbol interval
b variation with pulse transition time

this as the first of several alternative criteria for distortionless transmission.[6] For signals limited to a band below the frequency equal to the symbol rate, his first criterion requires the amplitude spectrum of a single equalised pulse to be antisymmetric within this band. On the basis that minimum bandwidth is desirable to avoid amplifying a great deal of noise at high frequencies, it is tempting to consider an amplitude spectrum that is flat up to half the symbol rate and zero thereafter. However, in practice, there must be a finite roll-off for the spectrum. Two more suitable spectra are shown with their corresponding pulse shapes in Fig. 7.8. These are known as 'raised cosine' or 'squared cosine' spectra because of the shape of the roll-off. The 'full' or 100% raised-cosine spectrum drawn with a continuous line generates little intersymbol interference in adjacent unit intervals, and it is relatively easy to approximate. The 75% raised-cosine spectrum shown by a broken line gives better immunity to near-end crosstalk (NEXT) and thermal noise because it requires less amplification, as illustrated in Fig. 7.9. This Figure assumes the transmitted pulse to be narrow and therefore to have a flat spectrum. Thus, in the case of a 50% duty-cycle rectangular transmitted pulse, the required gain of the equaliser plus amplifier must be further modified by the inverse of the transmitted pulse spectrum. The effect of this 'sin x/x' shaping is to increase the required gain by less than 1 dB at half the symbol rate and by about 2 dB at the frequency of maximum gain.

The effect of pulse shape on adjacent symbols can be seen in a sequence of pulses by superimposing the possible combinations onto one unit interval. For the popular full-raised-cosine spectrum this has been done in Fig. 7.10 using binary (two-level) and ternary (three-level) symbols. The opening around the decision instant has led such pictures to be known as *eye patterns.* The eyes shown are perfectly open at the decision instant; i.e. there is no intersymbol interference at this time, because the first Nyquist criterion is met. Although these eyes show all possible pulse combinations, some combinations may be excluded by the line-coding rules, so that eye patterns are often somewhat simpler.

The raised-cosine and similar equalisation characteristics are aimed at achieving zero intersymbol interference at the decision instant. However, it is also possible, and sometimes productive of an improved signal-to-noise ratio, to work with a large but precisely defined intersymbol interference. This is commonly known as *partial response equalisation* [7, 8]. In the simpler cases, a single pulse produces a response whose amplitude is only a fraction of that produced by a series of adjacent pulses. The result of this is that, for example, a two-level transmitted signal produces a multilevel 'eye' at the input to the next decision point. Fig. 7.11 shows in a stylised form how this arises as the equalised response is progressively narrowed and the intersymbol interference increases.

Obviously, this process will reduce the noise at the decision point, albeit at the cost of a gross reduction in the amplitude of the binary eye (see Fig. 7.11c). However, it is possible to recover the binary information by making a three-level decision on the perfectly open three-level eye which has appeared. The level separation is exactly half the amplitude of the original

raised cosine pulse spectrum definition:
A(f) = 1   for 2T|f| ≤ 1–a
A(f) = 1/2 {1+sin $\frac{\Pi}{2a}$ (1–2T|f|)} for 1–a ≤ 2T|f| ≤ 1+a
A(f) = 0   for 2T|f| ≥ 1+a

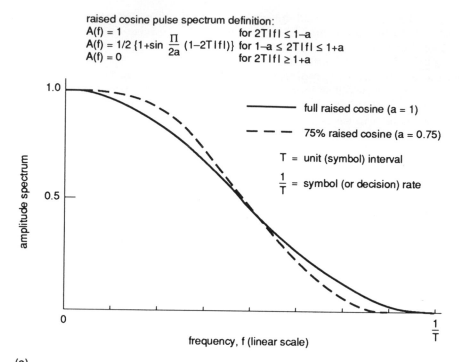

(a)

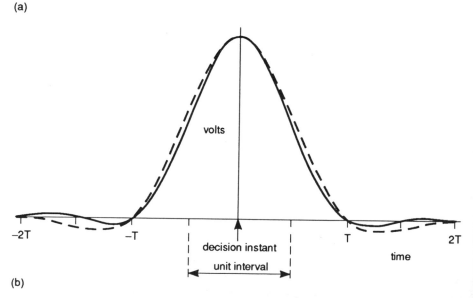

(b)

Fig. 7.8   Raised cosine spectra and corresponding pulse waveforms

a   Amplitude spectra
b   Pulse waveforms

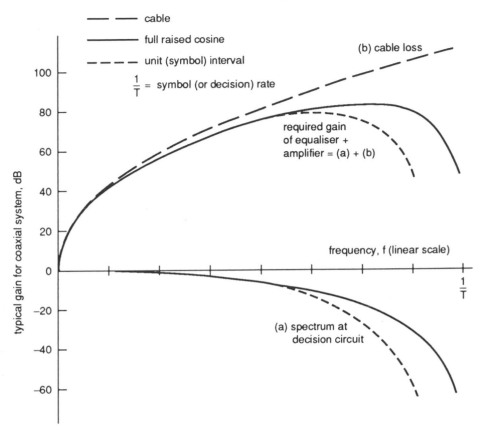

Fig. 7.9 Equalisation to obtain raised-cosine spectrum at decision circuit

binary eye. Thus, if the narrowing of the response to this extent has reduced the noise by over 6 dB, the signal-to-noise ratio will be better than in the more conventional case of zero intersymbol interference.

Recovery of the binary signal from the three-level decisions is required in each repeater before generation of the two-level output pulses. Although this requires only fairly simple logic operations, the power consumed by such logic may well be embarrassing. An alternative approach [9], known as *quantised feedback* or *decision feedback*, offers a potentially simpler solution. In this, a two-level decision only is made, and the output at each decision instant is used to subtract from the partial-response equalised signal the intersymbol interference which the pulse producing that decision may be expected to produce at subsequent decision instants. This effectively partially re-opens the closed binary eye of Fig. 7.11c and allows the single two-level decision circuit to operate satisfactorily. The technique may also be used to equalise or compensate for the effects of removal of the low-frequency components of the signal. In general, it is only limited by the problem of

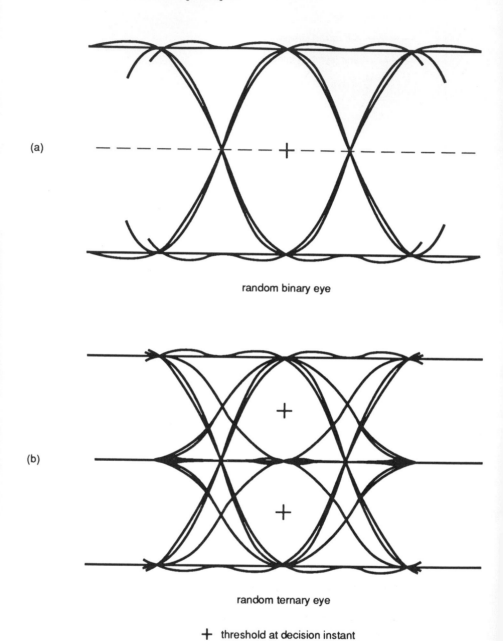

(a)

random binary eye

(b)

random ternary eye

+ threshold at decision instant

Fig. 7.10 Eye patterns for 100% raised-cosine spectrum
a random binary signal
b random ternary signal

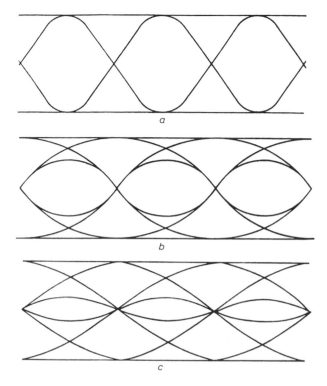

Fig. 7.11 Closure of binary eye caused by progressive increase in intersymbol interference

generating sufficiently precisely the intersymbol interference which is required to be subtracted from the signal.

7.4 Imperfections and eye margins

The theoretical error rate used as a basis for repeater design must be less than the value apportioned from the network requirements. The difference between the two, i.e. the *margin*, allows for:

(i)   Imperfections in the cable transmission characteristics.
(ii) Repeater performance that will fall short of the ideal.
(iii) The degree to which measurements may be unrepresentative and inaccurate.

System economics require these degradations to be kept to a minimum consistent with simple installation and reliable operation. The close relationship between error rate and signal-to-noise ratio (Fig. 7.2) allows error rate degradation to be expressed in terms of signal-to-noise degradation. The latter occurs when the noise is higher or when the effective signal amplitude is lower than expected theoretically. Higher noise is characterised by a noise-amplification factor *N*, usually expressed in dB; an example is the receiver noise figure in a coaxial system. Lowering of the effective signal amplitude can occur even in the absence of noise. An example is intersymbol interference due to non-ideal pulse shaping. Such impairments are liable to add on an amplitude basis. They can be marked upon an eye pattern, such as that in Fig. 7.10 for a three-level signal. Fig. 7.12 shows the peak amplitude of each possible impairment expressed as a fraction δ of the basic pulse amplitude. (Equivalently this is the peak-to-peak impairment as a fraction of the peak-to-peak signal.) The amplitude of individual impairments can then be summed to give a total δ. The total required margin in signal-to-noise ratio for a system using an *L*-level line signal is therefore:

$$\text{margin} = N + 20 \log_{10}(1 - (L-1)\delta) \text{ dB}$$

The characteristics of individual cable pairs can differ from those assumed for design and so give rise to noise amplification and reduction of effective signal amplitude as follows. Variations in the cable attenuation, due to temperature or variation in pair length, do not affect the curve shape for attenuation versus frequency. They can be equalised automatically, but they

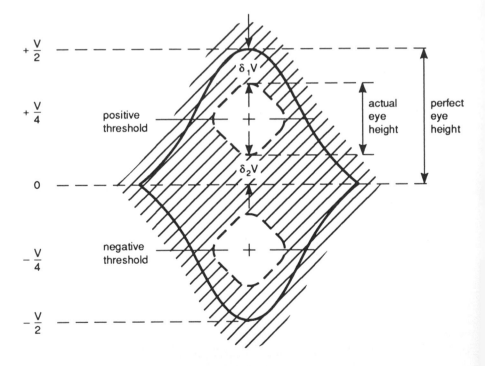

Nominal peak to peak signal = V. Eye imperfection parameter $\delta = \delta_1 + \delta_2$

Fig. 7.12   Eye degradations

give rise to a significant allowance in N. Allowances in N are also necessary in the case of symmetric paired cable for uncertainty in the crosstalk characteristics. Margins must also be allowed because a practical repeater cannot achieve its theoretical ideal performance. The effect of noise is amplified, as a result of imperfect equalisation and thermal noise in receiver circuits, requiring an N contribution. Contributions to δ arise from variability of the transmitted pulse, removal of low frequencies, non-ideal pulse equalisation, imperfect recovery of the sampling instants and the uncertainty region in any decision circuit. Usually, these effects cannot be made negligible without significantly affecting repeater cost, size, power consumption or reliability. Even with high-quality coaxial cable, practical equaliser tolerances restrict the optimum numbers of levels in the line signal to five or fewer.

7.5 Line coding

7.5.1 General

Apart from the repeater signal-to-noise ratio requirements which may indicate an optimum number of levels in the line signal, there are also 'system' features such as power feeding and quality-of-service monitoring which affect the form in which the information may be transmitted. This formatting and structuring of the information prior to transmission is termed line coding [10]. It should exhibit certain properties:

(*a*)   The line code should be transparent; i.e. it should not restrict the information which may be transmitted.

(*b*)   It should not require the transmission of DC or low frequencies, to facilitate power separation etc.

(*c*)   It should restrict the highest frequencies necessary for adequate transmission (commonly by multilevel coding) to maximise the repeater signal-to-noise ratio within the constraints of the cable characteristics and the repeater power requirements. Both (*b*) and (*c*) may be called spectrum shaping.

(*d*)   The coding should produce a sufficient density of transitions between code states to ensure adequate timing-signal extraction (timing information).

(*e*)   If the coding is carried out in blocks, producing blocks or 'words' of more than one line-code symbol, some block-alignment information must be present to ensure the decoding is carried out correctly.

(*f*)   It must provide an error-monitoring capability. The redundancy in the coding should be organised so that any transmission errors will produce easily detectable coding-rule violations.

(*g*)   Redundancy is also desirable to allow the signal to carry auxiliary service information, such as protection-switching controls.

(*h*)   The number of states (i.e. radix) of the code must be severely limited, as indicated above.

In the choice of a suitable line code for a specific system, it is instructive to consider first the deficiencies for this purpose of the original bit sequence (Fig. 7.13*a*). The problems relating to the regeneration of the signal arise

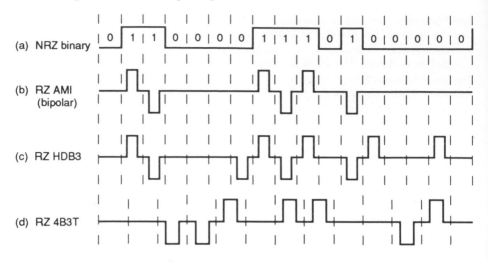

signals are shown as transmitted to line

Fig. 7.13   Examples of line coding

a Binary (non-return to zero)
b AMI (bipolar)
c HDB3
d 4B–3T

because it is likely to contain long runs of identical digits. When these occur, the retiming operation may lose track of the centre of each symbol interval. Also, removal of the DC component for power separation will cause the waveform to 'droop' and so drift relative to the decision threshold. These problems would be manageable if the source were a random bit sequence. In this case, the longer the run of identical digits the less frequently it occurs, and repeater designs have been achieved such that loss of timing or DC reference occurs sufficiently infrequently that the consequent error rate is acceptable. Where the source is not effectively random, as in telecommunications, the bit sequence can be encoded into one with virtually random properties without any increase in the number of symbols or the symbol rate. The encoding technique is to use a *scrambler* formed by a shift register with feedback from the output to intermediate shift points via exclusive-or gates [11]. Decoding is by a descrambler of similar construction. This could be said to use non-redundant binary (NRB) coding. Its most obvious shortcoming as a line code is this very lack of redundancy which precludes error monitoring.

The simplest means of providing an error-monitoring capability in such a system is by insertion into the digit stream of extra digits used to check the parity of the blocks defined by the insertion points. Any single error, or odd number of errors, in a block can be thereby detected and, for low error rates and where 'bursts' of errors do not occur, these may be used to monitor the average error rate in the system.

7.5.2 Low-disparity binary codes

For situations where the statistical approach to loss of timing or DC reference is unacceptable, problems in repeater design can be considerably eased by using a redundant code that ensures a balance of ones and zeros in the long term. If a 'one' is transmitted as $+\frac{1}{2}$ and a 'zero' as $-\frac{1}{2}$, the effective imbalance or *disparity* is the sum of a sequence of these values and is known as the *cumulative digital sum* (CDS). The variation between the extreme values of this sum is called the *digital sum variation* (DSV). If the DSV is small, the code is a *low-disparity code*. Some examples are explained in Section 7.6 and in References 12 and 13.

The main disadvantage of redundant binary codes for use over metallic pair cables is the increase in symbol rate, and hence repeater gain required, which such redundancy entails. An alternative method of adding redundancy is to increase the number of levels or states in the signal, instead of increasing the digit rate.

7.5.3 Alternate mark inversion

One of the simplest line codes with more than two levels, and one of the earliest used, is the *bipolar* or *alternate mark inversion* (AMI) code. This is a three-level or ternary code, where binary zero is represented by the centre '0' level of the ternary signal and binary 'ones' are represented alternately by the extreme values '+' and '−', as shown in Fig. 7.13b. Obviously, AMI has enough states to represent the binary information without restriction. It can be seen that this process constrains the CDS. If a '+' is given the value of +1 and a '−' the value of −1, then the coding rule limits the CDS to two states and hence the DSV to 1. Since each mark (+ or −) cancels out any imbalance caused by the previous one, it follows that this is equivalent to removing DC and low-frequency energy. The corresponding frequency spectrum is shown in Fig. 7.14a. Transitions between code states are synonymous with the presence of 'ones' in the binary information, so this code only provides adequate timing information for the repeaters if the original bit sequence contains a sufficient density of 'ones'. Alignment information is unnecessary, since the coding process only involves one bit at a time. Any single line error will cause a violation of the alternate-mark-inversion rule, which is readily detectable. AMI is the code most extensively used in 'first-generation' PCM systems [2, 14], operating at 1.5 Mbit/s over twisted-pair cable. Nevertheless, its inability to restrict the maximum time between transitions has led to the adoption for many 'second generation' systems of modified AMI codes.

7.5.4 High-density bipolar codes

The modified AMI codes [15] which guarantee timing information are called *high-density bipolar* (HDB) codes or *bipolar zero-substituted* (BZS) codes. These are basically bipolar codes but with extra marks inserted in the coded signal when more than a certain number of consecutive spaces ('O' levels)

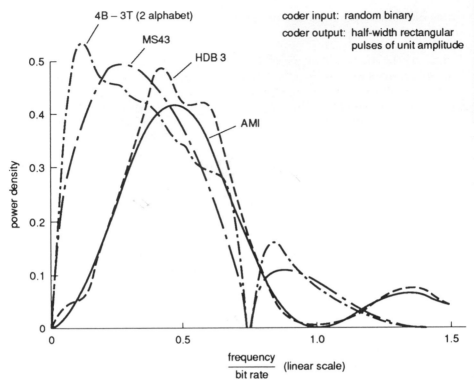

Fig. 7.14 Spectra of line codes

a AMI
b HDB3
c 4B–3T
d MS43

would otherwise exist. For example, in HDB3, when more than three spaces occur the next space is replaced by a mark. The inserted mark is recognisable because its polarity is chosen to violate the alternate-mark-inversion constraint. To avoid introducing any DC energy, the violations themselves are formed into an AMI sequence by adding a mark, if necessary, three symbols before any violation (see Fig. 7.13*c*). This requires a three-bit shift register in each coder or decoder, but it ensures that single line errors can still be readily detected. The frequency spectrum for HDB3 is shown in Fig. 7.14*b*.

The HDB3 code is extensively used in second-generation PCM systems, operating at 2 Mbit/s. It has also been adopted by the CCITT as the equipment interface code at 2, 8 and 34 Mbit/s.

The bipolar zero-substituted codes are very similar. The terminology BNZS indicates that *N* zeros are to be substituted. The run of zeros, which is replaced by a pattern of ones and zeros containing a bipolar violation is generally longer, so that the inserted pattern is always balanced in its own right and contains an even number of marks. For example, in the B6ZS code, six consecutive zeros are substituted by the sequence: mark, zero, violation, mark, zero, violation. These codes have been adopted by the CCITT as the equipment interface codes at the North American rates of 1.5, 6 and 45 Mbit/s, using B8ZS, B6ZS and B3ZS respectively. Because of the short substitution length, B3ZS has two alternative substitution sequences, and is actually identical to HDB2.

Interestingly, despite the apparent differences between these various HDB and BZS codes, they can all use an identical form of error monitor. This arises because the AMI constraint on the inserted 'Violations' limits the CDS to three states and hence the DSV to 2 for all these codes. Error monitoring can be performed by using an up/down counter which follows the CDS of the coded signal. The counter is limited to the number of CDS states of the code (three in this case). When an error occurs, it creates an offset of one state in the counter so, when the code reaches the appropriate CDS limit, counter overflow indicates the occurrence of that error. Limiting the counter to the number of valid code CDS states ensures both correct counter start-up and state correction after the occurrence of each error.

This technique is generally known as *DSV error monitoring* and can be applied to any code with a bounded DSV, simply by the use of a counter with the appropriate number of states.

7.5.5 4B–3T codes

Both bipolar and HDB codes are inefficient in conveying only one bit by each ternary symbol, whose potential information capacity is $\log_2 3 = 1.59$ bits. An example of a more-efficient ternary code is the 4B–3T class of codes. These are so called because they use three ternary symbols (information capacity 4.77 bits) to convey four bits of information. Table 7.1 defines such a code [16]. Of the 27 possible ternary words, 7 have no net digital sum and 6 of these are used to represent 6 binary words. The seventh word '000' is not used, as it contains no timing information. The remaining 20 ternary words form 10 pairs of opposite digital sum to represent the remaining 10 binary words. The choice of which ternary word of each pair to use at any instant is such as to counteract the *cumulative digital sum* (CDS) at that instant. In this code, any ternary word with positive digital sum is paired with a negative sum word for allocation to the same source word. If the accumulated digital sum is negative, the positive word is used. and vice versa. The frequency spectrum for 4B–3T is shown in Fig. 7.14c.

As the DSV for the 2-alphabet 4B–3T code is 7, rather than the 1 or 2 for AMI and HDB respectively, 4B–3T is more sensitive to low-frequency distortion. The MS43 code of Franaszek [17] is a 4B–3T code which reduces the DSV to 5 by employing three alphabets. These are titled A1, A2 and A3 in the code definition of Table 7.2. At the end of a word, if the digital sum is at its negative extreme, the choice is taken from A1. Similarly, for the positive extreme, A3 is used. Otherwise, A2 is used in order to minimise the word-end DSV itself. The reduced DSV improves both the low frequency spectrum shaping and the DSV error monitoring capability, compared with 4B–3T. The frequency spectrum for MS43 is shown in Fig. 7.14d. MS43

Table 7.1   Two-alphabet 4B–3T
Code

Binary	Ternary Alternatives	
	A1	A2
0000	0−+	0−+
0001	−+0	−+0
0010	−0+	−0+
0011	+−+	−+−
0100	0++	0−−
0101	0+0	0−0
0110	00+	00−
0111	−++	+−−
1000	0+−	0+−
1001	+−0	+−0
1010	+0−	+0−
1011	+00	−00
1100	+0+	−0−
1101	++0	−−0
1110	++−	−−+
1111	+++	−−−

Alphabet A1 contains only words of zero
or positive disparity and is used whenever
the sum is negative.
Alphabet A2 contains only words of zero
or negative disparity and is used whenever
the sum is positive.

is the first code so far mentioned which has sufficient flexibility to provide
for carrying auxiliary service information. This can be done by modification
of the alphabet boundaries within the DSV constraints. Both of the above
4B–3T codes ensure frequent transitions for retiming.

Since 4B–3T codes replace four AMI symbols with only three, they enable
30-channel (2 Mbit/s) PCM systems to operate with the same symbol rate
on the line as 24-channel (1.5 Mbit/s) systems. Thus, the same repeater
spacing can be used. 4B–3T codes were also adopted for use in the earlier
digital transmission systems operating over coaxial cables (e.g. 120 Mbit/s
systems in the UK).

7.5.6 The 6B–4T Code

The theoretical bandwidth efficiency of a ternary code can only be reached
by using statistical rather than deterministic limitation of the DSV. This
requires scrambling to ensure that the statistics are controlled. However,
compared with 4B–3T, an improvement can be made by using longer code
words, for example, the 6B–4T code. This uses 4/6 of a ternary digit per

Table 7.2 Three alphabet 4B–3T code (MS43)

Binary	Ternary alternatives		
	A1	A2	A3
0000	+++	−+−	−+−
0001	++0	00−	00−
0010	+0+	0−0	0−0
0100	0++	−00	−00
1000	+−+	+−+	−−−
0011	0−+	0−+	0−+
0101	−0+	−0+	−0+
1001	00+	00+	−−0
1010	0+0	0+0	−0−
1100	+00	+00	0−−
0110	−+0	−+0	−+0
1110	+−0	+−0	+−0
1101	+0−	+0−	+0−
1011	0+−	0+−	0+−
0111	−++	−++	−−+
1111	++−	+−−	+−−

Alphabet A1 contains words of positive disparity and is used when the cumulative sum is at its minimum.
Alphabet A2 contains words of unit or zero disparity and is used when the sum is not at an extreme.
Alphabet A3 contains words of negative disparity and is used when the sum is at its maximum.

information bit, whereas 4B–3T uses 3/4. Thus, the transmitted digit rate and the required bandwidth are only 8/9 of those for 4B–3T. For a '$\sqrt{f}$' cable, the attenuation in dB is thus reduced by 5.7%. For a thermal-noise-limited system, this can improve the signal-to-noise ratio by 5 or 6 dB.

The structure of the 6B–4T code is too large to tabulate here, as there are basically $2^6 = 64$ binary words to be represented by $3^4 = 81$ ternary words. Of these 81 ternary words, 64 are selected as follows:

(a) 19 have no net digital sum and 18 of these are used to represent 18 binary words. The 19th word '0000' is not used, as it contains no timing information.
(b) 16 words of +1 disparity and 16 of −1 disparity are used, despite this disparity, to represent 32 binary words.
(c) The remaining 30 words form 15 pairs of equal and opposite disparity and the lower-disparity pairs (10 of ±2, 4 of ±4) are used to represent the remaining 14 binary words.

The DSV is unbounded, but the probability distribution of the CDS may, nevertheless, be used to estimate probabilities of eye closure due to low-frequency (LF) cut-off. For low values of droop, say 1—2% per digit, the effect is to require about one octave lower LF cut-off compared with 4B–3T. This is a very reasonable penalty in view of the improvement at the top. The unbounded DSV of this code means, however, that an alternative method of error monitoring must be used [18].

The 6B–4T code is used in the UK in digital transmission systems operating on coaxial cables at 140 Mbit/s.

7.5.7 Partial response coding

In the context of crosstalk-limited systems, the precise spectrum produced by the line coding process is important. Fig. 7.14 shows the spectra of four common three-level codes described above. Bandwidth can be conserved by reducing redundancy; however, it is barely worth trying to obtain a redundancy of less than 2 or 3%. (There are techniques available, e.g. Carter coding [19], but the encoding and decoding processes are only just manageable.)

If the bandwidth is paramount, but the modulation method cannot support more than 2-level modulation, then coding techniques which concentrate on spectral shaping are more appropriate than trying to squeeze the last 1% out of the redundancy. This may be the situation in microwave radio, where the channel spacing places very hard limits on the bandwidth that can be used.

These techniques, not surprisingly, generally rely on filtering of the signal to constrain it. Hence, they give only a partial response, as outlined in Section 7.3. However, there is a whole class of line codes based on performing this function digitally. As might be expected, these are termed *partial-response codes*, and in this group of codes the primary emphasis is on spectral shaping.

The closest in similarity to the equalisation described in Section 7.3 is in the best known of these codes, the *duobinary code*. Just as the filtering effectively integrates the response of adjacent pulses to produce a 3-level eye, so the duobinary coding [20] adds the binary information in adjacent digit periods to produce a squared-up version of the same 3-level signal. The virtue of this is that it has a spectral null at half the symbol rate. This reduces crosstalk to other systems where the frequency of maximum gain usually occurs not far above half the symbol rate.

The failure of duobinary as a line code is its lack of a DC null. This is remedied in *modified duobinary*, which is also known as interleaved bipolar. The binary signal is not added but subtracted, and not to the adjacent digit but to the next one. This produces spectral nulls at DC and half the symbol rate. The nulls are narrower than in AMI or duobinary. Also, the system is more critical of the exact frequency of maximum gain. However, there is still a DSV of only 2.

A whole range of partial-response coding schemes has been devised. Even AMI may be described and generated in such terms. The most common, and so probably the most useful, are:

Class 1: Partial response — duobinary
Class 2: Partial response — bipolar
Class 4: Partial response — modified duobinary (also known as inter-leaved bipolar)

7.6 Codes for optical-fibre transmission

7.6.1 General

Section 7.5 described codes for transmission over metallic lines. In other transmission media, such as microwave radio or optical fibre, the different demands of the medium drive us towards a somewhat different type of line code.

In optical fibre, there is not the same rapidly-rising attenuation with frequency as there is in coaxial cable, and so not the same pressure for reduction of the transmitted symbol rate. Also, the linearity of the transmitting device cannot easily support a multi-level code. So, in optical fibre systems, 2-level on–off modulation of the light source is commonly employed, generating a binary signal. Just as in the ternary case, values can be assigned to the code states; if the 'one' or 'on' state is assigned $+1/2$, and the 'off' or 'zero' state $-1/2$, then these values may be used to construct the equivalent CDS and DSV of the code used.

Again, redundant coding can be employed, not only to provide a monitoring capability as in the case of simple parity-bit insertion, but also to overcome the problems of loss of timing or DC drift where a statistical approach to these is unacceptable. This leads to the concept of a family of rate-increasing 2-level codes, generally known as *mB-nB codes*. The rate increase and code properties can be traded off against code complexity, just as in the ternary codes.

The same principles of line coding are commonly employed:

(*a*)   'Balanced' code words to reduce low frequencies.
(*b*)   Transitions for timing extraction.
(*c*)   Restricted word selection for word alignment.
(*d*)   Cumulative digital-sum alphabet selection and error monitoring.

7.6.2 1B–nB codes

The first stage consists of the 'single bit' encoding schemes, where one binary information digit is converted to two or more line-code symbols. The simplest of these (1B–2B) is the diphase or Manchester code, also known as WAL1. A binary '0' is represented by the digit pair '01' and binary '1' is represented by '10'. This has the following properties:

(*a*)   Enough states.
(*b*)   Good LF cut off, DSV of 1.
(*c*)   Symbol rate: twice the information bit rate.
(*d*)   Timing: very good, at least 1 transition for 2 code digits.

(e)   Alignment: only statistical, requiring transitions in binary data. $1010 \equiv$ 0101 in code.

(f)   Error monitoring: by DSV or illegal words.

(g)   No redundancy allocated to auxiliary services.

(h)   Radix of 2.

An extension of this is the WAL2 code (1B–4B) given by:

$$1 \qquad 0110$$

$$0 \qquad 1001$$

This has even better LF cut-off. However, since it operates at four times the bit rate, it has more HF and it is no better on alignment. Despite the alignment problem, the simplicity of such codes is very attractive for the equipment-interface situation (i.e. back on copper cable again) where bandwidth is of fairly low significance.

7.6.3 CMI code

A marginally more-complex code, known as CMI, obviates the alignment problem. CMI stands for *coded-mark inversion* because of its derivation in some measure from the well-known AMI code, as follows:

Binary	0	1
AMI	0	-1 and $+1$ alternately
CMI	01	00 and 11 alternately

CMI has the following properties:

(a)   Enough states.

(b)   Good LF cut-off, DSV of 1.

(c)   Symbol rate: Still twice but less HF energy than diphase.

(d)   Timing: Very good, at least 1 transition for 3-code digits.

(e)   Alignment: negative transitions only at start of binary symbol. Thus, if timing is extracted only from negative transitions, it may be derived at the *binary* clock rate and used to decode directly.

(f)   Error monitoring: DSV or illegal words.

(g)   Again, no redundancy is allocated to auxiliary services.

(h)   Radix of 2.

One of the very convenient features of these 2-level codes concerns equalisation. They are incredibly tolerant of *over*-equalisation with no degradation whatsoever of the received eye. However, this is primarily of interest in the interface case where noise is no problem.

The CMI code has been adopted by CCITT [21] as the recommended interface between equipments at 140 Mbit/s, just as HDB3 or BZS codes have been adopted at all lower rates in the CCITT hierarchy. Thus, these are the two types of line code seen on the outside of all line systems and digital multiplexers for interconnection purposes.

CMI is also one of the codes popularly used on short-haul optical-fibre systems. However, its high redundancy make it less suitable for long-haul systems, where bandwidth needs to be conserved to get the maximum repeater spacing. For these applications, higher-order mB–nB codes are better suited.

7.6.4 5B–6B code

The 5B–6B code [13] is probably the most commonly-used code for long-haul optical-fibre transmission systems, being a reasonable compromise between an increase in symbol rate and coding complexity. The code structure is of the alphabetic type, as follows:

(a) Of the 64 possible 6-digit code words, 20 have no net imbalance and 18 are used to represent 18 of the binary 5-digit words.

(b) There are then 21 pairs of opposite-disparity words, of which 14 (all of unity disparity) are used to represent the remaining 14 binary words.

Omission of two of the zero-disparity words is *not* for timing content this time, as DC balance is synonymous (in the binary case) with transitions and hence timing content. Instead, it is to keep the DSV down. As in the case of MS43, some flexibility exists in this code for modifications to allow an auxiliary service channel capability.

Of course, 20% increase in the symbol rate is a lot at some frequencies, so more-efficient codes can be employed, e.g. 7B–8B or 9B–10B etc. However, a lower symbol rate means lower redundancy and longer coding-block lengths, with correspondingly poorer control of code characteristics. In particular, DSV error monitoring rapidly becomes unworkable because of the long times to reach CDS limits; thus, insertion of parity bits again becomes desirable.

7.6.5 SDH line coding

The new synchronous digital hierarchy (SDH), which is known in North America as SONET (synchronous optical network) is described in Chapter 9. In the SDH, all multiplexing is carried out with reference to a common frame rate of 8 kHz and the line coding is performed by adding 'overhead' digits within that frame structure.

Because of the very flat gain/frequency response provided by optical transmission and the impracticability of feeding power to repeaters over glass fibre, no attempt is made to structure the signal to remove DC or low frequencies. Adequate DC balance and timing content are provided, without recourse to redundancy, by scrambling. The overhead digits are grouped in 8–bit bytes and arranged in a very simple fashion to provide the other required code functions. The bytes are separately allocated to specific functions, such as: frame/block alignment, parity error monitoring, system-management communication channels, engineers' telephone circuits, protection-switching control, channel identification, etc.

7.7 Higher-order codes

The codes described in Sections 7.5 and 7.6 use only two or three levels. The required bandwidth can be reduced by adopting more than three levels in the code, thus reducing the transmitted digit rate. However, in line systems, signal amplitude degradations associated with practical cables and repeaters severely limit the number of levels that can be used. For example, a total degradation of $\delta = 0.25$ (which is the typical result of practical tolerances) is satisfactory for a ternary repeater, but it would completely close up an eye of five or more levels. Thus, codes of more than three levels are not very practicable, unless an adaptive equaliser is used, even on the highest-grade coaxial cable.

Multilevel signals are used in digital microwave systems, as described in Chapter 12. For example, a 64-level signal is used to fit 140 Mbit/s within a bandwidth of 30 MHz. This is done by using quadrature amplitude modulation (QAM). Systems using 256-level QAM are under development.

Optical-fibre transmission systems currently use binary codes. However, this situation may change with improvements in the linearity of lasers which can enable them to generate a range of levels of light intensity. The move to longer wavelengths (1.3 and 1.5 μm) results in PIN diodes having superior sensitivity to avalanche photodiodes (APD). The PIN diode, unlike the APD, can permit uniformly-spaced signal levels and decision levels. Consequently, future optical-fibre transmission systems may also use multilevel codes [22].

7.8 References

1 BYLANSKI, P., and INGRAM, D.G.W.: 'Digital Transmission', (Peter Peregrinus, 1980) 2nd edn.
2 BENNETT, G.H. (Ed.): 'PCM and digital transmission', (Marconi Instruments, 1983)
3 Bell Laboratories staff: 'Transmission systems for communications' (Bell Telephone Labs., 1971) 4th edn.
4 ROWE, H.E.: 'Timing in a long chain of regenerative repeaters', *Bell Syst. Tech. J.*, 1958, **37**, pp. 1543–1598
5 BYRNE, C.J., KARAFIN, B.J., AND ROBINSON, D.B.: 'Systematic jitter in a chain of digital regenerators', *ibid*, 1963, **42**, pp. 2679–2714
6 NYQUIST, H.: 'Certain topics in telegraph transmission theory', *AIEE Trans.*, 1928, **47**, pp. 617–644
7 LENDER, A.: 'The duobinary technique for high speed data transmission', *AIEE Com. Electron.*, 1963, **82**, pp. 214–218
8 KRETZMER, E.R.: 'Generalisation of a technique for binary data communication', *IEEE Trans.*, 1966, **COM-14**, pp. 67–68
9 WALDHAUER, F.D.: 'Quantised feedback in an experimental 280 Mb/s digital repeater for coaxial transmission', *IEEE Trans.*, 1974, **COM–22**, pp. 1–5
10 CATTERMOLE, K.W.: 'Principles of digital line coding', *Int. J. Electron.*, 1983, **55**, pp. 3–33

11 SAVAGE, H.E.: 'Some simple self synchronising digital data scramblers', *Bell Syst. Tech. J.*, 1967, **46**, pp. 449–497
12 CATTERMOLE, K.W.: 'Principles of pulse code modulation' (Iliffe, 1969)
13 GRIFFITHS, J.M.: 'Binary code suitable for line transmission', *Electron. Lett.* 1969, **5**, pp. 79–81
14 DAVIS, C.G.: 'An experimental pulse code modulation system for short-haul trunks', *Bell Syst. Tech. J.*, 1962, **41**, pp. 1–25
15 CROISIER, A.: 'Introduction to pseudoternary transmission codes', *IBM Jr. Res & Dev.*, 1980, **14**, pp. 354–367
16 WATERS, D.B.: 'Data transmission terminal'. British Patent no. 1156 279
17 FRANASZEK, P.A.: 'Sequence state coding for digital transmission', *Bell Syst. Tech. J.*, 1968, **47**, pp. 143–157
18 JESSOP, A.: 'Line system error detection without frame alignment', IEE Colloquium on data transmission codes, Digest 1978/S1, 10 November 1978
19 CARTER, R.O.: 'Low-disparity binary coding system', *Electron. Lett.*, 1965, **1**, pp. 67–68
20 KOBAYASHI, H.: 'Correlative level coding and maximum likelihood decoding', *IEEE Trans.*, 1971, **IT-17**, pp. 586–594
21 CCITT Recommendation G703: 'Interface at 139264 kbit/s'
22 BROOKS, R.M., and JESSOP, A.: 'Line coding for optical fibre systems', *Int. J. Electron.*, 1983, **55**, pp. 81–120

7.9 Acknowledgements

The author acknowledges the assistance of R. J. Catchpole (STL), with whom he prepared an earlier, unpublished version of this chapter, on which a large part of the present chapter is based.

Chapter 8
Plesiochronous high-order digital multiplexing
S.P. Ferguson

GPT Limited

8.1 Introduction

Soon after the introduction of 24- and 30-channel PCM systems in the 1960s and early 1970s, time division multiplexing was used to combine the serial streams of digits or bits (binary digits) from four such systems into one bit stream, for more economical transmission. This became known as second-order digital multiplexing. Subsequent developments have led to third, fourth and fifth-order multiplexers, as progressively more streams are combined. These are used in a hierarchy to assemble telephony, data or other traffic to the highest bit rate possible for economic transmission.

At each level in the hierarchy, several bit streams, known as *tributaries*, are combined or separated by a multiplexer/demultiplexer called a *muldex* (often abbreviated to mux). The steps in the hierarchy were chosen to allow flexibility in traffic planning and an economic balance between muldex costs and transmission costs. Fig. 8.1 shows a typical arrangement. In each case the streams are plesiochronous, meaning that their bit rates are close but not necessarily identical, giving the multiplex structure its name of *plesiochronous digital hierarchy* (PDH). The output from a multiplexer may serve as a tributary to a multiplexer at the next level in the hierarchy, or it may be sent directly over a line or radio link as shown in Fig. 8.2.

During the late 1980s a new multiplex hierarchy emerged, and was defined by the CCITT in 1990 as the *synchronous digital hierarchy* (SDH). It is based

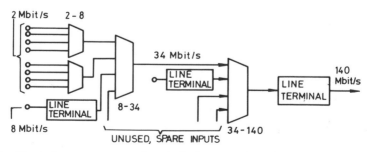

Fig. 8.1   Typical arrangement of a plesiochronous hierarchy

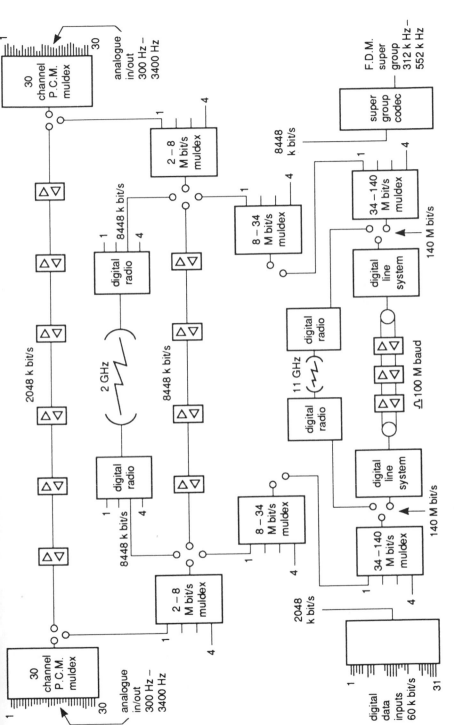

Fig. 8.2 Hierarchy in a digital network

on increments of traffic at 155·52 MBit/s, and on networks which are essentially synchronous, and originated as the 51·84 Mbit/s SONET standard in N. America (3×51·84 Mbit/s = 155·52 Mbit/s). The bandwidth properties of optical fibres have changed the balance between transmission and muldex costs, compared to the use of copper cables and radio, and network needs are better served by having an initial multiplex step of larger size. In addition, synchronous operation provides simpler multiplexing to very high bit rates and leads to lower switching costs for bandwidth management, together with new switched services. This development is described in Chapter 9.

The PDH is described in this chapter, alongside references to particular features of SDH where they relate in some way to techniques developed originally for PDH. However, SDH results in very different muldex equipment designs and associated problems, compared with PDH.

8.2 Plesiochronous muldex standards

There are three 'islands' or standards of plesiochronous digital multiplexing, centred on Europe, North America and Japan, without complete mutual compatibility but operating on similar principles. The most obvious differences between the three 'islands' are in the bit rates of the various orders, as shown in Figs. 8.3 and 8.4. Table 8.1 lists the bit rates in use and Table 8.2 lists the appropriate CCITT recommendations for muldex design. Other differences are mentioned in the text. Transatlantic digital transmission has been the spur to the definition of new muldexes which bridge between the 'islands'. These, however, do nothing to overcome the theoretical incompati-

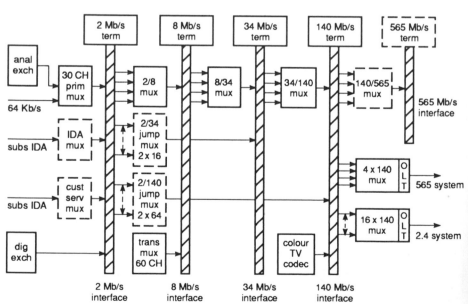

Fig. 8.3   European digital hierarchy

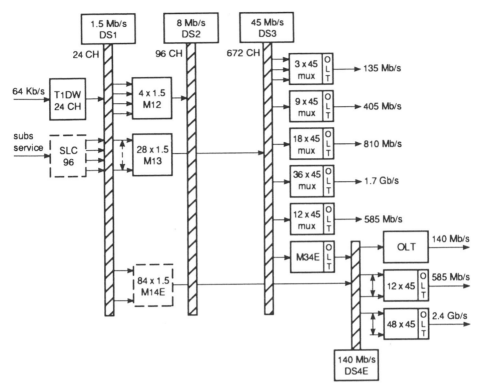

Fig. 8.4 USA digital hierarchy

Table 8.1 Digital muldex hierarchy 'islands' and interworking

Island	PCM used	High Order Hierarchy
A (Europe):	A-law 30 channel:	2048–8448–34, 368–139 264–(564 992) kbit/s
B (Japan):	mu-law 24 channel:	1544/3152–6312–32 064–97 728 kbit/s
C (N. America):	mu-law 24 channel:	1544–6312–44 736–139 264(new)/(274 176)(old)/–(564 992) kbit/s
Between A & B/C:	A-law 30 channel:	2048–6312–44 736–139 264 kbit/s
Between B & C:	mu-law 24 channel:	1544–6312–44 736–139 264 kbit/s

1 International links are to be at the 139 264 kbit/s level, corresponding with 1920 telephone channels for the A-law hierarchy, and 2016 channels for mu-law.
2 Rates in brackets are not CCITT standards.
3 In N. America, 1544 = 'DS1', 6312 = 'DS2', 44736 = 'DS3', 139 264 = 'DS4'.
4 The UK originally used a 1536 kbit/s 24 channel system, then adopted 2048 kbit/s when it became a CCITT standard. Also the next level above 8448 kbit/s was originally 120 Mbit/s in the UK (described in Brigham [1]), and was replaced by 34 and 140 Mbit/s when they were adopted by CCITT.

Table 8.2 CCITT recommendations for European and N. American high order muldexes

Bit rate (Mbit/s)	Recommendation	Bit rate (Mbit/s)	Recommendation
2—8	G742	1·5—6	G743
8—34	G751	2—6	G747
34–140	G751	6—32 or 45	G752
140—565	G954 Annex B (See Note 2)	45—140	G755

1 These are for positive justification (explained in Sections 8.3 and 8.6). Positive/zero/negative justification muldexes for plesiochronous networks are also recommended by CCITT (e.g. G745) but are not widely used; see Section 8.6 for technical description.
2 This is not a CCITT recommendation, but CCITT gives it for information.

bility at the PCM level. They merely allow transmission over common equipment; the PCM incompatibility is resolved by other means.

The examples of plesiochronous muldexes given in this chapter are based on the European muldex standards. Where N. American standards are significantly different, they are mentioned. The European standards are based on *A*-law encoded speech assembled to 2048 kbit/s from 30 channels, then grouped in fours to 8448, 34368, 139264 and 564992 kbit/s. The last is widely used, but it is not a CCITT standard. These rates are often referred to as 2, 8, 34, 140 and 565 Mbit/s respectively. The bit rates increase by factors greater than four because of the practicalities of the multiplexing process.

The grouping factor of four is common to the European standards and is largely a consequence of break points in predicted costs of technology at the time each order of muldex was designed. In particular, 2 Mbit/s was determined by cable-pair loss and crosstalk and by regenerator design with predetermined regenerator spacing (based on audio loading coils). the choice of 8 Mbit/s was based on TTL-device capability around 1970. 140 Mbit/s was based on cable loss and regenerator design for spacing already set by FDM systems on coaxial cable; this factor set the approximate rate. The exact rate was set to be a multiple of 2048 kbit/s, (with $68 \times 2 \cdot 048 = 139 \cdot 264$ Mbit/s), as a far-sighted but ultimately unused step towards synchronous multiplexing. 34 Mbit/s was chosen as a natural break point between 8 and 140 Mbit/s, giving two steps of four.

The rate of 565 Mbit/s was set by logic device limitations and by practical regenerator design on large-bore coaxial cables. A further step to 2·5 Gbit/s is now in place, with the rate again being set by semiconductor device limitations and by the precedent of four being a factor for all lower orders. The emergence of the SDH standard means that 2·5 Gbit/s system use synchronous techniques to carry plesiochronous traffic.

Some administrations refer to *skip-muldexes* or *jump-muldexes* (or -muxes). Taking 8 to 140 Mbit/s as an example, these would use the conventional

8—34 Mbit/s and 34—140 Mbit/s multiplexing methods but the 34 Mbit/s level would be inaccessible. A new assembly process from 8 directly to 140 Mbit/s could be designed, but it would require a new muldex standard, and no new plesiochronous muldex standards are expected to be approved by CCITT. Skip muxes for 2—34 Mbit/s are common, and are often combined with optical line systems.

The overall delay through a multiplexer–demultiplexer path is insignificant in network terms and is typically 25—30 bit periods at the tributary rate, for a single PDH multiplexing step. This corresponds to about 20 μs total, from 2 Mbit/s via 8, 34 and 140 Mbit/s and back to 2 Mbit/s.

8.3 Basic principles

If the inputs to a multiplex are assumed to be synchronous, that is they have the same bit rate and are in phase with one another, then multiplexing them is relatively simple; they can be interleaved by taking a bit or group of bits from each input in turn. This can be done by a switch which samples each input under the control of the multiplex clock. There are two main methods of interleaving digital signals, namely *bit interleaving* and *word interleaving*. These are illustrated in Fig. 8.5. In bit interleaving, one bit is taken from each tributary in turn. Accordingly, if there are N input signals each with a rate of f_t bit/s, then the comined rate will be Nf_t bit/s and each element of the combined signal will have a duration equal to $1/N$ of the duration of an input digit. In word interleaving, groups of bits are taken from each tributary in turn and this involves the use of storage at each

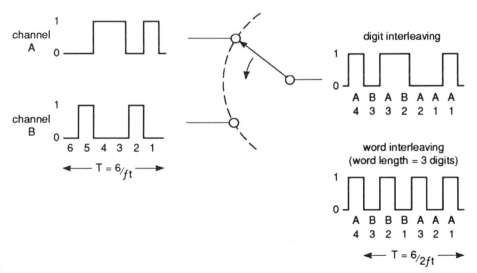

Fig. 8.5 Interleaving digital signals

a bit interleaving
b word interleaving

input to hold the bits while waiting to be sampled. A common form is *byte interleaving*, where one byte is taken in turn from each input.

Bit interleaving is more straightforward than word interleaving because it requires less storage during the multiplexing process. Also, it produces a more randomly distributed output bit stream, which aids timing recovery and reduces the probability of false alignment at the receiver due to framing-signal imitations. In applications where timing accuracy is important, it also has the advantage that the delivery of separate timing information for each tributary can be in the smallest element of time measurable in the system, i.e a single bit period.

Word interleaving is more appropriate for multiplexing signals comprising blocks of bits, as may occur in data transmission, where timing accuracy is not usually as critical as in the delivery of analogue information which has been encoded into digital form. It is also valuable where individual data blocks are to be accessed directly from an aggregate bit stream.

Bit interleaving is used for the PDH, and word interleaving is used for SDH and for data-oriented multiplexing techniques, which are more commonly used in applications such as local area networks (LANs). The SDH interleaves bytes, and under particular multiplexing conditions, these can be made to coincide with the boundaries of individual octets or speech samples in PCM coding.

When the tributaries have been interleaved, they must be correctly separated again into their correct digital streams at the receiving or demultiplexing end of the system. It is axiomatic in digital multiplexing that the data content of each input stream must be unconstrained, so no property of the tributary data may be used to aid recognition. Instead, a single marker is inserted at regular intervals to locate the starting point for interleaving, and this is used by the demuliplexer to align itself into the correct sequence.

The marker is introduced after dividing the aggregate stream into *frames*, each of which consists of a fixed number of consecutive bit positions or time-slots. A recognisable sequence of bits, called the *frame-alignment word* (FAW), is added to be the marker. The FAW may occupy a number of adjacent time-slots, i.e. bunched framing, or it may be allocated to a number of isolated time-slots distributed through the frame, i.e. distributed framing. The means for aligning to the FAW at the demultiplexer are described in Section 8.5.

Each bit time-slot within the frame is uniquely allocated either to information bits from the incoming tributaries or to control bits such as the FAW. The inclusion of control bits means that the multiplex rate is higher than the aggregate of the tributary rates.

In a transmission network which has not been designed for synchronous operation, the inputs to a digital multiplex will not generally be synchronous, particularly above 2 Mbit/s. Although they have the same nominal bit rate, they commonly originate from different crystal oscillators in the network and can vary within the clock tolerance, i.e. they are plesiochronous. To meet this situation, the multiplexer must make the inputs appear to the interleaver to be synchronous; this is done by using a process known as *justification*.

Justification is a process of changing the rate of a digital signal in a controlled manner so that it can accord with a rate different from its own inherent rate, usually without loss of information. This process is usually referred to as *stuffing* in the North American literature, with the same term also being used on both sides of the Atlantic to describe the insertion of frame-alignment words and other control bits. The word 'justification' is taken from the printing trade, where it is used to describe the process of adjusting the word spacing so that the initial and final letters of a line are vertically aligned at the margins. The justification process is explained in more detail in Section 8.6.

8.4 Outline of a plesiochronous muldex

A muldex consists of a multiplexer and demultiplexer. They are normally combined in one equipment, in fact on one plug-in card in the latest developments which use large-scale integration. The alarm circuits, which call and assist the maintenance staff in fault diagnosis, are usually common, as is the power convertor. The transmit multiplexer and the receive demultiplexer are functionally separate. The transmit and receive parts of a muldex are shown respectively in Figs. 8.6 and 8.7.

The transmit multiplexer (Fig 8.6) accepts a number of input tributaries, usually via coaxial or balanced-pair cable. Each cable carries the binary data and its associated clock in coded form. This is called the *in-station digital interface*. The inputs all have the same nominal bit rate, but the actual rates and phases will differ; i.e. they are plesiochronous. The multiplexer combines these inputs by justification (to align the streams of bits together), followed by cyclic bit interleaving as shown in Fig. 8.5a. In addition, bit patterns are added to align or synchronise the distant demultiplexer and

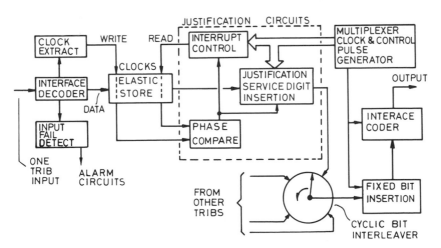

Fig. 8.6 Multiplexer

to permit the transmission of some alarm conditions. The overall bit rate is typically around 405 to 412% of the nominal tributary bit rate, and is defined by a local crystal oscillator.

The demultiplexer (often called a receive multiplex), shown in Fig 8.7, performs the reverse functions and produces the majority of the alarm conditions, such as 'loss of alignment' and 'distant alarm'. The receive multiplex also produces a less-welcome effect. The process of justification involves the multiplexer in estimating the tributary input bit rates periodically and sending that information to the receive multiplex to allow reconstitution of the original bit rates. The process is akin to the amplitude sampling process in PCM in that it produces an imperfection, just as quantisation distortion is produced in PCM. The imperfection is called *waiting-time jitter*; it is jitter on the demultiplexer tributary outputs, which has a peak–peak level sometimes as high as 0·3 unit intervals and includes components down to zero frequency.

A half-way level between a multiplexer and a demultiplexer is the *drop and insert* or *add-drop muldex*, shown in Fig 8.8. Such a muldex is used where access is needed at a station on a route between two muldexes, but only for a proportion of the tributaries. Equipment is saved by fitting hardware only for the relevant tributaries, while the remaining traffic goes straight through by means of synchronising the outgoing frames and bit rates to the incoming ones. The advantage of reduced hardware is not so apparent with modern miniature muldexes, while the reduction in waiting-time jitter for the 'straight through' tributaries is not significant in a network, according to both unpublished and published studies (see Section 8.10). The new synchronous hierarchy (SDH) is better suited to achieving a cost reduction by

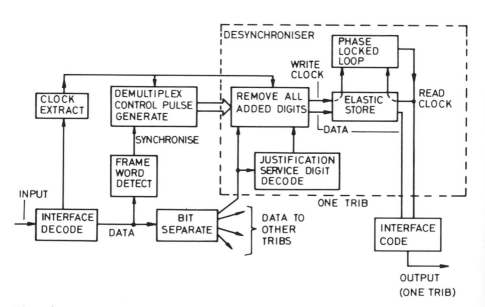

Fig. 8.7   Demultiplexer

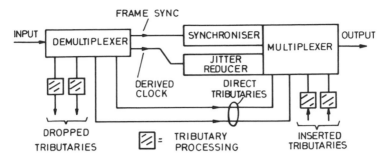

Fig. 8.8 Drop and insert muldex

the use of add-drop techniques, because it does not require processing at intermediate bit rates in order to access a small fraction of the 'through' traffic.

There is a trend for plesiochronous muldexes to be physically integrated with optical line systems. For example the combination of 2—8 Mbit/s and 8—34 Mbit/s muldexes and a 34 Mbit/s line system is described as a '16 times 2 Mbit/s optical system'. Total integration on these lines is the norm for equipments based on the SDH.

8.5 Multiplex frames

Framing is the basic assembly technique for a digital muldex. In each fixed-length group of bits, i.e. frame, the multiplexer inserts a unique group of bits onto which the demultiplexer can synchronise, i.e. the frame-alignment word (FAW). Since the multiplexer assembles the bits from its inputs in a known order by bit interleaving, demultiplexer synchronisation permits correct reconstruction of the tributary signals.

The frequency with which frame-alignment words are generated affects the resynchronisation time after a transmission break, as does whether the FAW is bunched or distributed. Generally, a bunched word gives more rapid alignment; it is used in the European hierarchy, with distributed FAWs being used elsewhere. The realignment time is also affected by the complexity of the frame-alignment circuits. Generally, fairly simple circuits are used, because the muldex specifications were defined before the widespread use of large-scale integrated (LSI) circuits. The results of the decisions taken are that European times are typically hundreds of microseconds, while North American times extend to tens of milliseconds. CCITT recommendations for European muldexes give the realignment times indirectly, by giving a 'reference' realignment strategy.

The generally adopted strategy is:

(*a*) Receipt of four incorrect FAWs constitutes 'loss of alignment'. This causes traffic output to be suppressed and replaced by the alarm indication signal (AIS) (for explanation of AIS see Section 8.11). A free search for the FAW then begins.

(b)   When a FAW is found (and this may be spurious, because of a simulation in traffic), the frame counters are reset and a further search for the FAW is inhibited until one frame period later.

(c)   If no FAW is then found, the free search is resumed.

(d)   When three correct consecutive FAWs have been found, the 'loss of alignment' condition ends and traffic output is restored.

A common criterion for alignment is that it shall be successful within the specified realignment time on 99% of occasions. To meet this, the above strategy requires between 10 and 11 frame periods, using the European CCITT frame structures in Table 8.2.

Within the European hierarchy, a guideline exists for defining the required realignment time as one goes up through the orders of the multiplex hierarchy. The rule is that on 99% of occasions the time to realign at one level should be less than the time taken for loss of alignment to be detected at the next lower level. Because the latter normally requires a confidence count of four incorrect FAWs to be found in sequence, this takes just over three frame periods. In theory, this rule helps to prevent the propagation of frame-alignment loss down through the demultiplex chain, for very short interruptions and provided that the detection of alignment loss was in fact spurious. The assumption here is that a burst of errors occurred, which filled the FAW out-of-alignment confidence counters in the demultiplexer but did not disturb its clock counters.

A similar rule is applied to the realignment-time specification of transmission systems between muldexes. For example, a 34 Mbit/s radio system should realign in less than four frames of an 8–34 Mbit/s muldex.

If a loss of alignment is caused by a break in transmission, then the receive circuits have no clock reference from the multiplexer and start to drift. Depending on their stability (e.g. whether a crystal clock-extraction circuit is used), restoration of the signal may not occur until one or more clock cycles have been slipped between transmitter and receiver. Such slippage inevitably leads to a loss of alignment, whether or not the break lasted long enough for four FAWs to be lost.

Listed below are the main considerations in the choice of frame length and of FAW and the numbers of sequential (in)correct FAWs deemed to constitute (mis)alignment. Owen [2], Bylanski [3] and Häberle [4] expand on some of these, and Jones [5] provides some interesting comparisons between alignment strategies:

(a)   Bunched words give faster realignment times than do distributed words, for a given complexity of digital processing.

(b)   Short FAWs have a high chance of simulation in the traffic, while long ones have a high chance of being corrupted by random errors. Both effects extend the realignment time beyond that achieved with an intermediate length.

(c)   A large confidence count to establish loss of alignment extends the detection-time component of the overall realignment time. A small count gives a high risk of instigating unnecessary searches, with their associated risk of temporarily locking to a simulated FAW in the traffic, and it therefore delays realignment.

(*d*)   A large confidence count to confirm the gaining of alignment extends the overall realignment time, but it guards against spurious loss of alignment at high error rates.

(*e*)   FAWs should be chosen so that when a part of a FAW is taken with adjacent traffic bits, with or without a bit error having occurred, the FAW is not reproduced in a new position.

(*f*)   FAWs are so chosen that no combination of bit-interleaved FAWs from lower-order systems can simulate the FAW of the higher-order muldex. The FAW in the European hierarchy is . . 1111010000. .

For considerations (*a*), (*b*) and (*c*), computer simulation is necessary to optimise the parameters for each frame length. If, for example, the realignment strategy described previously for European muldexes is used, then the only variables left are the FAW length and its distribution, and the frame length.

Depite consideration (*f*), there is an axiom that the data input port on each tributary of a multiplexer is bit-sequence independent; i.e., any pattern can legitimately be offered to a multiplexer. (Tributary input ports designed to accept the N. American DS3 or 45 Mbit/s are an exception to this). As the tributaries are usually not synchronous, even deliberate attempts to simulate a FAW are unlikely to succeed. Occasional simulations of a FAW are not significant because, once alignment is found, the demultiplexer looks for the FAW only in the exact position in which it occurred in the previous frame.

8.6 Justification

8.6.1 Basic operation

Digital multiplexers conventionally accept plesiochronous tributary inputs; i.e. the bit rates of the inputs are close to their nominal rate, but they have no other phase or frequency relationship to each other nor to the multiplexer output bit rate. Not only must the multiplex perform bit interleaving on these inputs, but it must also permit the reconstruction of the original tributary signals at the demultiplexer. This is achieved by justification. The European plesiochronous hierarchy uses a technique called *positive justification*. *Positive/zero/negative justification* systems also exist in less-used plesiochronous standards, but are a key part of the SDH. Positive justification will be explained first.

Before bit interleaving can occur, the bit rates of the various tributaries must be brought to a common figure. This is done by first writing the input data for each tributary into a separate first-in-first-out store (FIFO), using the clock derived from the tributary input. This *transmit elastic store* is shown in Fig. 8.6. Next, the data are read out from all the stores in parallel by a common readout clock which is derived from a local crystal oscillator. Bit interleaving can then occur.

In order to avoid store overflow, the readout clock is arranged to be faster than the fastest expected input clock. In order to avoid the stores being emptied, a pulse is occasionally removed from the read clock, for each tibutary individually. When this happens, no data bit is read out of the store; instead a 'dummy' or *justified bit* is transmitted, which will be removed again at the demultiplexer by means to be described in Section 8.7. (It is possible in principle to use this bit for other purposes, such as error monitoring or an auxiliary channel, but this patented technique is not covered in CCITT and so would not be compatible between equipments from different vendors.) Whenever sent, justified bits are always transmitted in the same frame time-slots, one per tributary, and they are sent just often enough to prevent the store from emptying. When a justified bit is not to be sent in this time-slot, known as the *justifiable time-slot*, then a data bit is sent as normal.

The justification decision is made as follows. At the start of each frame, within a window of a few bit periods, the phase of the clock derived from the tributary input is compared with the phase of the elastic-store readout clock, as shown in Fig. 8.9. If the readout clock advances beyond a defined point, justification occurs in that frame, together with the associated disabling of a single pulse from the store readout clock. The clock speeds are so arranged that justification at an average rate of rather less than once per frame is enough to make the average readout clock rate equal to the input bit rate.

The sequence of events is that a decision to justify or not is taken separately for each tributary near the start of a frame. Then the decision is signalled to the demultiplexer and finally the justifiable time-slot is justified or not. The decision to send a justified bit, i.e. to justify, is signalled to the distant demultiplexer by a distributed group of bits called the *justification service digits*, so that the justified bit can be removed again. The resulting positions

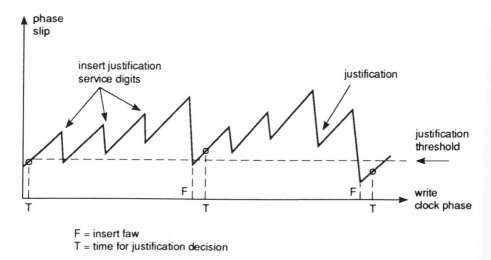

F = insert faw
T = time for justification decision

Fig. 8.9   Phase changes with bit stuffing and justification

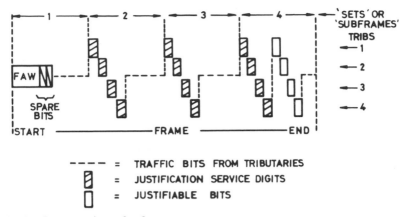

Fig. 8.10   Construction of a frame

of the justification service digits and the justifiable bits in the frame are shown in Fig. 8.10.

As well as being interrupted intermittently for the justification process, the data readout clock from the elastic store is interrrupted regularly for the creation of fixed gaps in the store output data; therefore its original speed is made correspondingly higher to allow for this. These gaps are then used for the insertion of the frame-alignment word, the justification service digits and spare signalling bits; this process usually precedes the bit interleaving and is commonly termed *bit stuffing*. Sometimes, especially in N. America, this term is also taken to include justification. Fig. 8.11 shows the circuit arrangement and Fig. 8.9 shows the phase changes. The original uninterrupted readout clock rate, when multiplied by the number of tributaries, gives the multiplexer output bit rate.

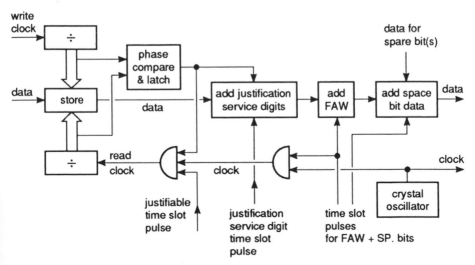

Fig. 8.11   Transmit bit insertion

The size of the transmit elastic store is chiefly based on the size of gaps created by the above processes (typically up to 4 bits) and by the need to absorb the tributary input jitter which is not successfully coded in phase and amplitude into the justification service digits. In practice, the latter corresponds to jitter at frequencies higher than about 1/4 of the frame rate. Typically, an eight-bit store is used. Section 8.10 also discusses storage considerations.

The above process is positive justification; i.e. a data bit is occasionally replaced by a dummy bit which does not contribute to the emptying of the elastic store. Negative justification is used where the read clock to the elastic store has not been arranged to be fast enough always to prevent store overflow. Instead, an extra data bit is occasionally removed from the store and transmitted in a spare time-slot and this action is signalled to the demultiplexer. Both positive and negative justification may be used in the same multiplexer, and the process is then called positive/zero/negative justification; it is used in the SDH. However, from 2 Mbit/s upwards in the European PDH, only positive justification is used.

8.6.2 Jusification ratio

The justification process can occur at frequencies between zero and the frame rate, depending on the proportion of frames justified, i.e. on the *justification ratio*. This range corresponds to the range of variation of input-tributary bit rates (including the effect of low-frequency jitter on the bit rate) which can be accepted before the justification process fails. This range can be found from:

$$f_t = (b - j) \times f_0 / B \qquad (8.1)$$

where

f_t = tributary bit rate

b = number of bits available to carry information per tributary per frame

j = justification ratio

f_0 = multiplexer output bit rate

B = total number of bits per frame

and the frame rate is given by f_0/B.

Taking the 2—8 Mbit/s muldex as an example and allowing for worst-case muldex output rates:

If all frames are unjustified (i.e. $j = 0$), then:

$$f_t(\text{max}) = 206 \times \frac{8448 \times 1 \cdot 00003}{848} = 2052 \cdot 2 \text{ kbit/s}$$

If all frames are justified (i.e. $j = 1$), then:

$$f_t(\text{min}) = 205 \times \frac{8448 \times 0 \cdot 99997}{848} = 2042 \cdot 3 \text{ kbit/s}$$

The CCITT-specified tolerance of the tributary bit rate is well within this range; at 2048 kbit/s it is ±50 ppm or ±102·4 Hz. However, low-frequency jitter contributes a varying frequency offset of up to ±11 310 bit/s (see Section 8.10.2 for derivation). This is beyond the tolerable range and it is accommodated partly by the justification process and partly by storage in the transmit elastic store. The latter is practicable because CCITT recommendations set limits on input jitter amplitude which must be tolerated.

In practice, it is necessary to restrict the justification ratio to a narrower range of variation than that which would cause justification failure, because of its effect on wating-time jitter. This restriction sets a minimum figure for the frame rate, for a given tolerance on the tributary rates, because justification is conventionally done no more than once per frame (or multiframe in N. American standards shown in Table 8.5). The restriction on frame rate is additional to that imposed by realignment-time requirements. An example of the range of operation is 0·4078 to 0·4407, as used in the 2—8 Mbit/s muldex. In order to cope with input jitter, it is conventional to set the nominal justification ratio near to 0·5 rather than near to 0 or 1, but never to exactly 0·5, for reasons explained in section 8.6.3.

The justification ratio varies with changes in tributary and aggregate bit rates. For assessment of consequential waiting time jitter, the effect of input jitter in increasing the variation of the instantaneous tributary bit rate is not considered. This is because input jitter to the multiplexer is found in practice always to reduce the waiting time jitter at the output of the following demultiplexer. The justification ratio can be calculated as follows.

Let

j = justification ratio

b = number of bits available to carry information per tributary per frame

f_t = tributary bit rate

B = total number of bits per frame

f_0 = multiplexer output bit rate.

Then the frame rate F is:

$$F = f_0/B \tag{8.2}$$

Now,

bits/second available for single tributary traffic = Fb

and

bits/second needed for single tributary traffic = f_t

so

bits justified per second = $Fb - f_t$

so

average number of justified bits/frame $= (Fb - f_t)/F$

(which is the justification ratio, j).
Substituting from eqn. 8.2:

$$j = b - Bf_t/f_0 \qquad (8.3)$$

Taking the 2—8 Mbit/s muldex as an example:

$$j = (206) - (2048 \times 848)/8448 = 0\cdot4242.$$

8.6.3 Waiting time jitter
Waiting-time jitter occurs at the demultiplexer outputs and tends to have noticeably high values when the justification ratio approaches that of two integers, particularly the ratio of small integers, such as 1 over 2. Depending on the circuit implementation, not all such ratios gave high jitter.

For one particular implementation of the justification circuits and associated elastic store, Duttweiler [6] has given a way of predicting this jitter. If the justification ratio is at or near an integer ratio, then the jitter amplitude in unit intervals is the reciprocal of the denominator of this ratio, as shown in Fig. 8.12. However, as shown by Cleobury [7], alternative circuit arrangements can produce very different, and often lower, jitter levels, as shown in Fig. 8.13.

The mathematical model by Duttweiler corresponds to the phase comparison being done once per period of the clocks. In fact, the comparison can be done as slowly as once per N clock periods, (see Chow [8] for analysis of this case), where N is the number of storage elements in the elastic store, or at any rate in between (for example, see Fig. 8.14). This will have consequential effects on the waiting-time jitter. Computer modelling of the

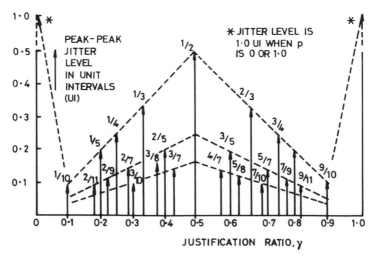

Fig. 8.12   Classical Duttweiler waiting-time jitter

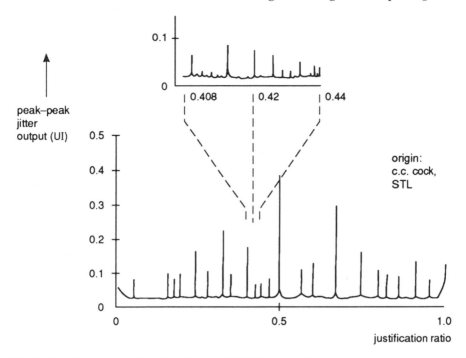

Fig. 8.13 Typical waiting-time jitter: 2–8 Mbit/s muldex

justification process is the only practical way of predicting the jitter amplitudes resulting from particular justification ratios, store sizes and phase comparators.

There is one case for which the analysis is simple and constant, irrespective of implementation. If the justification ratio equals 0 or 1·0, the waiting-time jitter is 1·0 unit interval peak-peak. This special case corresponds to positive/zero/negative justification, as used in the SDH.

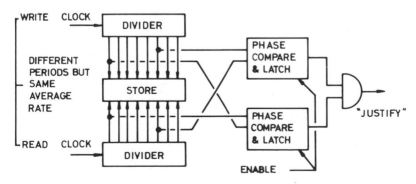

Fig. 8.14 Justification phase comparator with comparison done twice per eight bits

8.7 Dejustification

8.7.1 Bit removal

At the demultiplexer, shown in Fig. 8.7, after clock extraction, frame alignment and tributary bit separation on the incoming bit stream, the individual tributary data streams still contain all the added bits. Next, the justification service digits are decoded; then all the added bits, including the justified bits, are blanked out. The output of the blanking circuit is a clock and its associated data stream, with periodic gaps (up to 4 bits long, taking the 34—140 Mbit/s muldex as an example), as shown in Fig. 8.15.

The justification service digits must be protected against transmission errors. An incorrect decision about removing a bit at a tributary of the demultiplexer would cause both a bit error and, far more serious, a bit slip in the subsequent readout. This would casue a loss of alignment in any lower-order multiplexed traffic being carried through that tributary. The protection is achieved by spreading the service digits through a frame, as shown in Fig. 8.10, in an attempt to reduce the effects of burst errors, and by using repetitive transmission and majority voting at the demultiplexer to achieve limited error correction. Logic ones are sent for a justified frame, logic zeros otherwise. Voting is on a majority of 2 out of 3, or 3 out of 5, depending on the degree of error immunity provided by the particular frame structure.

Ideally, the criterion for error immunity should be that enough is provided so that random errors cannot cause dejustification slips more often than

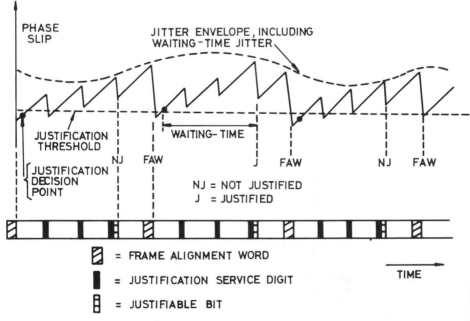

Fig. 8.15   Creation of jitter at input to jitter reducer

they cause loss of frame alignment. This is because even one such slip will cause transient loss of frame alignment in the next-lower-order muldexes. In practice, the probabilities of frame alignment loss and of dejustification slips have different characteristics when plotted against error rate. Therefore, in designing for error immunity, one must choose the error rate(s) and distribution at which the above criterion is to be satisfied.

Computer simulations (see Table 8.3) and practical tests show that this has not been done adequately for high error rates in the 2—8 and 8—34 Mbit/s muldexes, where 2-out-of-3 voting is used. Slips can occur here rather more often than losses of frame alignment; the 8—34 Mbit/s muldex is therefore the weak link in the European hierarchy. The transient alarm indications resulting from such justification errors will be from subsequent 2—8 Mbit/s demultiplexers and primary rate demultiplexers when they both lose frame alignment. This causes some difficulty in the location of transient faults.

In practice, these simulations are for extreme conditions, and an error rate of 10^{-4} is probably the worst acceptable before a system is taken out of service.

8.7.2 Jitter reduction

After the removal of the fixed and justified bits, the data content is identical to that of the distant tributary input and the clock has the same average rate, but with severe superimposed jitter corresponding to the gaps, as shown in Fig. 8.15. The clock is next passed through a jitter reducer consisting of a crystal-controlled phase-locked loop (see Byrne [9]). The corresponding data are written to an elastic store by the jittered clock, and read out a little later by the jitter-reduced clock, as shown in Fig. 8.7. The jitter reducer attenuates both the systematic jitter corresponding to the regular gaps and the justification jitter caused by the intermittent removal of justified bits. The latter has two components: one at the average justification rate and one (waiting-time jitter) which is the result of waiting for

Table 8.3 Results of calculation on 'survival' times of digital muldexes at high error rates (random errors)

E = equivalent binary error rate for 99% probability of survival for stated time in milliseconds (or other units shown);

E	First order (to 2 Mbit/s)		Second order (2–8 Mbit/s)		Third order (8–34 Mbit/s)		Fourth order (34–140 Mbit/s)	
	Justn.	Frame	Justn.	Frame	Justn.	Frame	Justn.	Frame
10^{-2}	—	8·0	0·8	12	0·36	5·4	5·4	1·23
10^{-3}	—	7400	84	100s	37·5	45s	5·3s	10s
M = mean time between losses of synchronisation at error rate of 10^{-3};								
M	—	735 s	33 s	2·8 h	14 s	75 min	35 min	17 min

The figures given are for a whole mux; for any one tributary they are four times greater.

the justifiable time-slot after the need for justification occurred. It is the second part which is broadband and troublesome.

The jitter reducer, shown in Fig. 8.16, acts as a low-pass filter to jitter. It has a 3 dB cut-off frequency in the range of a few tens to a few hundred hertz, depending on the muldex order. Because waiting-time jitter has spectral components down to zero frequency, it cannot be completely eliminated. (The SDH inherently generates high jitter levels in its framing and justification functions; consequently, the bandwidths of its jitter reducers are orders of magnitude narrower than those in the PDH.)

The CCITT specifications for jitter output on PDH demultiplexers allow for the difficulty of removing wating-time jitter, by permitting large amplitudes (e.g. up to 0·25 unit intervals (UI) on the 2–8 Mbit/s muldex) at low jitter frequencies where no phase-lock-loop attenuation occurs. In practice, wating-time jitter is not known to have caused network problems within the European hierarchy, despite trials which attempted to find these problems in long chains of muldexes; Section 8.10.3 gives further details.

The elastic store, as shown in Fig. 8.7, must absorb jitter from:

(a) Regular removal of FAW and other fixed bits (up to 4 UI),
(b) intermittent removal of justified bits (up to 1 UI).
(c) Jitter at the remote transmit tributary input, above the cut-off frequency of the jitter reducer (up to 1·5 UI).
(d) Jitter at the demultiplexer input (caused by the transmission path), scaled down by the number of tributaries (up to $1·5/4 = 0·4$ UI).
(e) Drift and tolerances in the jitter-reducer circuits (up to 2 UI).
(f) Overshoot of jitter-reducer phase error associated with system recovery after a loss of synchronisation (up to 2 UI).

Item (f) needs explanation. When a demultiplexer loses a usable input, a local standby crystal oscillator provides timing for the tributary outputs in order to keep subsequent systems on frequency, ready for a rapid recovery. For the best jitter-reducer recovery time, the oscillator signal replaces the clock normally offered to the jitter-reducer input as shown in Fig. 8.16.

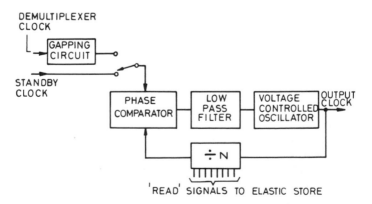

Fig. 8.16 Arrangement of jitter reducer

When frame synchronisation is restored, the jitter reducer's phase-locked loop has to change frequency slightly; if a second-order loop is used than a phase-error overshoot occurs. Normally, the loop is designed so that the step in frequency is within the range which the loop can accommodate without cycle skipping (see Byrne [9] and Gardner [10]) and with a minimum of overshoot. Section 8.10.4 discusses other points on jitter-reducer design.

Such overshoot is effectively the same as internally-generated jitter, and it moves the elastic store nearer to the limit of its operation. If the limit is exceeded, digital errors occur until the phase error recovers exponentially to within the limit. This recovery is relatively slow because of the loop's narrow bandwidth. A generous elastic store is a cheaper approach to rapid recovery than is a switched-bandwidth loop; it can give effectively instantaneous recovery once frame alignment is achieved. Since about 1985, European muldex designs have used receive elastic store sizes of at least 12 bits. Earlier ones used 10 or 8 bits and could not cope so well with item (*f*), leading to recovery times extending up to several milliseconds under severe jitter conditions.

Virtually all digital transmission systems use jitter reducers and have the same problem with phase overshoot and store capacity. Some European administrations are now requiring complete recovery to error-free operation, throughout the muldex hierarchy down to 2 Mbit/s, in around 1 ms, compared to the tens of milliseconds sometimes experienced. Breaks of 1 ms can affect data traffic, but not telephony or its associated signalling. Depending on the exchange signalling techniques in use, the threshold break length for telephony can vary from 20 ms to about 33 ms. Above the threshold figure, calls in process of being dialled have a rapidly increasing chance of dropping out.

8.8 Complete plesiochronous frame structures

There are several differences in the approaches to frame structure design between the European and other hierarchies, and these are described in the CCITT recommendations listed in Table 8.2. Summaries of some of the frame structures are given in Tables 8.4 and 8.5. Note that subframes are also known as *sets*.

8.9 Digital Interfaces

A typical repeater-station installation consists of racks of muldexes and racks of terminal equipment for transmission systems using optical and copper cable. Microwave radio systems traditionally have physically-separate arrangements because of the need to keep transmitters and receivers near their antennas on account of waveguide losses, and because of the advantage of hill-top sites.

A plesiochronous muldex has 10 or more traffic connections to other equipments (depending on which 'island' it is in). The design of such

Table 8.4   Frames for European high-order plesiochronous muldexes

Feature	Second order	Third order	Fourth order
Aggregate bit rate	8 448 kbit/s ±30 ppm	34 368 kbit/s ±20 ppm	139 264 kbit/s ±15 ppm
Number of tributaries	4	4	4
Tributary bit rate	2 048 kbit/s ±50 ppm	8 448 kbit/s ±30 ppm	34 368 kbit/s ±20 ppm
Type of justification	Positive	Positive	Positive
Bit numbering	*Subframe 1*	*Subframe 1*	*Subframe 1*
Frame alignment word	1–10 (1111010000)	1–10 (1111010000)	1–12 (111110100000)
Service bits	11, 12	11, 12	13–16
Tributaries (bit interleaved)	13–212	13–384	17–488
	Subframe 2 & 3	*Subframe 2 & 3*	*Subframe 2, 3, 4, 5*
Justification control (one bit per tributary)	1–4	1–4	1–4
Tributaries	5–212	5–384	5–488
	Subframe 4	*Subframe 4*	*Subframe 6*
Justification control (one bit per tributary)	1–4	1–4	1–4
Justifiable bits (one per tributary)	5–8	5–8	5–8
Tributaries (bit interleaved)	9–212	9–384	9–488
Justification signal (justified/unjustified)	000/111	000/111	00000/11111
Frame length	848 bits	1536 bits	2928 bits
Bits for tributary data	206	378	723
Frame rate	9 962 Hz	22 375 Hz	47 560 Hz
Nominal justification	0·424	0·436	0·419

interfaces therefore has a large effect on the cost of a muldex, especially as the cost of the associated digital processing has fallen with the greater introduction of LSI devices. The in-station traffic interconnections are generally made by 75 Ω coaxial cable, with 120 Ω balanced cable being used at 2048 kbit/s in some territories; this is particularly the case in French-influenced ones. Balanced 110 Ω cable is the norm for 1544 kbit/s interconnections in N. America. Each cable carries a composite of clock and data using simple codes. These coded interface signals do not contain information at zero frequency, as no DC path is guaranteed; this is because of concern over possible large 'earth' currents in a station via the cables.

Table 8.5 Frames for N. American high-order plesiochronous muldexes

Feature	Second order	Third order
Aggregate bit rate	6312 kbit/s ± 30 ppm	44736 kbit/s ± 20 ppm
Number of tributaries	4	7
Tributary bit rate	1544 kbit/s ± 50 ppm	6312 kbit/s ± 30 ppm
Type of justification	Positive	Positive
Subframes per frame	6	8
Frames per multiframe	4	7
Bit numbering	*Subframe 3, 6*	*Subframes 2, 4, 6, 8*
Frame alignment word	1	1
(distibuted)	(0 ... 1)	(1 ... 0 ... 0 ... 1)
	Subframe 1	*Subframe 1*
Multiframe alignment word	1	1
(distributed)	(0 ... 1 ... 1 ... X)	(X ... X ... P ... P ... P 0 ... 1 ... 0)
Service bit	Bit X in multiframe word	Bits X. X in multiframe word (2 identical bits)
Parity bits, even parity		Bits P, P in multiframe word (2 identical bits)
	Subframes 2, 4, 5	*Subframes 3, 5, 7*
Justification control (one tributary per frame)	1	1
	Subframe 6	*Subframe 8*
Justifiable bits (one tributary per frame)	2–5 (one per frame, for tribs. 1–4 respectively)	2–8 (one per frame, for tribs 1–7 respectively)
	All subframes	*All subframes*
Tributaries (bit interleaved)	2–49 (including justifiable bit in subframe 6)	2–85 (including justifiable bit in subframe 8)
Justification signal (justified/unjustified)	000/111	000/111
Frame length	294 bits	680 bits
Multiframe length	1176 bits (4 frames)	4760 bits (7 frames)
Bits for tributary data per multiframe (including justification)	288	672
Multiframe rate	5367 Hz	9398 Hz
Nominal justification ratio	0·334	0·390

The first code used, and still used on many 1544 kbit/s systems in N. America, is 'alternate mark inversion' (AMI), where zeros are sent as zero, and alternate logic ones are sent as marks of opposite polarity, so creating a three-level signal. Digital traffic with a high density of zeros gives poor clock information, so scramblers are used to improve this. Even so, scramblers do not help sufficiently with all possible traffic patterns and therefore digital errors can occur, so new classes of codes were created, as described in Section 7.5.

The code most commonly used in Europe is higher density bipolar 3 (HDB3) (see Croisier [11]). This is basically AMI but with an additional coding rule to add an extra mark in place of the last of any group of 4 zeros. The addition of the mark is obvious to the receiving equipment because it violates the normal rule that marks should alternate in polarity. HDB3 is used at 2, 8 and 34 Mbit/s and the early UK systems at 1 536 kbit/s and 120 Mbit/s. In N. America, similar codes are used, as listed in the CCITT recommendations in Table 8.2. The N. American standards are summarised in Table 8.6.

On 2 Mbit/s and 8 Mbit/s inputs to muldexes, equalisation for frequency-dependent losses in interconnecting cables is not required, because the maximum loss is specified to be relatively small (6 dB at half-bit rate). For line systems, the jitter caused by lack of equalisation of even this small loss is not always acceptable, since the input to the transmit terminal is the timing source for the line, and jitter on it narrows the 'eye' or retiming margin of the first regenerator. Equalisation is therefore often used at the inputs of transmit terminals in line systems. This does not apply where a muldex is combined with a line system, as described in Section 8.4.

The simpler arrangement of variable receiver threshold or gain is recommended for muldex inputs, because the combination of variable cable loss and worst-case permitted driver-pulse shapes leaves inadequate margin to cope with other degradations. The two important ones are jitter at the interface (see Section 8.10) and the interference caused by signal reflections at the input port or at cable discontinuities. These are re-reflected from the driver output port and appear on the input as interference. The CCITT has set a limit of 18 dB for tolerance to this interference.

At 34 Mbit/s and above, the maximum lengths of cable used in stations can cause significant pulse distortion and intersymbol interference, because

Table 8.6   Digital interfaces for N. America

Rate	Interface
1·5 Mbit/s	AMI for old systems, B8ZS* for new
6 Mbit/s	B6ZS* on balanced cable, or scrambled AMI on coaxial cable
45 Mbit/s	B3ZS*

* B3ZS, B6ZS and B8ZS are codes in the same class as HDB3, but with N. American terminology (binary *N* zero substitution), where *N* refers to the maximum number of consecutive zeros before coding. Examples of equivalent codes are: HDB3 = B4ZS, B3ZS = HDB2. Only the first of each pair of equivalent terms is normally used.

of the frequency-dependent cable loss caused by skin effect. The distortion is usually removed by variable automatic cable equalisers at equipment inputs. These use the overall cable loss as a guide to the 'tilt' required to correct the frequency-dependent losses. For this reason, simple resistive attenuators must not be used with interfaces based on variable automatic equalisers.

When 140 Mbit/s was introduced, only the very fastest and most expensive form of emitter-coupled-logic (ECL) devices could be guaranteed to code HDB3 properly, allowing for production tolerances. Therefore a new code was created, coded mark inversion (CMI) described by CCITT [12]. (This is patented and free use is available only as described by CCITT.) This is a two-level code. It codes logic zero as a half-width pulse, while logic one beomes a full-width pulse with alternate marks inverted. Unlike three-level codes, it has the advantage of allowing the use of a fixed cable equaliser to cater for the full range of in-station cable lengths.

An interface unique to the UK is 68 736 kbit/s, known as '68 Mbit/s'. This also works on CMI because of its need for ECL speeds and the pre-existence of 140 Mbit/s ECL CMI circuits. It is used for digitally-encoded broadcast-quality television and for a digitally-encoded FDM hypergroup, equal to 900 telephone channels. One 68 Mbit/s channel occupies the traffic capacity of two 34 Mbit/s streams. (Where coaxial cable interconnections are used with SONET or with SDH, the codes for the nearest plesiochronous rate are used, i.e. B3ZS for 51·84 Mbit/s and CMI for 155·52 Mbit/s. This eases planning constraints in station design.)

At 565 Mbit/s, even CMI becomes difficult to handle because of device-speed limitations. Therefore, the commonest commercial arrangment is scrambled AMI, despite its known limitations. No CCITT recommendation exists for an in-station digital interface at this level in the hierarchy. Generally, muldexes are combined with the associated line systems at this level, and no 565 Mbit/s interface is required in such cases.

The reference above to data errors with certian traffic patterns is not purely historical, nor is it limited to AMI interfaces. Digital transmission systems continue to be engineered and sold in which error-free performance occurs with pseudo-random bit sequence (PRBS) test patterns, but not with the output of multiplexers on which only some tributaries are carrying traffic.

This may appear surprising, since a PRBS contains all possible patterns of n bits, where $2^n - 1$ is the sequence length, and values of n up to 23 are commonly used. The explanation is that pattern-dependent or frequency-dependent phenomena are often related to resonances and may therefore need to be stimulated for some time to be fully revealed; a PRBS creates each pattern only fleetingly. The situation is similiar to that found when a spectrum analyser sweeps too rapidly and thereby fails to display a high-Q filter characteristic. The causes of pattern-dependent errors often lie in the analogue circuits, such as those for clock extraction and cable equalisation. In radio systems, they can lie in IF amplifiers or demodulators. Fortunately, many commercial pattern generators offer patterns which can reveal the problem, when passed through the offending equipment, without the use of actual multiplex frame patterns.

8.10 Jitter at interfaces

8.10.1 Approach to jitter specification

Jitter has a profound effect on the design of high-order muldexes. It is generated and filtered by all digital transmission equipment in a wide variety of ways. This very variety has prevented the use of simple all-embracing planning rules for jitter in a network. However, CCITT specifications exist and do help in the design of particular pieces of equipment. The specifications are based on simplified concepts of how transmission equipments work. Three properties of an individual equipment are defined:

(a)   Tolerance to jitter on digital input ports, for error-free operation
(b)   Maximum output jitter in the absence of input jitter (i.e. residual jitter)
(c)   The jitter transfer function between input and output ports.

(a) and (b) are specified in terms of amplitudes of single sine-wave components versus frequency. (b) is specified in terms of jitter amplitude created by the total received jitter energy, with optional bandwidth constraints. There are also some statistically-specified parameters applied to demultiplexers, in order to allow for the statistical nature of waiting-time jitter.

In high-order muldexes, item (a) requires that digital storage be provided in the multiplexer. This storage is an additional part of the elastic store already required for the justification process. Item (b) in a demultiplexer has waiting-time jitter as the main component, together with some contribution from the effects of data on the clock circuits. Crystal-contolled phase-locked loops are used as jitter reducers in demultiplexers; these limit the bandwidth and therefore the total energy of waiting-time jitter. For item (c), the dominant effect is the filtering action of these jitter reducers; their action also applies to any jitter components passing through the multiplex/demultiplex process.

8.10.2 Input tolerance

The CCITT input jitter tolerance specifications [13] apply equally to all digital input ports, and not just those of digital multiplexers. Those for 34 Mbit/s are summarised in line (a) of Fig. 8.17. Figures for other rates follow the general form of line (a). The tolerance for a typical commercial 34—140 Mbit/s multiplexer, on a 34 Mbit/s input port, is shown by curve (b) of Fig. 8.17.

Straight-line sections of the specification limits are numbered and correspond to the features below:

1   This is the effect of wander, caused by slow changes in transmission-system propagation delay. Plesiochronous muldexes handle this via the justification process and are effectively transparent to it.
2   This is the cumulative effect of low-frequency jitter-generating mechanisms in transmission equipment. The size of the muldex elastic store limits the tolerance here, particularly when the jitter frequency is greater than about 1/10th of the frame rate (e.g. as in the 2–8 Mbit/s specifications).
3   This slope is the result of filtering action intrinsic to most jitter sources.

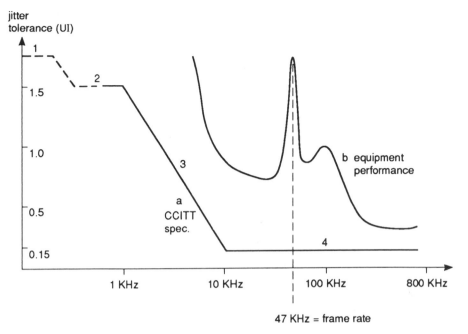

Fig. 8.17 Input jitter tolerances

It corresponds to a peak frequency shift of:
$$\Delta f = m \times a$$
where

Δ_f = peak frequency shift, Hz

m = jitter modulating frequency, Hz

a = peak jitter amplitude, rad.

For a 2—8 Mbit/s muldex, $\Delta_f = 2400 \times 0.75 \times 2\pi = 11,310$ Hz.

4 This jitter is the residual output of a transmission equipment, after internal filtering. It corresponds to the width of the 'eye' available for retiming in the input clock-extraction circuit, often dominated by the speed of the retiming-bistable device. For good jitter tolerance, the bistable device must have 'set-up' and 'hold' times very much less than the clock period, i.e. it must be capable of operating at a speed much higher than the clock rate.

The tolerance of a commercial multiplexer is shown as curve (b) in Fig. 8.17. At jitter frequencies related to the muldex frame rate, some interesting effects occur. Once per frame, the phase of the clock for each incoming tributary is measured and used in the justification decision process. Incoming jitter at frequencies less than half the frame rate will be reproduced. to a limited extend at the demultiplexer output. The limits to the effectiveness of this process are the Nyquist sampling theorem and the finite bandwidth of the jitter reducer. As far as the sampling process is concerned, jitter at frequencies less than half the frame rate will be reproduced.

However, higher-frequency components will suffer aliasing effects, and reappear at the demultiplexer output, shifted in frequency.

One consequence of sampling is that jitter at the frame rate is sampled always at the same point in the jitter waveform. It is therefore not recognised by the phase measurement for justification, and so is also not reproduced at the input to the demultiplexer elastic-store. Consequently, there is a large tolerance to jitter at this frequency, as shown in Fig. 8.17.

A second consequence, which is potentially more serious, is that jitter at half the frame rate is effectively stored twice in the transmit elastic store, requiring extra storage capacity if digital errors are not to occur. This double storage occurs because justification is decided on at the start of a frame but not carried out until near the end, by which time the phase excursion caused by jitter is in the opposite direction. However, the original jitter amplitude is correctly assessed, it is signalled in the justification service digits and it is recreated by the dejustification process. As a result, the receive elastic store has to provide storage only for the original jitter amplitude. Fortunately the CCITT jitter specifications require tolerance only to relatively small jitter amplitudes at the critical frequencies.

Because HDB3 guarantees that marks of the same polarity never occur in adjacent clock periods, it is possible to create clock-extraction circuits with a theoretical high-frequency peak–peak jitter tolerance approaching 1 unit interval (UI). A peak–peak tolerance of 0·7 UI is practicable at 2 Mbit/s clock rates, falling to around 0·3 UI at 34 Mbit/s because of component limitations. CMI does not allow such circuit arrangements, and the theoretical best case is 0·5 UI with perfect components. At 140 Mbit/s with practical components, even a tolerance of 0·15 UI is not easy to arrange. Jitter tolerance specifications in N. America are more demanding than those in Europe.

Levett [14] describes the features of a muldex which limit jitter tolerance at various frequencies.

8.10.3   Residual jitter

Waiting-time jitter specifications are based on the Duttweiler [6] model of the jitter creation mechanism (Section 8.6.3), together with the attenuation expected of practical jitter reducers. The largest values of waiting-time jitter occur for specific steady-state relationships between the clock rates of the input tributary and of the multiplexer, as described in Section 8.6.3. These conditions cannot be sustained in the presence of even small amounts of input tributary jitter. The consequence is that waiting-time jitter in PDH systems very rarely approaches the theoretical worst case allowed for in specifications, even under laboratory conditions. It has even been suggested that jitter or 'dither' be artificially added to control waiting-time jitter in commercial systems, but this has not been specified by CCITT.

Accumulation of waiting-time jitter in muldex chains has been shown to be on a power basis [15], which accords with the fact that the jitter sources are uncorrelated. Accordingly, the jitter amplitude builds up as the fourth root of the number of demultiplexers. There had previously been suggestions, but no supporting evidence, that the relationship might be as severe as a square-root law. This would have led to problems in long muldex chains. One unpublished European trial sought such problems and confirmed their absence. However, there have been suggestions that some early designs of N. American muldex suffered from excessive jitter in long muldex chains.

8.10.4 Jitter transfer function

For property (*c*) of Section 8.10.1, the jitter-transfer function of a muldex is normally dominated by the crystal-controlled phase-lock loop in each tributary of the demultiplexer. This provides a low-pass-filter effect. In theory, cut-off frequencies can be arbitrarily low; however, in practice, costs tend to rise rapidly with Q values above about 100 000. Second-order loops are commonly used in order to allow separate specification of loop band-width and of the drift of loop phase error (see Gardner [10] and Byrne [9]). A consequence of such a loop is that jitter gain is almost unavoidable at some jitter frequencies. Practical damping factors can hold this gain down to around 0·3 dB or less, per muldex.

The apparent jitter-transfer function can be affected by the level of input jitter used for the measurement. If the phase-lock loop is overloaded with jitter, its linear response will cease and subsequent behaviour will depend on the loop design. In some designs, the attenuation is deliberately made to increase with increasing input jitter level. In general, if gross overloading occurs, then the elastic store overflows and digital errors will also result.

Above the 3 dB cut-off frequency of a loop, the roll-off rate is generally asymptotic to 20 dB per decade. (It is initially asymptotic to 40 dB in a N. American specification, leading to suspected 'pulse edge overshoot' problems in jitter waveforms for chains of early muldex designs.) The ultimate

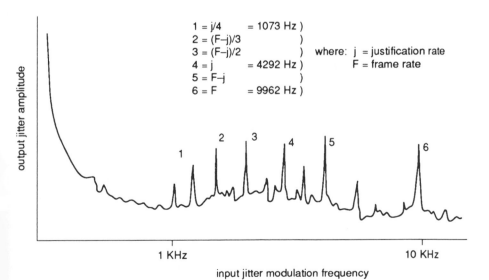

Fig. 8.18 Measured jitter–transfer function of a 2–8 Mbit/s muldex

attenuation is typically in the range 30—70 dB, depending on the care taken in design and construction. The measurement of attenuation factors greater than 20 or 30 dB is practicable only by frequency-selective techniques, because of the presence of relatively-large levels of residual jitter. Details of layout, decoupling, shared use of integrated circuits etc. can affect this attenuation and can also affect residual jitter levels. Lack of care in these areas can also introduce a non-linear jitter-generating mechanism, whereby a steady phase movement on the input signal causes one or more phase jumps at the output. This effectively translates tolerable low-frequency jitter to less-tolerable high-frequency jitter. Intermodulation effects can also occur between the various jitter components. Fig. 8.18 shows the results of this effect in a 2—8 Mbit/s muldex.

The frequencies at which the CCITT requires that attenuation starts were first based on practicalities of circuit design for a reasonable cost, rather than on any theoretical network studies. This point accords with the general approach for CCITT jitter specifications, which are the result of compromises based on the peformance of various early equipment designs. Reference 13 gives an excellent summary of the accumulation of a jitter in a network, while Kearsey [16] covers more aspects of jitter in a network.

8.11 Use of muldexes to monitor performance in a network

Unlike radio systems and many line systems, muldexes do not contain features which push circuits to the limit of their ability to extract bits from distorted or weak signals. Accordingly, error monitoring on muldexes is not as comprehensive as on line or radio systems. Most of these systems allow for the detection of errors in almost any bit; in PDH muldexes, only the frame-alignment word is monitored for errors.

In the SDH, the combining of features from line and muldex terminals has led to the monitoring of all received bits, based on byte-interleaved parity (BIP). Reference 17 gives a good background to the characteristics of various error-detecting codes. Prior to the acceptance of SDH for major new muldex applications, there were proposals to improve the coverage of error monitoring in European PDH muldexes, by encoding existing spare bits. The use of simple parity bits is a feature of the N. American 45 Mbit/s DS3 system, as shown in Table 8.5.

Several administrations are implementing network-monitoring systems which gather and process centrally the error reports of many demultiplexers and transmission links. Ideally, this information would be used to give a measure of circuit quality. However, there are several hazards in, for example, attempting to use error-rate measurements on a 34-140 Mbit/s demultiplexer to derive the quality of a 64 kbit/s circuit carried on it.

One problem is that the occurrence of only three errors on the justification-control bits of one tributary in the 140 Mbit/s frame structure will lead to a loss of justification synchronisation, which in turn will cause at least a single-bit slip. This will cause loss of frame alignment on a following 8—34 Mbit/s demultiplexer, and this loss will be extended down to the primary PCM frame. The resulting error burst could last for a millisecond or more. On the other hand, errors which occur at other points in the 34–140 Mbit/s frame may have little or no effect on the PCM traffic, depending on where they occur. In effect, the 'error extension' of the transmission path, i.e. the multiplying factor for errors which occur in the path, varies over many orders of magnitude.

Studies in the CCITT are addressing this problem, and related ones in SDH, with the intention of completing a new recommendation (G82X: 'Network performance objectives for end to end connection occurring at and above the primary rate'). This aims to provide more-comprehensive guidance than CCITT recommendation G821, and to be independent of transmission media, protocols or bit rates.

When transmission performance becomes unacceptable, muldexes can initiate several alarms similar to those in radio or line transmission terminals. These are: 'power failure', 'demultiplexer loss of input/loss of alignment', 'loss of a multiplexer tributary input', 'distant terminal alarm', and 'receipt of alarm-indication signal' (AIS). Each hierarchical level of demultiplexer can detect and broadcast the AIS, and gives demultiplexers a unique ability to extend alarms through a network. AIS (known as 'blue signal' in N. America) is an all-ones pattern put out by any equipment which has raised a major traffic failure alarm. In response, the demultiplexer extends that signal to all of its tributary outputs. As a result, AIS (and its associated 'AIS received' alarms) can flood across a network from a fault.

There are differing views as to how rapidly this response should occur, i.e. whether or not time should first be allowed for all of the preceding high-order demultiplexes in a hierarchical chain to find frame alignment after a transmission disturbance, before AIS is introduced. This response of demultiplexers to AIS is one of several complications for the designers of network-management systems.

8.12 References

1 BRIGHAM, E.R., SNAITH, M.J., and WILCOX, D.M: 'Multiplexing for a Digital Main Network', *PO Elect. Engrs. J.*, 1976, **16**, pp. 93–102.
2 OWEN, F.F.E.: 'PCM and digital transmission systems' (McGraw Hill, 1982)
3 BYLANSKI, P., and INGRAM, D.G.W.: 'Digital transmission systems' (Peter Peregrinus, 1980)
4 HÄBERLE, H., 'Frame synchronisation PCM systems', *Electrical Communication*, 1969, **44**, pp. 280–287
5 JONES, E.V. and AL-SUBBAGH, M.N.: 'Algorithms for frame alignment; some comparisons', *Proc. IEE* 1985, **132**, Pt. F, pp. 529–536
6 DUTTWEILER, P.L.: 'Waiting time jitter' *Bell Syst. Tech. J.*, 1972, **51**, pp. 165–207
7 CLEOBURY, D.J.: 'Characteristics of a digital multiplex equipment employing justification techniques' IEE Conference on Telecommunication Transmission, 1975, IEE Conf. Pub. 131 pp. 83–86
8 CHOW, P.E.K.: 'Jitter due to pulse stuffing synchronisation', *IEEE Trans.* 1973 **COM–21**, pp. 854–859

9 BYRNE, C.J.; 'Properties and design of the phase controlled oscillator with a sawtooth comparator', *Bell Syst. Tech. J.*, 1962 **41**, pp. 559–602
10 GARDNER, F.M.; 'Phaselock techniques', (John Wiley, New York, 1979), 2nd edn.
11 CROISIER, A.; 'Introduction to pseudoternary transmission codes', *IBM J. Research & Development*, 1970, **14**, pp. 354–367
12 CCITT Recommendation G703: 'Interface at 139,264 kbit/s'
13 CCITT Recommendation G823: 'The control of jitter and wander within digital networks which are based on the 2048 kbit/s hierarchy'
14 LEVETT, F.A.W.: 'Jitter in digital multiplex equipment'. IEE Colloquium on Jitter in Digital Communication Systems, 1977
15 FRANGOU, G.J., and PATEL, N.: 'Jitter accumulation in digital networks'. IEE Conference on Measurements for Telecommunication Transmission Systems, 1985, IEE Conf. Pub. 256 pp. 27–28
16 KEARSEY, B.N.: 'Jitter in digital telecommunication networks', *Brit. Telecom. Eng.*, 1984, **3**, pp. 108–116
17 MICHELSON, A.M., and LEVESQUE, A.H.: 'Error control techniques for digital communication' (J. Wiley, 1985)

Chapter 9
Synchronous higher-order digital multiplexing
M.J. Sexton and S.P. Ferguson
STC and GPT Limited

9.1 Introduction

The need for improved quality of service, with its implications for customer responsiveness and service availability, and the need to contain or reduce network operating costs are driving forces felt, in some degree, by all telecommunication network operators. The pressures are particularly acute in the more-developed countries with liberalised regulatory regimes; however, they are also felt, albeit indirectly, in those countries which still have significant regional monopolies. One of the more-important mechanisms available to a network operator responding to both these needs is to upgrade the capacity and quality of the transport infrastructure by progressively replacing it with optical fibre terminated in facilities that enable 'hands off' operation and management. Such networks can be designed to be 'self healing' and to allow greater flexibility in responding to change.

Recent developments in VLSI and optical-device technology now allow a much greater degree of functional integration in equipment design, with consequent potential cost benefits. To realise these cost benefits, it will become necessary to interconnect telecommunication network elements directly via high-capacity optical interfaces. This also gives an important benefit in network reliability, because most reliability hazards are associated with the metallic interconnections between equipments. Optical interface standards are required before such developments can be widely adopted.

It is helpful in this context to view the network in layers, as illustrated in Fig. 1.4. The lowest layer represents the transmission or transport network. It is made up of copper and optical-fibre cables, line transmission systems, radio systems, multiplexers and a range of distribution frames and protection systems dealing with various capacities and providing a degree of flexibility in the allocation of the resources of the transport network.

The transport layer supports a number of logically-independent service-layer networks. A service layer is made up of service entities, both servers and users. The transport layer provides interconnection between service entities to form a service network. The most important of these service networks is the PSTN, where the servers are the switching machines, large and small, while users gain access via handsets, modems etc.

Rivalling the PSTN in importance as service networks are the private analogue leased-line networks. The packet-switched network and the digital leased-line networks are others. The ISDN aims to integrate a number of these service networks into a single integrated service network and network operators are already planning for a broadband service network. This may integrate and absorb the existing service networks or it may just operate as yet another logically-independent service network to be supported on the transport layer.

Networks are becoming fully digital and the balance of costs favours networks which use high-capacity optical transmission in local networks as well as in trunk networks. It is advantageous that the multiplexes used for these links should be compatible with the switching systems used at the network nodes; i.e. they should be synchronous rather than plesiochronous. The 30-channel and 24-channel primary groups already use synchronous multiplexing. The introduction of time-division digital exchanges provided a large cost saving by handling traffic in primary groups instead of requiring it to be demultiplexed onto individual audio circuits for space-division switching. A further cost saving is possible for high-capacity routes if higher-order multiplexed traffic can be switched directly, instead of being demultiplexed to primary multiplex groups. This necessitates the higher-order multiplexes transmitting in synchronism with the clocks of the digital exchanges.

Not all the traffic carried to a network node by a high-capacity transmission link may be switched there; some of the capacity may be required for PSTN traffic routed through to other nodes and for private circuits. Also, the introduction of broadband services will necessitate the provision of channels carrying traffic at much higher digit rates than 64 kbit/s. In order to accommodate these different needs flexibly, a *digital cross-connect system* (DCS), also known as a DXC, is required at each end of a high-capacity link. The requirements for a DCS and an example realisation are considered in Section 9.4. A further example of a DCS used in a local network is provided by the Flexible Access System (FAS) of British Telecom, which is described in Chapter 15. In FAS, a single optical fibre provides a large business customer with a range of services, including telex, telephony and digital private circuits. At the exchange, the private circuits are routed through a DCS which can be managed remotely from a control centre. These private circuits may also pass through a DCS at one or more intermediate nodes on the route to their destination.

Synchronous multiplexes may be used to provide a managed transmission infrastructure to give flexible allocation of transport capacity to various service networks. The general principles of the network architecture are illustrated in Fig. 9.1. This shows a small section of a network comprising three service nodes each having a digital cross-connect system (DCS). As examples of service-layer network components, two switches, a remote concentrator unit (RCU), a 64 kbit/s cross-connect and a flexible-access system are shown. The DCSs are used under the control of the network-management function to allocate circuit paths between service-layer entities to form service networks.

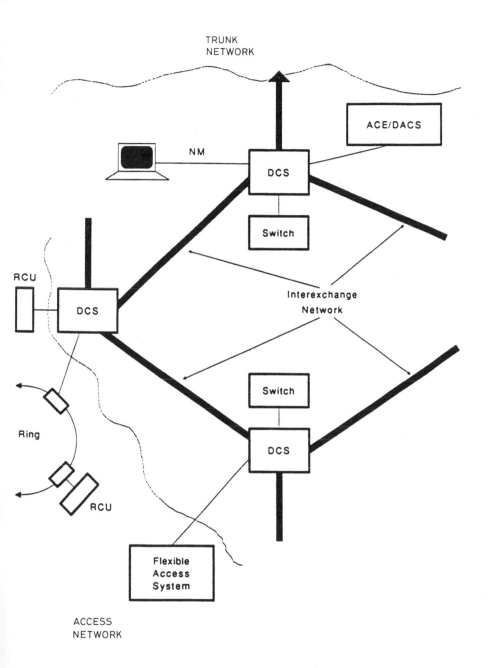

Fig. 9.1 A managed synchronous transmission network

Such networks are now being designed. While the general concept is clear, there remain many unresolved issues such as: dimensioning and functionality of the nodes, the number of layers in the cross-connect hierarchy for optimum connectivity, deployment strategies and compatibility with existing or planned management structures and services.

Different synchronous multiplex systems were developed for trials in the USA, UK and Japan. Then, the newly-formed Bellcore specification authority in the USA defined a standard known as SONET (synchronous optical network) [1, 2] which replaced an earlier and less-flexible synchronous multiplexing proposal known as SYNTRAN (synchronous transmission) [3, 4].

SONET was initially compatible only with the North American hierarchy of bit rates. However, it was modified in conjunction with the CCITT in order to provide options compatible with the European hierarchy. The outcome was the *synchronous digital hierarchy* (SDH) [5, 6, 7]. The preferred European options [8] were defined by the European Telecommunications Standards Institute (ETSI) and those for the USA [9, 10] by the American National Standards Institute (ANSI). By an appropriate choice of options, a subset of the SDH is compatible with both American and European standards; thus, interworking is possible. The CCITT defined the SDH for use as a network node interface (NNI). It is also defined as an option at the user/network interface of the proposed broadband integrated services digital network (BISDN).

The European and North American plesiochronous digital hierarchy (PDH) standards can be defined on two pages, as in Tables 8.2, 8.4 and 8.5. In contrast, SDH incorporates many complex features, and its definition requires many times that number of pages. This chapter therefore concentrates on the key points of SDH.

The PDH was defined in a way which minimised both the transmission bandwidth and the complexity of digital logic required, with around 100 000 equivalent digital gates being enough to multiplex and demultiplex between 64×2 Mbit/s and 140 Mbit/s. The priority in SDH is features rather than simplicity, as silicon and optical technologies have greatly advanced since PDH was defined. The nearest equivalent option in SDH to the PDH function of multiplexing 64×2 Mbit/s to 140 Mbit/s, is multiplexing 63×2 Mbit/s to 155 Mbit/s. In SDH, the muldex (combined multiplex and demultiplex) for this requires nearer a million equivalent gates.

9.2. Main features

SONET and SDH are complete transport systems in which muldexes and optical line terminals are combined, using frame structures that are optimised for use by digital switches. This contrasts with plesiochronous muldexes, whose frame structures were designed to minimise muldex costs and make efficient use of bandwidth. The key features of the synchronous digital hierarchy are as follows:

(*a*) The basic modules are synchronous to each other.
(*b*) Multiplexing is by byte interleaving.
(*c*) Any of the existing CCITT plesiochrous rates can be multiplexed into a common transport rate of 155.52 Mbit/s.
(*d*) This rate includes a generous allocation for a range of 'overheads'. These include multiplex overheads, such as tributary identity, and trans-mission-support services, such as error-performance checks, management telemetry and engineers' order-write (EOW) speech channels.
(*e*) Higher transport rates are defined as integer multiples of 15 552 Mbit/s in a $n \times 4$ sequence, for example, giving 622.08 Mbit/s and 2488.32 Mbit/s. Multiplexing to higher rates is fully defined. (Thus, upper limits are set by technology, rather than by lack of standards as was previously the case.)
(*f*) Since the modules and their contents are nominally synchronised, it is easy to 'drop and insert' and to cross-connect complete basic modules in a switch. This allows drop and insert and cross-connection down to 2 Mbit/s (in Europe) and 1.5 Mbit/s in (North America).
(*g*) The modules are based on a frame period of 125 μs. This allows cross-connection down to 64 kbit/s. Moreover, high-order multiplexed traffic can be fed directly into the switch core of a digital exchange. This gives both considerable economies and allows broadband switched services to be provided to customers.
(*h*) The technique makes allowance for non-synchronous operation and for wander. This allows common equipment to be used both in fully-synchronised local networks and in imperfectly-synchronised national networks.
(*i*) Synchronous multiplexers can accept traffic from any of the interfaces standardised for plesichronous multiplexes [11].
(*j*) The interfaces on the multiplexed-traffic ports can be optical. All signal processing required for optical transmission is included in the multiplexer framing process.
(*k*) The standard includes line coding, which is solely in the form of scrambling, except that CMI may be used in addition for copper cable.
(*l*) The allocation of bytes within a frame is flexible. This is achieved by having a carefully 'layered' specification.
(*m*) SDH includes management channels within the interface rate. These have a standard format for the structure of network-management messages, independent of vendor or operator.

9.3. SDH frame structure

9.3.1. The STM–1 synchronous transport module
The SDH has a repetitive frame structure with a periodicity of 125 μs, the same as that of primary-rate signals. This allows for cross-connection of components within primary-rate signals carried in the SDH frame. The

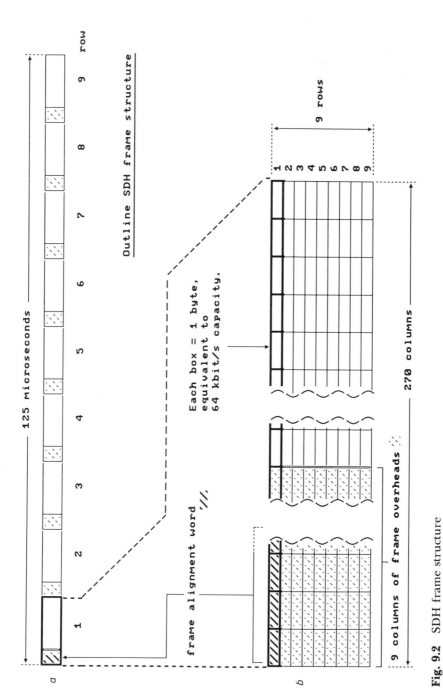

Fig. 9.2   SDH frame structure

(*a*) outline frame structure
(*b*) frame structure shown in rows and columns

basic SDH signal, termed a *synchronous transport module at level 1* (STM–1), is shown in Fig. 9.2*a*. This consists of nine equal segments with a burst of 'overhead' bytes at the start of each. The remaining bytes contain a mixture of traffic and overheads, depending on the type of traffic carried. The overall length is 2430 bytes, with each overhead occupying nine bytes. The aggregate bit rate is thus 155 520 kbit/s, which is usually called '155 Mbit/s'.

Fig. 9.2*b* shows how the frame is conventionally represented as 9 rows and 270 columns of 8 bit bytes, derived from the representation in Fig. 9.2*a*. The first 9 columns are mostly for *section overheads* (SOH), such as frame alignment, error monitoring and data communications. The remaining 261 columns comprise the *payload*, into which a variety of signals can be mapped.

The 155 Mbit/s interface of STM–1 offers a flexibility similar to the current 1.5 Mbit/s and 2 Mbit/s interfaces, but with greater richness of features. With growth in demand for bandwidth and falling costs of optical equipment, 155 Mbit/s interfaces are expected to replace multiple primary-rate interfaces in many applications.

The STM–1 frame can hold payloads at the existing PDH rates of 1.5, 6 and 45 Mbit/s or 2, 8, 34 and 140 Mbit/s. It can also accommodate new traffic structures, such as that of asynchronous transfer mode (ATM) systems or that proposed in IEEE 802.6 for metropolitan-area networks (MANs). ATM is a very-fast packet-switched system and MANs are intended for fast data transfer over public networks.

Rates below 155 Mbit/s can be supported by using a 155 Mbit/s interface with a payload area that is only partially filled. An example would be a radio system whose spectrum allocation limits it to a capacity of less than a full STM–1 payload, but whose terminal traffic ports are to be connected to the SDH through a cross-connect unit.

9.3.2. Tributary units and groups

Each tributary to the multiplex has its own payload area, occupying a group of columns, which is known as a *tributary unit* (TU). Since the tributary signals are normally continuous and the rows occur in sequence, the columns occupied by one tributary are relatively evenly spaced through the frame. Each column contains 9 bytes (one from each row), with each byte representing 64 kbit/s of capacity, or a total of 576 kbit/s per column. Three columns (i.e. 27 bytes) can hold a 1.5 Mbit/s PCM signal, with 24 time slots together with some overheads. Four columns (i.e. 36 bytes) can similarly carry a 2 Mbit/s PCM signal with 32 time slots. In North America, a TU is known as a *virtual tributary* (VT).

Several TU may be combined into a *tributary unit group* (TUG). There are several sizes of TUG. As an example, TUG–2 has 12 columns, containing 3×2 Mbit/s. The equivalent in the North American hierarchy (VT6) carries 4×1.5 Mbit/s. No overheads or extra processing are applied in forming a TUG from several TU. Its path must therefore be tracked by network management.

9.3.3. Pointers

At every level in a digital hierarchy, there is a need to locate overheads so that demultiplexing can be done. In the plesiochronous hierarchy, a unique frame-repetition rate is used at each level and a frame-alignment word identifies the start of each frame. In contrast, SDH and SONET use a system of *pointers*. These are located in known places in the frame and their arithmetical values show where the overheads of the next level down begin. Once the frame-alignment word has been found in the transport frame, by similar techniques to those used in plesiochronous multiplexes, all further demultiplexing is done by following the pointers.

A tributary unit traverses SDH networks end-to-end. It may be transferred from one STM frame to another which may not be completely in synchronism. The pointer of each TU has a value which corresponds to the time disparity between the STM frame and the payload of the TU before it is multiplexed into this frame. Cross-connection functions can be performed in the network without demultiplexing, simply by recalculating the pointer value of the TU when it is re-routed.

9.3.4. Higher-order transport modules

Network growth and the demand for broadband services are leading to very-high-rate optical transmission systems at 622 Mbit/s, 2488 Mbit/s and possibly beyond. These should ultimately replace 140 Mbit/s and 565 Mbit/s systems. Consequently, there is a requirement for synchronous transport modules operating at higher rates than 155 Mbit/s.

Higher-order synchronous transport modules can be assembled by further multiplexing. At each stage, four tributaries are combined by extracting the payload from each, recalculating their pointer values, then phase aligning and byte interleaving them and finally adding a new section overhead. The resulting digit rates are at $4n \times 155.52$ Mbit/s. STM–N is the generic term for these higher-rate modules. For example, STM–4 is at 622.08 Mbit/s. STM–16 is at 2488.32 Mbit/s and can carry 16 times the payload of STM–1.

SONET differs from SDH in that its transport entities are known as *synchronous transport signals* (STS) and range from 51.84 Mbit/s (i.e. one third of 155.52 Mbit/s) upwards. SONET nomenclature therefore defines 155 Mbit/s as STS–3, compared with the SDH term STM–1, and all other rates are scaled accordingly. At the optical interface, SONET defines *optical carriers* (OC), so that 51.84 Mbit/s is OC–1, 155.52 Mbit/s is OC–3, and so on.

9.4 The multiplexing process

9.4.1 Assembling the transport module

As an example, the multiplexing of 2 Mbit/s signals will be described. This is illustrated in Fig. 9.3. When plesiochronous traffic enters the SDH, a mapping process takes place in which each stream is adapted to the SDH network timing by justification (as described in Section 8.6). Single-bit justification is used and the mapping is said to be *asynchronous*.

If the traffic is at the primary rate and is already locked in frequency to the network clock of the PSTN, correspondence between the PSTN and

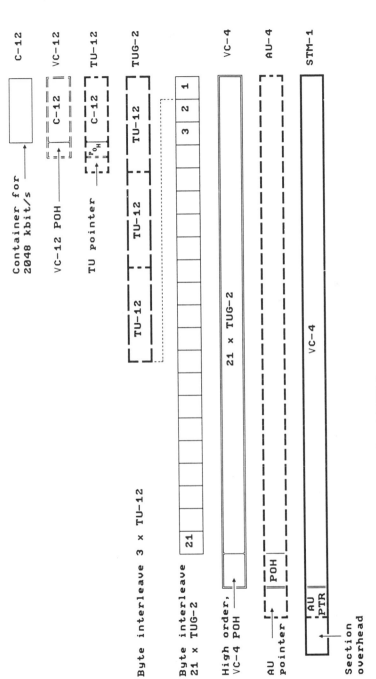

Fig. 9.3 STM-1 structure based on AU-4 and TUG-2

SDH clocks avoids the need for justification. *Byte-synchronous mapping* may be used, which relies on the primary-rate frame being at 8 kHz. This preserves visibility of that frame within the SDH frame and allows individual 64 kbit/s channels to be accessed directly in the SDH. Alternatively, *bit-synchronous mapping*, which simply disables the single-bit justification and takes no advantage of any structure in the primary-rate traffic, may be used.

The tributary signal has now been incorporated in a *container*, which can be handled in the SDH. In order to allow justification, the bit rate of the container is slightly higher than that of a plesiochronous tributary. Fig. 9.3 shows the 2 Mbit/s signal in container C–12.

Next, a *path overhead* (POH) is added to the container to form a basic *virtual container* (VC). This travels through the network as a complete package until demultiplexing occurs. Virtual container sizes range from VC–11 for 1.5 Mbit/s (where '11" denotes order 1, level 1) and VC–12 for 2 Mbit/s, through VC–3 for 45 Mbit/s to VC–4 for 140 Mbit/s. Thus, the example in Fig. 9.3 uses VC–12. In North America, the VC is known as a *virtual tributary synchronous payload envelope*.

The VC itself need not be fully synchronous with the STM–1 frame. For example, a VC could be cross-connected from one SDH to another with a slightly different frequency. Instead, the start point of each VC is indicated by a pointer and the VC together with its pointer constitute the tributary unit (TU). Fig. 9.3 shows a pointer added to VC–12 to form TU–12. As VCs travel across networks, being exposed to network timing variations, each VC is allowed to float between frames, within its TU columns, constantly tracked by its pointer, new pointer values being calculated whenever TUs are multiplexed together. Thus, it is the TU which is locked to the STM–1 frame.

Tributary units may then be assembled to form a tributary-unit group (TUG). In Fig. 9.3, three TU–12s are assembled to form a TUG–2. These are then further assembled, by byte interleaving and addition of path overheads, to form a higher order VC. Fig. 9.3 shows 21 TUG–2s assembled to form a VC–4. The higher-order VC is located within the STM by means of a further pointer. The VC and its pointer constitute an *administrative unit* (AU). Thus, addition of the AU pointer to VC–4 forms the administrative unit AU–4.

Administrative unit AU–4 forms the entire payload of a transport module STM–1. However, other multiplexing routes are possible. For example, Fig. 9.5 shows the use of VC–3 and AU–3 which contain only seven TU–2. Consequently, a further stage of multiplexing is needed to combine three AU–3 to form the payload of the transport module STM–1. In contrast, if the input to the SDH is a 140 Mbit/s digit stream, this provides the entire payload of STM–1 and no intervening multiplexing operations are required; it uses the entire traffic capacity of AU–4. A number of different multiplexing routes, for different PDH tributary bit rates, are shown in Fig. 9.7.

Finally, the STM–1 frame is completed by adding the *section overhead* (SOH) to the assembly of AUs. This is required for transmission purposes, as described in the next section. It is conventional to illustrate the complete composition of the STM by means of diagrams of the form shown in Fig. 9.4

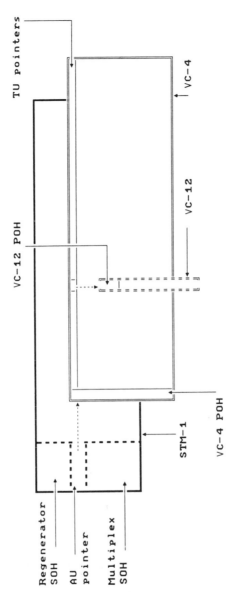

Fig. 9.4 Synchronous transport module using AU-4 and TUG-2

for STM–1 containing one VC–4 and Fig. 9.6 for three VC–3s. These correspond to Figs. 9.3 and 9.5, respectively.

Provision is made in the standard for forming a hierarchy of optical signals by multiplexing N synchronous transport modules STM–1 to form STM–N. This hierarchy is extendable to arbitrarily-high bit rates. The SDH has the implied layer structure shown in Fig. 9.8. The basic-VC path layer is defined for each signal type and the mapping variants described above occur at the high-order-VC path layer. Accommodation of differently-mapped VC is trivial, because of the homogeneity of the structure.

9.4.2. Transmission provisioning

For transmission, each higher-order VC is located within its synchronous transmission module (STM) by use of the AU pointer. This is associated with the common overhead which is attached for transmission, i.e. the section overhead (SOH). The AU pointer operates in byte steps for VC sizes up to VC–3, and in triple-byte steps for VC–4. The VC and its pointer together constitute an administrative unit (AU), which is considered as the unit of provision of bandwidth in the main network. The capacity of the section overhead increases in proportion with the order of the transport frame, N. It is divided into two: the *multiplex SOH* or MSOH and the *regenerator SOH* or RSOH. The former remains intact between end terminals on a line or radio system, while the latter can be re-used between regenerative points in a transmission path. Individual bytes in the SOH are assigned for various purposes, including some which are unique to radio. The SOH of STM–1 is shown in Fig. 9.9. However, there are differences between SDH and SONET in the way that the SOH is named and allocated.

The SDH frame alignment word (FAW) has been chosen to be a sequence of two bytes, $\ldots A_1 A_1 A_1 B_1 B_1 B_1 \ldots$, of total length $6N$ bytes for an STM–N signal. The key part of the sequence is the transition in the centre, and this remains unchanged in form as bit rates increase from 155 Mbit/s upwards and the total FAW length increases. This composite FAW can be produced at high bit rates through successive byte interleaves of identical aligned FAW sequences at lower bit rates, allowing low-speed logic to be used for all but the final interleave.

All of the synchronous transmission module is scrambled, apart from the first part of the SOH, which contains the frame-alignment word, and the signal is then sent to line. The scrambler is a seven-stage reset scrambler, with the frame-alignment word supplying the reset synchronisation at the demultiplexer. No other form of line coding is used. This arrangement is not as common as the use of conventional line coding, and puts a heavy demand on the performance of the clock-extraction circuits at the receiver. However, it has been proven in a number of transmission systems.

For digital transmission, timing information for the receiver has been traditionally provided by line coding, augmented by scrambling which also has other benefits. Where only scrambling is used, as in SDH, demands on clock-extraction circuits are increased because the pulse-transition density has a lower minimum level and a wider variation than with the use of typical line codes.

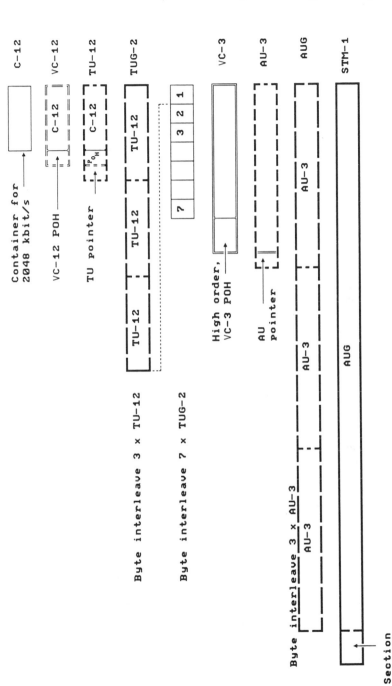

Fig. 9.5 STM-1 structure based on AU-3 and TUG-2

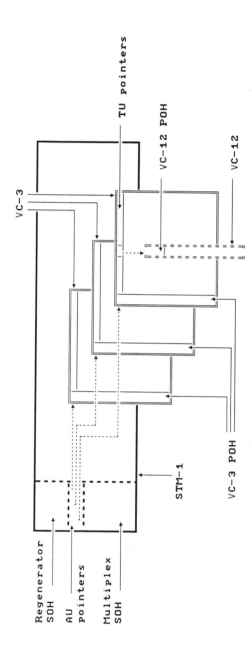

Fig. 9.6   Synchronous transport module using AU-3 and TUG-2

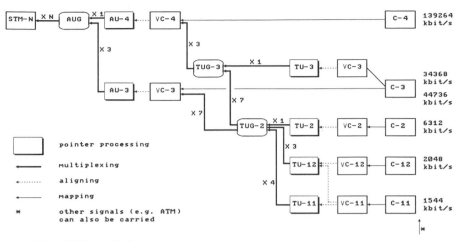

Fig. 9.7 SDH multiplexing routes

In addition, the use of scrambling alone exposes the transmission system to the risk that particular payload bit sequences may cancel the scrambler sequence and prevent transmission by stopping the recovery of a system clock at the receiver. In practice, experience has been satisfactory with various radio and line systems which rely only on scramblers as a form of line coding. The interleaving of many component signals and overhead bytes in order to form an STM–N signal for line transmission reduces the probability of such cancellation to vanishingly low levels.

However, ATM payloads are allocated a continuous burst of bytes in SDH, and this introduces the possibility of undesirable bit sequences which

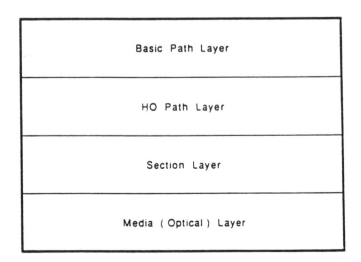

Fig. 9.8 Layered structure of STM-1

	A1	A1	A1	A2	A2	A2	C1 label	national use	
RSOH	← frame alignment word →								
	B1 parity	* note 1	* note 1	E1 EOW	* note 1		F1 user	national use	
	D1 regen. supy.	* note 1	* note 1	D2 regen. supy.	* note 1		D3 regen. supy.		
	← ADMINISTRATIVE UNIT POINTERS →								
MSOH	B2 parity	B2 error	B2 check	K1 note 2			K2 note 2		
	D4 DCC			D5 DCC			D6 DCC		
	D7 DCC			D8 DCC			D9 DCC		
	D10 DCC			D11 DCC			D12 DCC		
	Z1	Z1	Z1	Z2	Z2	Z2	E2 EOW	national use	
	← growth/future functions →								

Assignment of bytes in STM-1 section overhead

1. * = bytes whose use is specific to the medium, e.g. radio or optical fibre
2. K1/K2 = end-end signalling for protection control in line transmission
3. D4 - D12 = management channels, 576 kbit/s from Ethernet or X.25, for example

Fig. 9.9 Overhead-bytes assignment in the STM–1 frame

could disrupt clock extraction at the receiver. ATM therefore has its own scrambler in order to prevent the accidental or deliberate creation of 'clock-killer' cell sequences when ATM is carried over SDH transmission bearers. This scrambler has been chosen as a 43-stage sequence, which cannot continuously align destructively with the 7-stage scrambler sequence in SDH.

9.4.3 Additional options

Where byte-synchronous mapping is used for primary-rate traffic, a *floating mode* preserves 64 kbit/s visibility within each lower-order VC, while *locked mode* preserves it within a higher-order VC. The latter offers some simplification in digital processing; however, it is of value only where all the traffic in a higher-order VC is for the same type of service, e.g. PSTN. The SDH structure does not allow for floating and locked lower-order VCs to be mixed in the same higher-order VC. It does allow mixtures of floating-mode lower-order VCs with different mappings, so it can support a mix of services on the same bearer.

Lying between a higher-order VC and a TUG in concept is a *concatenated VC*. This is a collection of an arbitrary quantity of VCs which are associated one with another to form a combined payload capacity that can be used as a single container. Two types of concatenation are currently standardised: *contiguous* and *virtual*. In contiguous concatenation, the pointer in the first VC carries a normal pointer value, while the pointers of the rest of the concatenated VCs have a concatenation indication marker and share the use of the pointers in the first VC. In virtual concatenation, normal pointer values are used throughout, together with messages on network-management channels within the bearer which carries the concatenated VCs. For example, a concatenated $N \times$ VC–2 (known as a VC–2-mc) contains m virtual containers of the size originally defined to carry 6.336 Mbit/s PDH traffic, each having a nominal traffic capacity of 6.9 Mbit/s. The whole is used as a continuous channel of nominal bandwidth $N \times 6.9$ Mbit/s.

On STM–N transmission systems, several VC-4 may be concatenated into a single entity. For transport of ATM traffic for example, the provision of a very-large transmission entity can provide significant statistical gain in reducing the peak bandwidth demands of bursty traffic.

In a similar manner to the creation of a TUG, AU can be notionally combined, for the purposes of planning and routing broadband traffic, into an *administrative unit group* (AUG). As with a TUG, no overheads are attached to create this item, so its existence relies on network management tracking its path. For example, in North America, three AU–3s form an AUG which occupies all the payload area of an STM–1 at 155 Mbit/s.

9.5 Transiting a network

The terms VC, TU and AU are often used fairly loosely as being near-equivalent ways of describing a unit of bandwidth plus its overheads. Strictly, VC and TUG traverse the network intact, while TU and AU are manipulated at each routing point to achieve this.

As VCs traverse the network, their differential time delays result in pointer values being changed at each multiplexing or routing point In addition, jitter and wander can cause these values to change with time, corresponding to justification on a byte or triple-byte basis, depending on the VC size. Further contributions to the rate of pointer-value change are the possibility of synchronisation loss between different parts of a network traversed by a VC and, for international transmission, the slow drift which occurs between different national clocks.

Restoration of traffic to its original PDH form involves a *desynchroniser*, which introduces *pointer jitter* as it accommodates these changes. Hysteris buffers are used at places such as add-drop multiplexes and cross-connects where pointer values are changed. Therefore, any complementary pointer-value changes can be absorbed, thus reducing their contribution to overall pointer jitter at the desynchroniser. Only under very restrictive conditions is there no possibility of pointer jitter from demultiplexing after SDH transmission, irrespective of the mappings used.

With asynchronous mapping, waiting-time jitter also occurs. It is at a higher level than in PDH networks, because negative–zero–positive justification (as described in Section 8.6.3) is used. The theoretical jitter is then one unit interval peak-to-peak. The reason for this form of justification being used is that most payloads are assumed to be synchronous in the future; thus, justification decisions will be infrequent except when network synchronisation problems occur.

As a consequence of these two jitter sources, SDH network elements have jitter reducers which are orders of magnitude more powerful than those used in PDH networks.

9.6 Interfaces

9.6.1 Open interfaces

An important aspect of SDH for a managed network is its use of open interfaces, so that equipment from different vendors can be mixed with much greater freedom than in the past. Prior to the definition of SDH, common interfaces in transmission networks existed only at traffic level. They applied up to a maximum bit rate of 139 264 kbit/s, and then only on intra-station copper-cable interfaces. Interfaces for management and for other traffic, such as those for optical fibres, were proprietary. This situation acted as an obstacle to the growth of managed networks.

9.6.2. Interfaces for traffic

The SDH is based on a common signal structure at 155.52 Mbit/s ('155 Mbit/s'), with interface variants for copper cable and optical fibre; the former are for intra-station use and the latter for both intra- and inter-station use. Inter-station transmission may also be on radio, via the same intra-station interfaces [12]. SDH radio systems are being developed, using new frequency bands and using new modulation schemes in order to contain the bandwidth for compatibility with existing frequency plans.

Interfaces at successively higher rates, based on ×4 integer multiples of 155.52 Mbit/s, are defined in principle. However, they are unlikely to be used between equipments from different vendors above STM–4, nominally 622 Mbit/s.

All of these are supported by a wide definition of facilities for use at a network-node interface. The facilities include generous overhead channels; these are defined for many applications in line transmission and network management, such as embedded management-message channels, user channels, engineer order wire and channels for controlling end-to-end protection switching. In the past, these facilities were vendor-specific and so increased the cost and complexity of network operations as the number of vendors increased.

9.6.3 Interfaces for management communications

The SDH standard includes an interface for management communications. It provides for management messages which use a common profile independent of vendor or operator. The long-term aim is to have a core of common messages. To achieve this, network elements in SDH have a core set of common functions, effectively defining a set of common network-management objects and their attributes to which messages relate.

During the early stages of SDH managed networks, messages will be proprietary. Nevertheless, the existence of common objects with common attributes should allow simple conversion between the core messages of different vendors and operators [13]. This conversion will occur in the element-level controller. However, elements may have features beyond the defined core set, or the core set may not fully accord with standards (perhaps because elements were developed ahead of standards); in these cases conversion may be more complex.

The management interface takes two physical forms, both supporting data rates up to 576 kbit/s. The first is the CCITT Q interface, which has options of Ethernet and X.25 variants. The former allows a LAN (local area network) connection of multiple elements on the same site, and the latter allows connection of an element to a packet switch or other networking facility. The second form is the Qecc, functionally corresponding to the Q interface but carried within the composite SDH signal. Within that signal it is carried via the data communications channel (DCC).

9.7 Synchronous transmission network components

The principal component in the new synchronous network is a large digital cross-connect system installed at major network nodes. The traditional transmission multiplexer can be regarded as a special case; it is a small cross-connect with fixed port allocations. Between these extremes lies a range of multiplexer and cross-connect functions with varying degrees of flexibility.

The generic synchronous network element is illustrated in Fig. 9.10. The port-termination function can be defined to terminate G703-style network signals [11], mapping them into the appropriate synchronous containers. Alternatively, it may be defined to terminate SDH signals, in which case section layer overheads will be terminated and the VCs constituting the payload recovered.

The VC cross-connecting (VCX) function may be defined to reflect fixed allocations between the port terminations, as in a simple multiplexer/de-multiplexer. It may allow flexible allocation of VCs between certain ports, but not others (as in the add/drop mux), or it may allow fully-flexible allocation of VCs between any ports as in a non-blocking cross-connect. Furthermore, the VCX function may either be comprehensive enough to allow VCs of different order to be cross-connected simultaneously, or it may be optimised for cross-connecting one particular VC size such as VC–12 or VC–4.

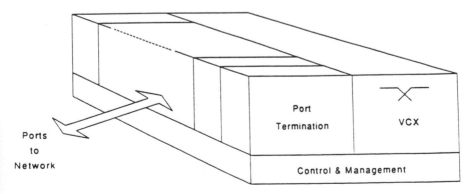

Fig. 9.10 General synchronous-network component

The control function will reflect the functionality of the element. This functionality will be modelled in data, both locally and in the data base of the element controller. Port provisioning and allocation of capacity between ports are among the functions controllable by centralised network-management facilities [14, 15].

It is expected that, in the course of the next few years, a range of these network components will become available, from which synchronous managed networks will be constructed. Under the general heading of synchronous multiplexers, one can expect to find access multiplexers, where the tributary interfaces are to G703 and the aggregate ports are to STM–1 or STM–4. These may be configured as terminal multiplexers or as add/drop multiplexers, as shown in Fig. 9.11. It is possible to construct rings or linear-bus networks from the latter. Similar functionalities will be available from SDH multiplexers in which both tributary and aggregate ports will be SDH signals (e.g. STM–1 to STM–4 and STM–4 to STM–16). These are illustrated in Fig. 9.12. Subscriber-loop carrier equipment or flexible-access multiplexers with SDH network port are also likely to appear.

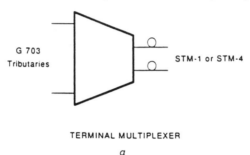

TERMINAL MULTIPLEXER

a

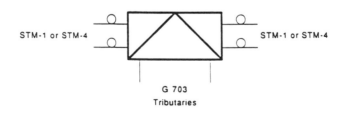

ADD/DROP MULTIPLEXER

b

Fig. 9.11 Synchronous-network access multiplexers

(*a*) terminal multiplexer
(*b*) add/drop multiplexer

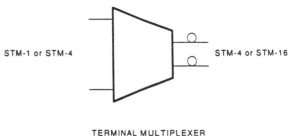

STM-1 or STM-4   STM-4 or STM-16

TERMINAL MULTIPLEXER

a

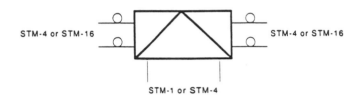

STM-4 or STM-16   STM-4 or STM-16

STM-1 or STM-4

ADD/DROP MULTIPLEXER

b

Fig. 9.12   SDH multiplexers

(*a*) terminal multiplexer
(*b*) add/drop multiplexer

Digital cross-connect systems in various sizes with different capabilities will also be available, from which synchronous network nodes can be constructed. The general structure of such a node is illustrated in Fig. 9.13. Its interface units are functionally equivalent to the multiplexer functions mentioned above. The SDH cross-connect core can be a high-capacity multilevel non-blocking switch or its capabilities may be optimised for the traffic mix at a particular node.

9.8 Applications and deployment strategies

9.8.1 General

Strategies for the deployment of synchrous multiplexes will be viewed differently by each network operator according to the pressures and constraints of the local situation. There are many stand-alone applications for managed synchronous networks which yield benefits but are fairly self-contained and can be deployed with minimal impact on the evolution of

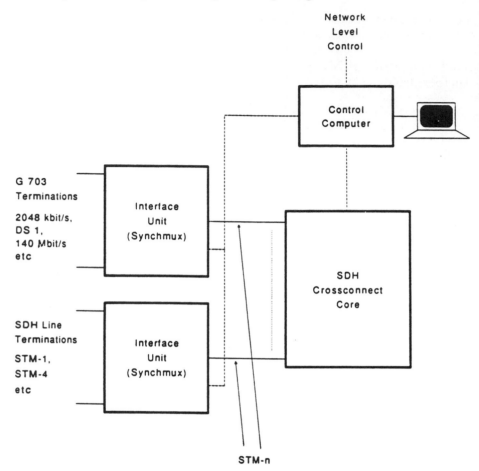

Fig. 9.13 Synchronous digital cross-connect system at a network node

the whole network. On the other hand, it is recognised that maximum benefit accrues when these principles are applied to a large contiguous network [16, 17]. Some deployment strategies are described below. They can be used, alone or in combination, to obtain significant direct benefit, while at the same time being a first step towards a fully-managed network.

9.8.2 Synchronous overlay
Most pressure is likely to be felt from private (leased-line) network users. Hence, it might be considered appropriate to install a high-quality managed network with wide-area coverage as an overlay specifically for this application. This has been the driver for the Flexible Access Systems (FAS) in the UK. Such an overlay network may, if appropriate, be extended to take an

increasing share of other service-network traffic until a large part of the core and access network has been penetrated.

9.8.3 Synchronous islands

Another deployment strategy uses the concept of 'synchronous islands'. A synchronous island is here defined as a contiguous area of the network which uses synchronous structures internally, yet interfaces to the surrounding network (an asynchronous sea, to extend the metaphor) via standard G703 series interfaces [10]. It can be deployed on its merits in many existing applications, without any implied commitment to general synchronous networking. As the network grows and new transmission links are required, the network operator has the choice of basing such new extensions on traditional transmission technology or on new synchronous technology. A synchronous extension to such an island will cost significantly less than a plesiochronous extension of equivalent capacity. This is because the format used internally is directly suitable for use in the extension and therefore can be implemented with minimum tandem processes. This same property contributes to higher reliability and smaller size. In this way, a synchronous island can be extended, merging with other adjacent islands until eventually the whole network becomes upgraded.

Examples of such synchronous islands are:

(*a*)   Network nodes based on a synchronous cross-connect core having plesiochronous interface units (including SDH mapping functions)
(*b*)   Small sub-networks of synchronous multiplexers (e.g. a junction ring)
(*c*)   Flexible-access cells using synchronous transmission from feeder node to remote.

Fig. 9.14*a* illustrates a synchronous island based on a synchronous cross-connect core connected via G703 interfaces to existing facilities. Fig. 9.14*b* shows this expanded by the addition of various transmission facilities which have the benefit of interfacing more directly with the core facility.

9.8.4 Cross-connect systems at network nodes

Cross-connect systems can be envisaged to operate either as network nodes in an exclusively synchronous network or, at the other extreme, as network nodes within an exclusively-asynchronous network to current standards. Needless to say, hybrid elements can be envisaged capable to terminating a combination of synchronous and asynchronous links. Such elements can act as gateways between the synchronous network and the vestiges of a plesiochronous network.

A basic system is illustrated in Fig. 9.14. The core cross-connect may be optimised to cross-connect at a single VC level or may make provision for cross-connecting at various levels. It may exploit traffic statistics (e.g. the proportion of terminated to transit paths at a node) to provide a restricted flexibility and it will certainly need to be expandable over a range of sizes. The interface units terminate the various network signals and, if required, map them into synchronous virtual containers according to G708/G709. Interface units may, in some cases, be required to provide a consolidation function to better load the core cross-connect ports.

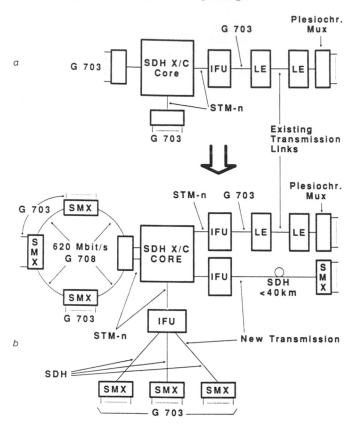

Fig. 9.14   Growth of a synchronous island

(*a*)   SDH cross-connect interfacing to existing facilities
(*b*)   addition of new transmission facilities

9.8.5   Junction network upgrade

In many European countries where digitalisation has started, the junction network consists largely of fairly-elderly copper cables carrying 2048 kbit/s 30-channel voice signals. This was the first part of the network to be digitalised and may now be regarded as unreliable and costly to maintain. There is some penetration of fibre systems at 8 M/bit/s and 34 Mbit/s levels, but the sections are short and multiplexing to route 2 Mbit/s signals between service points (switches etc.) is a disproportionate part of the cost.

The physical topology which has often developed has significant linkage between small local exchanges in addition to links up to the next level in the switching hierarchy. The application is therefore ideally suited to a transmission ring topology exploiting the add/drop capability of the synchronous mux. The application is illustrated on Fig. 9.15, where traffic is hauled between four remote switching units (RSU) and between each RSU and a principal local exchange (PLE). In the UK, this coincides with a

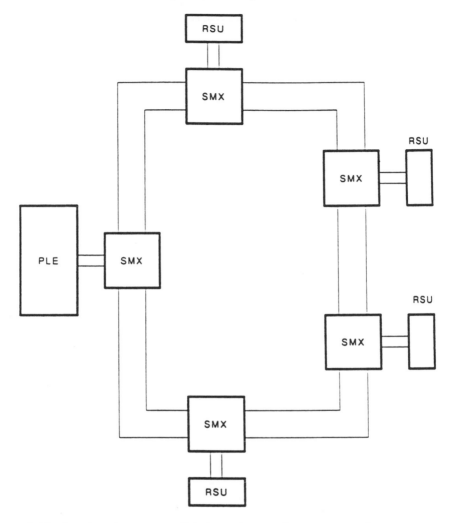

Fig. 9.15   Local exchange consolidation using synchronous mux ring

local-exchange consolidation programme, where local exchanges are replaced by RSUs which are connected to their parent PLE (but not to each other).

This topology has the advantage of high resilience, as there are always two physically-alternative routes available to route traffic. A synchronous ring will heal itself autonomously on failure, without any high-level network management intervention.

9.8.6 Local access host

The future evolution of flexible-access systems envisages that, in many cases, service access will be most effectively attained by remote line modules based on flexible multiplexers. These will generally be favoured over the RSU or small LE, because they will be more cost-effective both for low-density areas and where there is a requirement for non-telephony service. It is also consistent with a drive towards fibre to the customer and positioning for broadband services.

Such new-generation flexible-access systems may be based on a synchronous mux as host terminal. The tributaries facing the access network will terminate either point-to-point fibre connections to large business customers, or low-capacity rings or passive optical point-to-multipoint networks

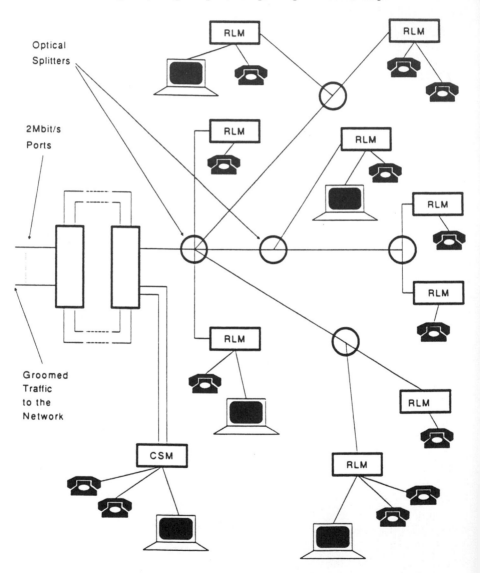

Fig. 9.16 Synchronous mux in flexible-access application

based on time-division multiple access. In the latter two cases, the signal formats may be equivalent locked-mode tributary units with channel-associated signalling, thus allowed B-channel 'grooming' in what is effectively a distributed multiplexer. Fig. 9.16 illustrates a possible arrangement.

9.9 Conclusions

The first high-order digital muldexes were plesiochronous because they were considered more flexible in the context of networks which were still mainly analogue. In particular, muldex levels or orders could be added as appropriate to reduce transmission costs, without concern over security of synchronisation clocks.

Networks are becoming increasingly digital, and the balance of costs favours networks which use high-capacity optical transmission in local as well as national links. These links are between synchronised exchanges and therefore invite synchronous multiplexing. In addition, the prospects for widespread use of broadband ISDN (which requires high-capacity links to individual customers) invite an increase in network capacity, and require that it be in a form suitable for switching, i.e. synchronous.

A managed transmission network concept, based on synchronous multiplexing, emerged from this background of market and technology pressures and the necessary standards support has been forthcoming from CCITT. The technical principles have been described above. Some transitional applications and deployment scenarios have also been described. There are many others, from the mundane replacement of plesiochronous equivalents to the probable future broadband applications. In the meantime, much work still needs to be done in the network-operating community in definition of network architectures, node capabilities and control interfaces before such networks become a reality.

9.10 References

1 BOEHM, R.J., and SHERMAN, R.C.: 'SONET (Synchronous Optical Network)'. IEEE Globecom Conf., 1985, pp. 1443–1450
2 'Synchronous optical networks (SONET)'. Bell Communications Research Document TA–TSY–000253
3 BALLART, R. *et al.*: 'Restructured DS3 format for synchronous transmission (Syntran)'. IEEE Globecom Conf., 1984, pp. 1036–1042
4 'Synchronous DS3 format interface specification'. Bell Communications Research Technical Reference TR–TSY–000021, June 1984
5 CCITT Recommendation G707: 'Synchronous digital hierarchy bit rates'.
6 CCITT Recommendation G708: 'Network node interface for the synchronous digital hierarchy'
7 CCITT Recommendation G709: 'Synchronous multiplexing structure', 1990 issue
8 ETSI: European Telecommunications Standard DE/TM–301

9 ANSI: American National Standard for Telecommunications T1. 105—
   1988: 'Digital hierarchy optical rates and formats'
10 ANSI: American National Standard for Telecommunications T1. 106—
   1988: 'Digital optical interface: single mode'
11 CCITT Recommendation G703: 'Physical/electrical characteristics of
   hierarchical digital interfaces'.
12 DAVEY, L., and SHAFI, M.: 'The application of radio relay systems in
   a synchronous digital network', *Telecom. J. Australia*, 1990, **40**, pp. 11–16
13 HEAD, K.: 'A SONET milestone for the history books', *Telephony*, 10
   Dec. 1990, pp. 24–27
14 MARSHALL, J., GALLAGHER, R., and COLE, R.: 'Managing flexibility
   in synchronous digital hierarchy networks'. IEE Nat. Conf. on Telecom.,
   1989, IEE Conf. Pub. 300, pp. 358–363
15 BERGH, S., GUNNARSSON, S., and JAKOBSSON, M.: 'New structure
   will facilitate telecom network management', *Tele* (magazine of Swedish
   Telecom), 1990, **2**, pp. 6–10
16 WRIGHT, T.C.: 'Deployment strategy for transmission networks based
   on the synchronous digital hierarchy', *Brit. Telecom. Eng.*, 1990, **9**,
   pp. 109–111
17 Special issue on 'Global deployment of SDH-compliant networks', *IEEE
   Communications Magazine*, Aug. 1990

Chapter 10
Line systems
P.J. Howard and R.J. Catchpole

BNR Europe Ltd.

10.1 Introduction

The aim of this chapter is to provide an overview of line systems. These systems are designed to transmit signals faithfully over cable from one part of the network, e.g. an exchange, to another part of the network. A system consists of terminal equipment, cable and, where necessary, intermediate repeaters, as shown in Fig. 10.1. The prime function of the terminal equipment is to convert signals at a standard equipment interface to and from a form more suitable for line transmission. In addition it provides Operation and Maintenance (O&M) features such as performance monitors, alarms and other supervisory signals. Usually, an Order Wire (OW) is also provided to allow communication between staff working on the system without interrupting service. The cable is buried, laid in ducts or hung overhead along a convenient route between the terminal equipments. The intermediate repeaters are located where signal access is needed or near enough together to avoid undue distortion or noise interfering with the transmitted signals.

Telecommunications line systems must be characterised by outstanding traffic performance, high reliability and maintainability, and long life. These characteristics have resulted in a network containing a wide variety of many generations of equipment. Some use analogue and others digital signals; some use twisted-pair or coaxial cables and others optical cables; some carry only telephony circuits and others a variety of services. The number of circuits can range from one to ten thousand. These types of systems will be summarised, and their relative places in the full spectrum of systems will be shown. More detail on some design aspects can be found in Reference 1.

10.2 Cables

Multichannel FDM systems were originally developed for open-wire pole routes and underground balanced-pair cables. The transmission characteristics of both these media are limited to two or three hundred kilohertz or less; consequently, system capacity never exceeded more than a few 12-channel groups. It was the introduction of coaxial-pair cables that made

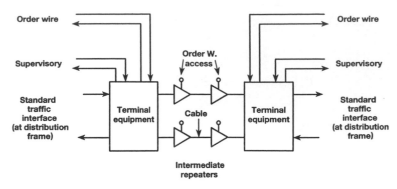

Fig. 10.1   Basic line system

high-capacity FDM systems possible. Two cable sizes have been standardised by the CCITT: one of 9·5 mm diameter (large core) and the other 4·4 mm diameter (small core). The inner central conductor is supported on widely-spaced polyethylene discs, so the cables are effectively air-spaced.

The key characteristics of these cables are impedance, attenuation, temperature coefficient and crosstalk. These are specified as follows [2, 3]:

(a)   Large-core cable
    Loss: $2·355 \sqrt{f} + 0·006f$ dB/m at 10°C where f is the frequency in MHz
    Impedance: $75 \pm 3 \ \Omega$ at 1 MHz
    Temperature coefficient of attenuation: 0·2% per °C
    Far-end crosstalk attenuation: greater than 130 dB
(b)   Small-core cable
    Loss: $5·32 \sqrt{f}$ dB/km at 10°C
    Impedance: $75 \pm 1 \ \Omega$ at 1 MHz
    Temperature coefficient of attenuation: 0·2% per °C
    Far-end crosstalk attenuation: greater than 130 dB

It will be seen that the loss increases with the square root of the frequency. The linear element is very small. A useful approximation is that small-core cable gives 2·25 times more loss than large-core cable. Hence, an equipment designed for small-core cable will operate on large-core cable at a repeater spacing 2·25 times longer. The crosstalk attenuation between adjacent coaxial pairs is very high and is not a limiting factor in system design.

10.3 Analogue system design

10.3.1 12 MHz system
The system design considerations pertaining to an FDM line system are illustrated in the following paragraphs by reference to a 2700-channel system operating on coaxial cable. As explained in Section 5.5, this requires transmission of a band of frequencies from about 300 kHz to 12·4 MHz.

The design objective is to provide a stable end-to-end transmission path, with a flat signal-level characteristic across the frequency band and with a maximum limit for the noise in each channel. This is accomplished by providing repeaters at regular intervals along the cable.

The basic features of an intermediate repeater are shown in Fig. 10.2. The loss/frequency characteristic of the cable is compensated by shaping the gain of the amplifier and by passive equaliser networks. In order to avoid adjusting the gain and equalisation to suit repeater sections of differing lengths, a *line build-out network* is included. This is chosen to make the characteristics of shorter lengths of line the same as those of a section of maximum length, for which the repeater is designed. Since the attenuation of the cable varies over time, automatic gain regulation is provided. Power-separation filters enable the amplifier to receive its DC power supply from the same cable pair that carries the signal. It is also necessary to protect the electronic circuits of the repeater from voltage surges that may occur on the line. Arrangements for power feeding and protection are discussed in Section 10.6.

The basic parameters of these systems were standardised by the CCITT. Thus, the repeater spacing and other key parameters are standardised, but the method of implementation is left to the individual manufacturers. The main recommendations relating to 2700-channel, or 12 MHz, line systems will be found in recommendations G332 and G345 [4, 5]. In the following sections, 'recommended' implies a CCITT internationally-agreed recommendation.

10.3.2 Gain shaping

The recommended repeater spacing for a 12 MHz system on 4·4 mm cable is 2 km, corresponding to a loss of about 40 dB at 12·5 MHz and 6 dB at 300 kHz. The usual design technique is to compensate for some of this slope with a shaped negative-feedback (NFB) network; i.e., minimum NFB is provided at the highest frequency, but not less than 15 to 20 dB in order to preserve the virtues that NFB brings.

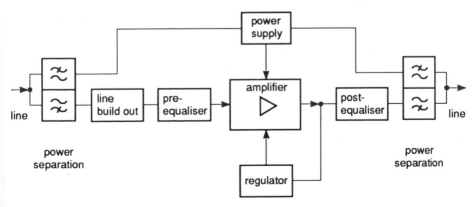

Fig. 10.2 Block schematic of intermediate repeater (protection arrangements not shown)

It is not practical to put the whole cable shape in the feedback path and additional fixed equaliser networks, with zero loss at the top frequency, are used. By splitting these between input and output, good return losses are obtained at low frequencies and buffering to lightning surges from the cable is provided.

A long tandem connection of amplifiers also requires equalisation at the terminal stations to compensate for any systematic error accumulating along the line.

10.3.3 Noise contribution

The CCITT noise objectives for a 2500 km Hypothetical Reference Circuit (HRC) [6] equate to 3 pW0p/km from the line plant (pW0p is the psophometrically-weighted noise in picowatts at a point of zero relative level). Thus, the noise contribution of each amplifier at 2 km spacing must never exceed 6 pW0p in any voice channel and a design target of 4 pW0p is set to ensure this. This will comprise, say, two thirds basic or thermal noise and one third intermodulation noise. Thus:

allowed thermal noise contribution (per channel) $= 2 \cdot 66$ pW0p

$$= -90 + 4 \cdot 25 \text{ dBm0p}$$

$$= -86 \text{ dBm0p}$$

But

amplifier noise output power (per channel) $= -139 + F_n + \text{gain dBm0p}$

$$= -99 + F_n \text{ dBm0p}$$

where -139 dBm is the thermal noise power in 4 kHz bandwidth at the operating temperature, F_n is the noise figure of the amplifier and the gain at the highest channel is 40 dB.

Hence, if the amplifier were operated at zero relative level, at a gain of 40 dB, the amplifier noise figure F_n would need to be better than $99 - 86 = 13$ dB. The amplifier cannot be operated at such a level because intermodulation would be too high. For an output level of -10 dBm, the noise figure would need to be 3 dB.

The intermodulation performance is determined by loading the amplifier with two or more sinewave signals (of frequencies A, B, C) at a high level, e.g. $+10$ dBm. Measurements of the resultant second-order intermodulation products, at $2A$, $A + B$, $A - B$ etc., and third-order products, at $3A$, $A + B + C$ etc., give the intermodulation coefficients for the amplifier. Thus, knowing the intermodulation performance and noise figure, it is possible to select line levels to give optimum performance. The calculation is complicated because some intermodulation products in a chain of amplifiers add on a square-law basis [8]. Hence, the calculation is most conveniently done with a computer and the form of a typical print out is indicated in Table 10.1.

The optimum line levels required at the line-amplifier outputs can be set by introducing an appropriate network in the transmit terminal (the pre-emphasis network) and taking this slope out at the receiving terminal by a de-emphasis network. The computer program can also be used to test the sensitivity of the design to level errors.

Table 10.1 Calculated noise performance on a 500 km route (9·5 mm cable)

Frequency MHz	Rel. level dBm	Thermal	A+B	A−B pW0p/km	A+B−C	Total
0·4	−22	0·092	0	0·003	0·001	0·096
2·3	−22	0·177	0	0·003	0·013	0·193
5·0	−20	0·280	0	0·002	0·029	0·311
7·6	−18	0·396	0	0·001	0·088	0·485
10·4	−15	0·487	0·001	0·001	0·223	0·712
12·3	−12	0·501	0·003	0	0·154	0·658

10.3.4 Overload Performance

In addition to fulfilling the requirements for noise performance, the amplifier must also tolerate the equivalent sine-wave power P_{eq} of the peak of the composite multiplex signal [9]. In accordance with CCITT Recommendation G.223, and this is given by [10]:

$$10 \log_{10} P_{eq} = \left[-5 + 10 \log_{10} N + 10 \log\left(1 + \frac{15}{\sqrt{N}}\right) \right] \text{dBm0} \qquad (10.1)$$

where N = number of channels.
In this case, $N = 2700$; therefore, $P_{eq} = +30·4$ dBm0.

On a pre-emphasised system, this would apply at the frequency of mean power. This is typically about 8 MHz, where the line level is −17 dBm. Thus, the overload requirement becomes +13 dBm. However, several decibels of margin must be added to allow for misalignments and a better specification would be +20 dBm.

It is important to recognise that both the noise and overload requirements have to be met with any secondary lightning protection components connected. These typically comprise high-speed diodes connected across the signal transistors and can contribute to the intermodulation products.

10.3.5 Level regulation

The coaxial-cable attenuation is subject to a temperature coefficient of 0·21% per °C. Cable is buried at a depth of about 80 cm and statistics gathered over many years show a mean annual temperature variation at this depth of ±10°C. Thus, the loss of a 2 km repeater section of small-core cable will change by:

$$40 \times 0·0021 \times 10 = ±0·8 \text{ dB at } 12·5 \text{ MHz}$$

and by

$$6 \times 0·0021 \times 10 = ±0·12 \text{ dB at } 300 \text{ kHz}$$

After a tandem connection of several repeaters this could accrue to an unacceptable level. It is therefore necessary to compensate for this effect. One solution is to include a temperature sensor in the repeater, which controls an appropriately-shaped network. This assumes that the local temperature is representative of the whole cable length and that the sensing element has very stable characteristics. Another technique is to send a pilot signal over the whole length of the system, and send back a command signal which adjusts the gain of the intermediate amplifiers. The most adaptable solution, however, is to use the line pilot signal to control directly the intermediate amplifiers. Typically, only every fourth amplifier is regulated and the rest operate on fixed gain.

A block schematic of a regulated repeater is shown in Fig. 10.3. A line pilot frequency of 12 435 kHz is allocated for this purpose and transmitted at 10 dB below the traffic level. It is picked off by a crystal filter in the regulator, amplified, rectified and smoothed. The DC level is compared with a reference voltage and the difference signal is amplified to control the current into a thermistor.

The resistance of this thermistor, which changes with current, determines the loss of a network inserted in the amplifier interstage network, or in the negative-feedback path. When the thermistor resistance is nominal, this network has its nominal loss. If the resistance falls, the loss is reduced; if it rises, the loss is increased. The network configuration is such that its loss/frequency characteristic is identical to the cable shape.

10.3.6 Secondary regulation

A second pilot, at 308 kHz, is available to control networks compensating for other errors. The use of such networks is generally restricted to about every hundredth repeater. The difficulty in practice is determining the real requirement and then controlling it, when the maximum error may indeed not be at 308 kHz. For telephony applications, additional automatic-regulating pilots are commonly transmitted within each group or supergroup, as described in Section 5.4.6. These pilots control flat-gain networks, which

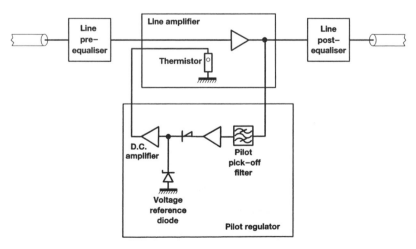

Fig. 10.3   Block schematic of pilot-regulated repeater

are reasonable to compensate the gain drift that is likely to occur across a limited bandwidth.

Another method is to readjust periodically the residual equalisers provided at the ends of a route, which notionally compensate for the systematic fixed equalisation error. Equipment is available to do this automatically by means of inter-supergroup pilots, which are transmitted for maintenance surveillance in the frequency gaps between adjacent supergroups. The solution adopted depends on the lengths of route involved and the maintenance policies of the operating administration. Up to route lengths of about 300 km on small-core or 650 km on large-core cable, modern coaxial line equipment has sufficient stability to require no maintenance attention in this respect.

10.3.7 Crosstalk

The most stringent crosstalk requirement stems from the requirements for the transmission of broadcast-programme material over coaxial systems. In this case, three or four telephony channels are replaced with a 12 kHz programme modulation equipment.

Assuming unidirectional programme circuits are operating in opposite directions over a 300 km coaxial link, an 86 dB crosstalk margin can be allocated between the two directions of transmission [11]. It is known that crosstalk can add in phase; i.e., over narrow bands voltage addition can occur. Thus, in the 300 km route, with 2 km-spaced repeaters, there are 150 crosstalk sources with voltages adding; i.e. an addition of $20 \log 150$ or 44 dB. The crosstalk requirement per repeater is then $(86 + 44)$ or 130 dB. Referring to Fig. 10.4, this means that the signal loss from the output of one amplifier to the input of the other needs to be 170 dB. Achievement of such a performance consistently in a production environment requires a design in which this need has been recognised from the outset. By careful attention to screening and by minimising common-path earth currents,

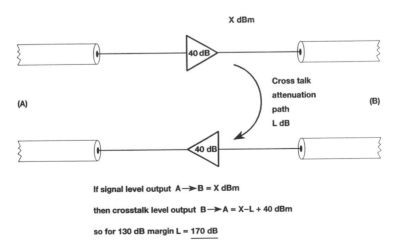

If signal level output A—►B = X dBm

then crosstalk level output B—►A = X–L + 40 dBm

so for 130 dB margin L = 170 dB

Fig. 10.4   Crosstalk attenuation in a repeater

Table 10.2   12 MHz system-Performance data

Parameter	Requirements	Measured
Equalisation against cable	0·5 dB	0·25 dB
Noise figure at 12·5 MHz	5·5 dB	3·5 dB
Intermodulation margin		
2A − B into 12·5 MHz	70·5 dB	74·7 dB
A − B into 0·3 MHz	70·0 dB	82·9 dB
A + B into 12·0 MHz	56·0 dB	62·0 dB
(Fundamentals at +10 dBm)		
White noise loading, nominal	66 dB NPR	68 dB NPR
White noise loading, +6 dB	66 dB NPR	69·7 dB NPR
Control network shape error,		
For ±4·0 dB excursion at		
12·435 MHz	±0·12 dB	±0·06 dB
Calculated system noise on		
4·4 mm cable	2 pW0p/km	1·05 pW0p/km
2-way repeater voltage drop		
at 110 mA	—	21·5 V
Crosstalk margin at 12·5 MHz	130 dB	134 dB

crosstalk attenuations of this order can be realised. The crosstalk attenuation between adjacent coaxial pairs in a cable is very high at the frequencies involved and it is not a limiting factor in system design.

10.3.8 Performance data

The data in Table 10.2, relating to a 12 MHz design [7] are indicative of the results achievable.

10.4 FDM transmission systems currently in use

10.4.1 Underground-cable systems

The principal characteristics of some of the analogue line systems commonly used in Europe and elsewhere [7, 12] are given in Table 10.3. In the USA, the most important systems are the Bell L4 and L5 systems [13, 14], operating at 18 MHz and 60 MHz respectively. Proposals for a 200 MHz system with a capacity of 30 000 channels were made. However, in view of the worldwide change to digital transmission, this system was never developed.

10.4.2 Cable television systems

Cable television (CATV) systems use very similar techniques to distribute multichannel TV to the home. The design trade-offs are very similar, although lesser emphasis is placed on reliability and maintainability. Loss of service only affects a few homes for CATV, but it is more serious in telecommunication networks. In the future there will, however, be even less distinction as CATV operators begin to carry telephony and other access services as a consequence of deregulation.

Table 10.3   FDM systems currently in use

Bandwidth	Channel capacity	Cable	Repeater spacing	Notes
4 MHz	960	9·5 mm	9·1 km	
12 MHz	2700	9·5 mm	4·55 km	Also TV transmission
18 MHz	3600	9·5 mm	4·55 km	Also TV transmission
60 MHz	10 800	9·5 mm	1·5 km	Also TV transmission
4 MHz	960	4·4 mm	4 km	
12 MHz	2700	4·4 mm	2 km	
18 MHz	3600	4·4 mm	2 km	
40 MHz	7200	4·4 mm	1 km	Not standardised by CCITT

10.4.3 Other configurations

Economic studies show that the combination of multi-core cables with 4-wire repeaters, as shown in Fig. 10.5a, is the best solution for land-line buried cables.

A different situation prevails where the cable is aerially-suspended and uses a single core for both directions of transmission. The single-amplifier configuration of Fig. 6.4b is then used. The directional filters provide separation between the two directions of transmission. Some operating administrations have extensive well-maintained pole routes. Examples are

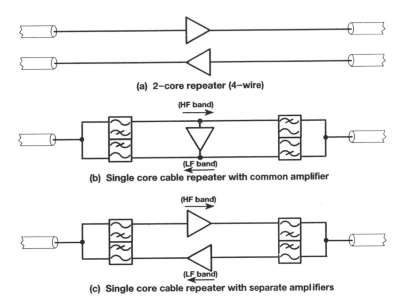

(a) 2–core repeater (4–wire)

(b) Single core cable repeater with common amplifier

(c) Single core cable repeater with separate amplifiers

Fig. 10.5   Repeater configurations

a   Two-cable-core repeater
b   Single-cable-core repeater with common amplifier
c   Single-cable-core repeater with separate amplifiers

railway authorities and the sparsely-populated rural areas in some parts of the world. It can then be economic to string a single coaxial cable on existing poles. Typical of these systems is a 120-channel system in which SG1 and 2 are transmitted in one direction and SG4 and 5 in the other [15]. The vacant SG3 provides a loss window for the directional filters and the single line amplifier is essentially that designed for a conventional 3000-channel system (SG1–SG5). As the single amplifier serves both directions of transmission it has a bandwidth of twice the traffic capacity.

10.4.4 Submarine cable systems

Submarine cable systems use a single-core cable with submerged repeaters. The cable has a solid dielectric. Systems operating up to 5 MHz use the configuration of Fig. 10.5b in their submerged repeaters [15]. However, systems with greater channel capacity use separate amplifiers for the two directions of transmission, as shown in Fig. 10.5c. Thus, the bandwidth required for the amplifiers is halved. The STC NG system [17] provides up to 4000 circuits, depending on the terminal equipment used, and the amplifiers operate up to 40 MHz.

10.5 Digital systems for copper cable

10.5.1 General

The improvement in quality of long-distance telephone calls over the past decade is largely due to the introduction of digital transmission. With digital transmission, the distortion and noise are effectively confined to the A/D and D/A conversion, and do not accumulate whatever the length of cable or number of intervening repeaters. Therefore, telecommunication networks are moving as rapidly as is practical to Integrated Digital Networks (IDN).

The frequencies involved in digital transmission are much greater than for analogue transmission. For example, a 4 kHz telephony channel is typically coded into 8 bits at an 8 kHz rate, thus generating a 64 kbit/s signal that requires at least 32 kHz for its effective transmission. Cable attenuation follows the square-root-of-frequency law and is therefore much higher than at 4 kHz. However, the digital signal is far more rugged and can tolerate a noise power within 30 dB of that of the signal. This effect approximately cancels the effect of increased attenuation. Consequently the achievable repeater spacings for digital systems are very similar to those of the equivalent analogue systems. The principle of digital transmission as explained in Chapter 7.

10.5.2 Digital systems on balanced-pair cables

Balanced-pair cables were originally installed for audio-frequency transmission between telephone exchanges. They were well maintained and so have an effectively infinite life. As the demand for telephony grew, installing new cables was avoided by achieving 'pair gain' by means of FDM systems. Since the crosstalk between cable pairs increases with frequency, the bandwidth which can be used is limited and this restricted balanced-pair FDM systems to 12 and 24 channels.

PCM systems have now made analogue systems obsolete in these short-haul applications. In addition to providing the advantages of digital transmission, they employ cheaper terminal equipment. Although digital signals have greater immunity to crosstalk, they contain much higher frequencies at which the crosstalk attenuation is low. The number of channels that may be transmitted over balanced-pair cables is thus still restricted by crosstalk considerations. 24-channel and 30-channel systems are commonly employed, operating at about 1·5 Mbit/s and 2 Mbit/s respectively. These use the primary multiplexers described in Section 6.10.

The cables had been installed with loading coils approximately every mile (1·83 km) to reduce audio-frequency attenuation. These coils are replaced by repeaters that amplify and regenerate the digital signals before they become vulnerable to errors caused by the cross-talk. These repeaters therefore must be small and power needs to be fed to them down the copper pairs. It is fortunate that the end loading sections on a route were of half length. As a result, the replacement repeaters have good immunity to the impulsive noise which is often present near telephone exchanges.

The systems are vulnerable to the near-end crosstalk paths shown in Fig. 10.4. Consequently, systems carrying traffic in one direction are separated as far as possible from systems transmitting in the opposite direction. Unless there is a separate cable for each direction, the pairs for the two directions are chosen at opposite sides of the cable, with a large number of unused pairs acting as a screen between them. New cables have been installed progressively, either as replacements or to cater for growth. These cables include a transverse metal screen to separate the two directions of transmission. Consequently, the crosstalk limitation is that due to far-end crosstalk rather than near-end crosstalk. Some of these new cables are also of lower capacitance to reduce attenuation at high frequencies. These low-capacitance cables are used in North America and Japan to enable systems to operate at 6 Mbit/s and carry 96 channels.

The first digital telephone transmission system used commercially was the Bell T1 system [18] in the USA. Experimental systems installed in 1961 proved so successful that the basic design of current junction systems in the USA is virtually unaltered. Systems of the same general type are also in widespread use in other parts of the world. The T1 system uses the 24-channel primary multiplex described in Section 6.12.3. It operates at 1·544 Mbit/s and uses the AMI (or bipolar) line code described in Section 7.5.3. The same cable is used for transmission in both directions, so the cable pairs used must be separated by at least one layer or unit in order to contain near-end crosstalk (NEXT) to an acceptable level. This results in a 'cable fill' (i.e. the proportion of pairs which provide an acceptable performance for digital transmission) of about 40%.

30-channel systems [19], which were developed later in Europe operate at 2·048 Mbit/s. These systems were usually implemented to give one ternary line digit for each input bit, using the HDB3 code described in Section 7.5.4. However, the UK had already employed a 24-channel system and it was found that the new 30-channel system was vulnerable to the increased crosstalk caused by the higher digit rate. As a result, a 'new generation' system was developed using the 4B–3T code described in Section 7.5.5.

This enables a 2 Mbit/s system to transmit ternary digits at the same rate as previous 1·5 Mbit/s systems, so that the same cable fill can be obtained [20].

This was followed in the USA by the introduction of the T1D 48-channel system, with the same repeater spacing as the earlier T1 system but with tighter restriction on the choice of cable pairs. The T2 system [21] provides 96 channels, but it requires the use of low-capacitance cable pairs.

The CCITT has issued recommendations for both the 24-channel primary multiplex group (using μ-law companding) and the 30-channel primary multiplex group (using A-law companding). There is now a family of systems based on each and some examples are shown in Table 10.4. Low-capacity systems use the primary group directly. High-capacity systems use higher-order multiplexing to build up assemblies of these groups, as described in Chapters 8 and 9.

It is now possible to extend digital working right up to the premises of individual subscribers over their copper access pairs from local exchanges. Transmission is at 144 kbit/s, using the techniques described in Chapter 15. These techniques provide the 'basic access' needed for an Integrated Services Digital Network (ISDN).

10.5.3 Digital systems for coaxial cable

Digital line systems operating at 140 Mbit/s were designed for operation on either 4·4 mm or 9·5 mm coaxial cable, at the same repeater (regenerator) spacing as 12 MHz analogue systems. Thus, they operated on spare tubes in existing cables alongside existing analogue plant, or replaced 12 MHz equipment to provide high-speed digital links with a capacity of 1920 voice

Table 10.4   Example digital landline systems on copper

Bit rate	Channel capacity	Line code	Cable	Repeater spacing	Example country
1·5/2 Mbit/s	24/30	AMI/HDB3	Audio quads	1·8 km	US(T1) Europe
3 Mbit/s	48	AMI/PR	Audio quads*	1·8 km	US(T1C) /US(T1D)
6 Mbit/s	96	B6ZS	Low cap. pairs	1·8 km	US(T2)
8 Mbit/s	120	HDB3	0·7/2·4 mm coax.	4 km	Italy
34 Mbit/s	480	MS43	0·7/2·4 mm coax.	2 km	Italy
140 Mbit/s	1920	6B-4T	1·2/44 mm coax.	2 km	UK
140 Mbit/s	1920	MS43	2·6/9·5 mm coax.	4·5 km	FRG

* In the AMI case, separate cables are needed for the two directions of transmission.

channels [22]. The equipment is physically compatible with the analogue system and operates with the same power-feeding system.

The design considerations are of course very different, as explained in Chapter 7. The output stage of the repeater combines separately-generated positive-going and negative-going pulses. The pulses are of 6 V amplitude and are nomially restangular in shape. They have a duration of half width, which in a 6B-4T line code means 5·4 ns. (The 6B-4T line code uses 4 ternary digits to carry 6 message bits, as explained in Section 7.5.6. Thus, the transmitted digit rate is only two thirds of the traffic rate, i.e. 140 Mbit/s are conveyed at 93·3 Mbauds.)

An overall block schematic of the regenerator is given in Fig. 10.6. The input stage is an analogue amplifier providing a maximum gain of 80 dB at 70 MHz. The overall transfer characteristic from line output to the regenerator decision point is determined by the loss of the cable, the input equaliser and the gain of the amplifier. It is shaped to be 5 dB down at 45 MHz, which leads to the pulse spreading to full width. This shaping is an optimum trade off between amplifier noise at the decision point and inter-symbol interference due to excessive pulse spreading. Compensation for cable-loss change with temperature is achieved by an automatic gain control (AGC) circuit activated by peak power at the decision point. The AGC control provides an adjustment proportional to the square root of frequency in accordance with the cable shape. The regenerator shown in Fig. 10.6 includes a post-equaliser, whereas these are not used in regenerators for twisted-pair cables. On coaxial systems, a high-pass post-equaliser helps in overall design refinement without affecting the thermal-noise performance.

Although the 140 Mbit/s regenerator incorporates familiar analogue elements, the amplifiers and equalisers can be far less accurate than in the

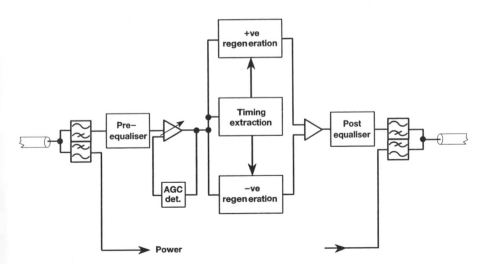

Fig. 10.6   Block schematic of 140 Mbit/s regenerator

analogue system. This is because regeneration occurs at each repeater and impairments do not accumulate.

One drawback with the 140 Mbit/s system described above was that it had less voice-traffic capacity than the 2700-channel FDM system it replaced. This led to an interest in 565 Mbit/s systems offering 7680 voice channels. In some countries, notably Germany, this was developed to span the 60 MHz repeater spacing of 1·5 km on large-core cables. Work in the UK was directed to a longer span, so that only one intercept would be necessary on existing 12 MHz routes, i.e. 2·25 km on large-core and 1 km on small-core systems. Development models of such a design were produced, but interest in the system waned with the advent of optical-fibre technology.

10.6 Remote repeaters

10.6.1 Accommodation

We have seen that coaxial cable repeaters for FDM or digital working are spaced a few kilometres apart, perhaps only 2 km. Similarly, 2 Mbit/s PCM regenerators on pair cable are spaced at around this distance. In very few instances are these intermediate repeaters housed in surface buildings and it is generally advantageous to provide underground accommodation. British Telecom practice is a good example of the type of solution generally adopted. A resin-impregnated cast-iron case approximately 600 mm long, 400 mm across and 400 mm high houses the equipment. The lid is removable for installation and maintenance and is provided with a gasket to ensure an airtight seal between the machined surfaces. Moisture ingress is prevented by pressurising the case with dry air from a foot pump to about two thirds of an atmosphere. A pressure-sensitive switch in the case applies an earth to one of the interstice wires in the cable to signal to control when pressure is lost. Thus, the case is usually protected by static pressure alone and not coupled into the cable-pressurisation system.

Cases of this type can withstand continuous immersion in water several metres deep and can therefore be installed directly in deep cable chambers. However, there is operational merit in installing the repeater case in its own footway box adjacent to the cable manhole with tail cables extending from a joint in the cable chamber. Footway boxes are simple brick-built or precast enclosures, typically less than 1 m deep and fitted with removable lids. The tail cables are terminated in an air-tight gland at the cable entry hole to the case. This termination can also incoporate the connectors to which the repeater equipment is connected.

10.6.2 Power supplies

The power consumption of wideband analogue repeaters and high-bit-rate digital repeaters is in the range of 2W to 4W and power is fed along the inner of the coaxial cables. Separation of the DC and HF signals is achieved in a power-separating filter as shown in Fig. 10.7a. It will be noted that the voltage conversion and stabilisation is confined to the terminal station or power-feeding point. The prime purpose of the zener diode shown in Fig.

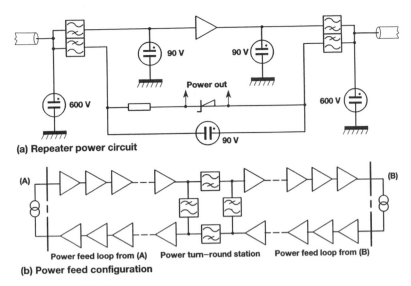

(a) Repeater power circuit

(b) Power feed configuration

Fig. 10.7 Power feeding arrangements

10.7*a* is not stabilisation, but to divert power surges. The commonly-adopted solution is to pass a constant DC along the inner of each coaxial pair so that the equipments are powered in series. The connection can be end to end, or in a loop as shown in Fig. 10.7*b*.

Considerable operational cost benefit is obtained if the system can be inherently safe, i.e. if a cable gang can work safely on multicore cables without having to shut down every system. A coaxial tube tends to be inherently protected, as an earthed sheath surrounds the live conductor. Consequently, higher voltages are permitted than on balanced-pair cables. In the UK, the current is limited to 50 mA maximum and the voltage to earth must not exceed 250 V; i.e. the power-feeding supply is limited to 500 V and 50 mA. The system can then be considered inherently safe. Taking into account the resistance of the inner conductor, the number

Table 10.5 Power feed station spacings

System	Current	Voltage	Power feed spacing
12 MHz/9·5 mm	50 mA	250—0—250	67 km
12 MHz/9·5 mm	110 mA	325—0—325	245 km
12 MHz/4·4 mm	50 mA	250—0—250	30 km
12 MHz/4·4 mm	110 mA	325—0—325	95 km
60 MHz/9·5 mm	300 mA	600—0—600	130 km

The 12 MHz data also apply to the 140 Mbit/s (digital system).

of repeaters that can be powered can be calculated. The corresponding spacings between power-feeding points are given in Table 10.5.

In other countries, personnel safety is assured by adopting safe working practices [14], e.g. by not working on live cables, by the use of key-locked switches, by providing automatic switch-off on open-circuit, by earth-current trips etc. [7]. Larger currents and voltages can then be used to give increased power-feeding spans as shown in Table 10.5. For the 60 MHz FDM systems in the UK, a power-feeding scheme using guard tones and handshake protocols permits the use of the high current and voltages which this system demands. Limiting factors are the voltage rating of the coaxial cable, cable connectors, the power-separating filter components and the surge arrestors discussed below. The problem with the latter is that high power feeding can maintain an arc in a gas discharge tube after it has been struck by a power surge.

The biggest power feeding schemes of this kind will be found on transatlantic submarine cables where several kilovolts are applied to each end of the cable [17].

The same constant-current feeding technique can be adapted to pair-cable systems like the 2 Mbit/s PCM system. However, in this case the DC loop is provided by the phantom circuit of the 'go' and 'return' copper pairs. Optical-fibre systems present a different challenge as the cable can be metal-free as discussed in Section 10.7.1.

10.6.3 Surge protection

Metallic cables are subject to induced currents arising from lightning, electric traction and high-voltage transmission lines. Precautions must therefore be taken to protect the telephone cable and equipment from damage. Obviously, the magnitude, duration and repetition rate of induced surges are very variable and the subject has been studied very fully by the CCITT, drawing on the experience of many countries. The two main documents are the recommendations for the installation of telephone plant which cover cable [23] and Recommendation K.17 which gives the test surge conditions which equipment should survive [24].

Immunity to surges normally involves at least three levels of protection. Firstly, a gas-discharge tube is connected between coaxial inner and outer. Secondly, a network, usually part of the cable equaliser, is connected to either side of the amplifier/regenerator. Thirdly, fast-acting diodes are connected around vulnerable transistors. Longitudinal protection is provided by zener diodes, backed up with gas-discharge tubes and shunted by a bypass capacitor. An example of repeater protection is given in Fig. 10.7a. It will be noted that the transmission path is interrupted while the protective devices are operating.

Most of these protection components are directly in the signal path and their influence must be considered at all stages of the design. Relevant component surge ratings are seldom catalogued and considerable test time is involved in establishing base data and alternative sources of component supply.

10.6.4 Supervisory systems

It is a basic need for maintenance staff to be able to locate rapidly and unambiguously a faulty repeater, so that it can be replaced and service restored. If the equipment can also signal partial failure or give early warning of potential failure, so much the better.

A variety of solutions have been adopted on analogue coaxial systems. The early designs used the interstice pairs laid up between the coaxial tubes to signal status information by DC conditions. However, such systems are limited in length and information detail. A more versatile system was designed when 12 MHz was first applied to 4·4 mm cable. The line-amplifier gain shaping is extended to 13·5 MHz and a 13·5 MHz oscillator is provided at each repeater. An interrogation signal is transmitted over the incoming coaxial pair to the first repeater out. This pulse is delayed for 250 μs and then serves to connect the 13·5 MHz oscillator to line for 100 μs, whilst at the same time regenerating the interrogation signal and passing it on to the next repeater [7]. Thus, the interrogating terminal receives a sequence of bursts of 13·5 MHz signal, each one originating at a different repeater. The interrogation sequence is repeated automatically and either the absence of, or a level error in, a pulse gives an alarm. The system can be readily extended to include miscellaneous alarms from intermediate power feeding stations. It can also be used for detecting partial failures and short breaks in transmission. The route information is displayed by direct readout and on a chart recorder.

Another example is the system adopted on several 60 MHz and 18 MHz systems. Here an individual frequency is allocated to each repeater and the oscillator runs continuously. A tunable receiver at the terminal is used to examine each repeater in turn. Again, the terminal equipment can be either manually or automatically controlled to various levels of complexity [12].

For digital systems, there has been considerable debate on whether fault location should be accomplished in or out of traffic. 'Out of traffic' techniques, using progressive loop backs between send and receive paths, are simpler to implement but less useful to maintenance staff. Most administrations now demand methods which can be used in the presence of traffic.

The Bell T1 system uses the AMI code, which has an error-monitoring capability, as described in Section 7.5.3. However, error monitoring is not carried out within the line transmission system. It is carried out during demultiplexing, by detection of errors in the frame-alignment word. Repeater faults are detected out of traffic. A special sequence of digits ('trios') containing an audio-frequency component is transmitted along the chain of repeaters and the audio component is detected at the terminal station. All repeaters in a housing share a service monitor, consisting of a bandpass filter of unique frequency connected to a supervisory pair. The faulty repeater is detected by transmitting different audio frequencies and monitoring the signals received from the supervisory pair.

Cable faults are detected by measuring the line voltage, both with normal and reversed polarity. The voltage reversal connects a resistor between the pairs at each repeater site. Finally, a speaker circuit, accessible at each repeater site, is provided over a spare cable pair.

The main features of the system developed in the UK are as follows. The line code has redundancy which allows the insertion of parity bits i.e. the transmitted signal has no DC content. A monitor at each repeater measures the mean DC level of the signal, which will no longer be zero if errors are present. The supervisory system is arranged to interrogate each repeater monitor in turn and the error rate is determined. The repeater supervisory unit also stores other information, for example loss of input signal, and many additional conditions can be signalled from intermediate power feeding points. The coaxial tube is used as the telemetry bearer and the information is transmitted in a formalised coding structure. This data stream can be recognised by the terminal supervisory receiver and analysed in a software-controlled microcomputer. Thus, status information can be presented to maintenance staff in a logical summary form by alpha-numeric display and by paper printout.

This concept of the automatic diagnosis of line fault conditions is a powerful one. There is a tendency to establish maintenance control centres able to monitor route performance remotely and dispatch repair teams directly to where they are needed. It is also expected that supervisory systems will become increasingly employed to collect and analyse miscellaneous data for onward transmission to remote maintenance centres, and to be part of the telemetry circuit linking maintenance centres.

Power-feeding faults are one class of fault for which these telemetry systems do not cater. A characteristic of the power-feeding loop shown in Fig. 10.7b is that a break anywhere cuts power to all stations. This might be within a repeater, a power-separating filter say, or along the length of the coaxial cable. The technique almost universally used is to have normally-reversed-biased diodes connected between the inner conductors at each repeater. If the power-feeding supply is now reversed in polarity, the longitudinal and transverse diodes are all forward biased and the loop resistance indicates how far away is the break [7]. A cable break can then be located from the indicated repeater by pulse-echo methods.

These techniques can be adapted to permit accurate location over very-long power-fed sections. However, they are not fully compatible with the concept of remote maintenance control centres.

10.7 Optical-fibre systems

10.7.1 General

The major change in new line systems over the past decade has been the fibre-optic revolution. Attempts at microwave waveguide systems were overtaken by dielectric waveguides for higher-frequency (and hence higher-bandwidth) signals, i.e. optical signals. These optical fibres were rapidly developed for lower attenuations than metallic media to achieve greater repeater spacings and hence lower system 'cost of ownership'. Furthermore, they were lighter, more flexible, immune from EMI and, eventually, cheaper than copper. Second-generation systems, using single-mode fibres, overcame the bandwidth limitations from the waveguide dispersion of multimode fibres. Regarding cable fault location, pulse reflectometers were developed that could pinpoint a fibre break to within 0·1% of a repeater section.

Optical-fibre digital line systems have been developed at all the bit rates available on copper systems. Also, 565 Mbit/s and even higher rate systems have been introduced for commercial service. These systems are characterised by very-long repeater sections and the technology which has made this possible is described in Chapter 11.

The impact of optical analogue systems has not been so dramatic. They are finding a role in cable TV systems as multichannel bearers; however, conventional VHF multi-channel TV on coaxial cable is currently more economic overall. Current optical transmitters and detectors tend to be non-linear devices more appropriate to digital transmission.

10.7.2 Intermediate repeaters

One of the outstanding features of optical fibre systems is the long distance between repeaters and the consequent economic advantage in both first cost and operational maintenance. Thus, the majority of junction systems can be realised without an intermediate repeater. The first generation of 140 Mbit/s optical trunk systems [25] operated on multimode fibre in the 850 nm window. Repeater spacings were of the order of 9 km and two bothway repeaters were accommodated in the repeater case described in Section 10.6.1. Optical repeaters require more power rails than conventional repeaters and these are supplied from a power convertor at each repeater. The total power consumption is then in the 10 W to 20 W range. It has been found convenient to provide this via copper pairs (0·9 mm) either incorporated into the optical cable or via a separate cable using a standard 50 mA constant-current scheme. Other copper pairs can be used to transmit telemetry information from the remote repeaters.

With the introduction of single-mode optical technology these methods of power feeding are not so attractive [26]. The repeater spacing of 140 Mbit/s trunk systems is already in the range of 30—45 km. Thus, either large conductors or a number of small conductors are needed to overcome the voltage drop. The transmission quality for pulsed supervisory signals on copper pairs over such distances is poor and unpredictable. In some applications, for example in regions of high lighting activity and along electrified railways, a metal-free cable is preferred.

In a densely-populated country like the UK, reliable power and accommodation is normally available in existing exchanges and repeater stations within the 30 km range of search. In less-populated countries, underground accommodation is commonly adopted, powered by a spur cable from the nearest telecommunication building. Alternatively a local power source can be provided in the form of solar cells, wind power, thermoelectric power etc. All these are able to provide the 100 W or so required for a multi-repeater installation more economically and reliably than a small diesel installation.

The particular solution adopted and the size of the associated secondary battery will depend on the geographic location, accessibility and maintenance philosophy.

10.7.3 System supervisory

The remaining problem of the remote optical-fibre repeater is the transmission of supervisory information back to the control terminal. This is even more essential if the site is powered from a small primary generator with only a few days reserve of secondary-battery capacity. Options which have been used include:

(a)   A separate fibre carrying a low-bit rate system (e.g. 2 Mbit/s) allocated for telemetry
(b)   Using redundancy in the line code to transmit telemetry information
(c)   By an additive low-frequency modulation below the spectrum of the traffic signal, or by modifying the traffic signal in some way.

The last method is attractive if it can be implemented within an acceptable reduction of the traffic-path capability, and if it can provide sufficient telemetry capability.

10.7.4 Submarine optical fibre systems

Optical-fibre technology has had a particularly strong impact on the design of submarine transmission systems and it is interesting to compare these with equivalent land-line systems.

The world's first international underwater optical fibre system was installed in 1986 between England and Belgium [27], followed by the world's first transoceanic optic fibre system (TAT–8) in 1988 between Europe and the USA. Key data are given in Table 10.6.

An essential difference between land-line and submarine system design is the attention given to reliability. Whilst some improvement can be achieved by optimising system configuration, the real advance has to be at component level in order to achieve the system design life expectancy of 25 years. [26].

The extensive prequalification and reliability programme on all major components has followed the general philosophy of identifying major failure modes, redesigning active elements and package configuration where

Table 10.6   Example submarine optical fibre systems

	UK—Belgium	TAT–8
System Length	113 km	9360 km
Capacity per fibre pair	3840 ch	3840 ch
Line rate	324 Mbit/s	296 Mbit/s
Operating wavelength	1300 nm	1330 nm
Number of repeaters	3	~140
Line current	1·55 A	1·6 A

necessary [28]. Repeated accelerated life testing and the development of suitable screening limits, to isolate infant mortality occurrences with respect to the predominant failure distributions, has followed. For example, the PIN–FET optical-receiver modules are tested up to 200°C. The maximum activation energy of the special integrated-circuit uncommitted transistor arrays has been determined by many thousands of hours testing at 145°C. The design and manufacturing procedures for the laser diode were directed to ensure 25 years operational life. By the end of 1988, this claim was supported by 10 million life-test hours. The recognised end-of-life criterion is when threshold current has increased by 50% from its starting value. Tests on a batch of these high-reliability lasers showed that half would change less than 1·6% in 25 years and 98% less than 50%.

The underwater optical fibres are encased in a protective copper tube which forms the hydrostatic pressure barrier and a physical barrier to the ingress of hydrogen. It also provides the metallic path for power feeding repeaters.

A comprehensive supervisory and telemetry facility provides for measurement of error ratio, received light level and laser bias current. Supervisory access to each regenerator in the repeater case is possible from any fibre. The telemetry path is obtained by modulating a carrier derived from parity violations to the even-mark-parity line code (24B–1P, i.e. one parity bit for every 24 traffic bits).

10.7.5 Penetration into the access network

Access from a subscriber's site to the local exchange is normally by means of a copper pair, typically 1 km in length. At first sight there would seem to be no need to introduce electro-optics and fibre cables into this area. However, there is growth in traffic, often in a startling and unpredictable manner when business sites are concerned. There are problems with ageing copper cables and there are needs for wideband services, e.g. CATV and eventually interactive visual services. The growth in traffic can be met economically by subscriber carrier systems that use multiplexing from the local exchange for most of the way to the subscriber premises, with an individual copper pair then only needed for the last few metres of distribution. This has been the practice for the past decade in North America, initially using copper systems, and later fibre-optic systems. The UK is now leading Europe by the introduction of similar access systems based on fibre-optic systems right up to major sites in business districts, as described in Chapter 15. Such systems are characterised by modest bit rates, in the region 2—140 Mbit/s, with sophisticated management systems.

The reduction of the cost of fibre-optic technology over the past decade has been dramatic, and the cost/performance of the terminating electro-optics will continue to improve over the next decade. This will allow the concepts to be applied to progressively-smaller business sites, and eventually to domestic subscribers. The first step is likely to be the provision of remote multiplexers on the premises of smaller businesses, or housed in a street cabinet in their neighbourhood. These multiplexers would be connected to the main network via fibre optics, and to the terminations via copper pairs.

10.7.6 Planning for growth

A fascinating aspect of telephone cable history is the way which it has proved possible to exploit existing cables for higher frequencies than prevailed at the time they were installed. This obviously is to the credit and far sightedness of the original designers. It is also evident that the original cable system must have been economically viable at say 600 circuit capacity even if it has been subsequently upgraded to 2700 circuits. Similarly, junction cables originally installed for audio service can be unloaded and equipped with 30-channel PCM systems. Today, the same questions are being asked of optical-fibre cables. What options will a user have to increase circuit capacity in the years ahead and in what way should the optical fibre be specified now to ensure that this will indeed be possible?

For multimode fibres the way ahead is not clear because of bandwidth limitations. The bandwidth limitation of single-mode fibres is far less, and is extendable by improvements to the characteristics of the source. This explains the early and rapid adoption of single-mode fibre cable as a 'future proof' investment, even though the elements of single-mode systems are only recently cost competitive with their multimode equivalents. Today, single-mode technology is the usual solution for all junction and trunk installations [26].

'Upgrading' can be by bit rate. The new 565 Mbit/s systems have the same repeater spacing as earlier 140 Mbit/s systems (45 km). Upgrading can also be by increasing repeater spacing. Using narrow-spectrum lasers at 1550 nm allows 90 km spacing on existing fibres. Recently, both in the USA and the UK, plans have been announced to introduce Wavelength-Division Multiplexing (WDM) onto single-mode fibre systems on a commercial basis [29]. This permits either two-way working on a single fibre, or two systems to be independently transmitted over the same fibre.

The ease of transmitting wide bandwidths over optical fibre also encourages the use of multiple terminations on a single fibre. With the low attenuations achieved with fibre, such bus and ring structures can easily extend over considerable distances. Thus, the technologies of local-area networks (LANs) can be extended to wide-area networks (WANs). For the telecommunications access networks, the natural layout of roads and services is a 'tree and branch' structure. Fibre can follow this pattern because of the ease with which branching couplers can be spliced in. This will lead to a Passive Optical Network (PON) arrangement using Time-Division Multiple access (TDMA) to achieve communication with all the terminations sharing the one fibre [29].

10.8 Conclusions

The evolution of line-system technology continues at a dramatic pace. The inventions of the thermionic valve, transistor and integrated circuit each caused orders of magnitude falls in the per-kilometre cost of a circuit. The invention of pulse-code modulation assisted this trend. It also realised dramatic improvements in speech-circuit quality and allowed the introduction of an integrated digital network.

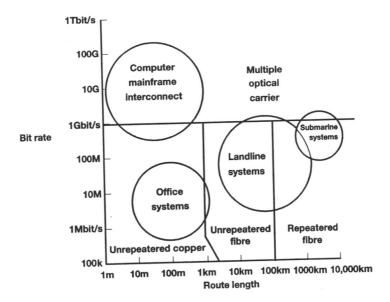

Fig. 10.8 Line system trends

The invention of fibre optics has caused further falls in costs. We can now think in terms of communications that are effectively distance-independent, and even hold out the prospect of being effectively bandwidth independent. Consequently, the importance of dependent repeaters, in terms of reliability, surge immunity and low power consumption, is becoming less.

All new long-haul line systems are now fibre-optic. Also, as costs decrease, this technology is penetrating closer to the subscriber's premises. Nevertheless the most cost-effective way to provide telephony and CaTV services into the home is still to use analogue transmission, over balanced-pair and coaxial cable, respectively.

Therefore, a wide variety of system designs will continue to exist on the network into the next century, as shown in Fig. 10.8.

10.9 References

1 Bell Laboratories Staff: 'Transmission systems for communications' (Bell Telephone Laboratories, 1971) 4th edn
2 CCITT Recommendation G622: 'Characteristics of 1·2/4·4 mm coaxial cable pairs'
3 CCITT Recommendation G623: 'Characteristics of 2·6/9·5 mm coaxial cable pairs'
4 CCITT Recommendation G332: '12 MHz systems on standardised 2·6/9·5 mm coaxial cable pairs'
5 CCITT Recommendation G345: '12 MHz systems on standardised 1·2/4·4 mm coaxial cable pairs'

6 CCITT Recommendation G222: 'Noise objectives for design of carrier-transmission systems of 2500 km'

7 HOWARD, P.J., ALARCON, M.F., and TRONSLI, S.: '12 MHz line equipment', *Elect. Comm.*, 1973, **48**, pp. 27–37

8 BENNETT, W.R.: 'Cross modulation requirements on multichannel amplifiers', *Bell Syst. Tech. J.*, 1940, **19**, pp. 587–610

9 HOLBROOK, B.D., and DIXON, J.T.: 'Load rating theory for multi-channel amplifiers', *ibid.*, 1939, **18**, pp. 624–644

10 CCITT Recommendation G223: 'Assumptions for calculation of noise on hypothetical reference circuits for telephony'

11 CCITT Recommendation J22: 'Performance characteristics of 10 kHz type sound-programme circuits'

12 BECKER, L.: '60 MHz line equipment', Elect. Comm., 1973, **48**, pp. 38–46

13 KLIE, R.H., and MOSHER, R.E.: 'The L–4 coaxial cable system', *Bell Lab. Record.*, 1967, **45**, pp. 211–217

14 Various authors: 'L5 coaxial-carrier transmission system', *Bell Syst. Tech. J.*, 1974, **53**, (10)

15 CRANSWICK, H.B., and JOLLY, C.W.: 'Coaxial cable system for developing countries', *Elect. Comm.*, 1969, **44**, pp. 24–32

16 BENNETT, A.J., and HEATH, G.A.: '1840-channel submerged-repeater system', *Ibid.*, 1971, **46**, pp. 139–156

17 Various authors: 'The NG–1 submarine cable system', *ibid.*, 1978, **53**, (2)

18 FULTZ, K.E., and PENICK, D.B.: 'The T1 carrier system', *Bell Syst. Tech. J.*, 1965, **44**, pp. 1405–1451

19 WHETTER, J., and RICHMAN, N.J.: '30-channel pulse-code modulation system. Part 2: 2·048 Mbit/s digital line system', *PO Elec-Engrs. J.*, 1978, **71**, pp. 82–89.

20 CATCHPOLE, R.J., DYKE, P.J., and USHER, E.S.: Planning and implementation of PCM systems on symmetric pair cables', *Elect. Comm.*, 1982, **57**, pp. 180–186

21 DAVIS, J.H.: 'T2: A 6·3 Mb/s digital repeatered line', IEEE Int. Conf. on Coms. Record, 1969, pp. 34·9–34·16

22 REEVES, H.S.V., HEMSWORTH, A.D., and SHEPPARD, D.: 'Design and field trial of a 140 Mbit/s coaxial line system'. IEE Transmission Conference, 1981, IEE Conf. Pub. 193, pp. 154–158

23 CCITT V series: 'Recommendations' for the installation of telephone plant'

24 CCITT Recommendation K17: 'Tests on power-fed repeaters using solid-state devices in order to check the arrangements for protection from external interference'

25 THOMAS, D.S.: '140 Mbit/s multimode optical system', Elect. Comm., 1982, **57**, pp. 201–207

26 RADLEY, P.E. *et al.*: 'Single-mode transmission systems, present and future'. Proc. 4th World Telecom. Forum, Geneva, 1983

27 BURVENICH, J.P. *et al.*: 'The first European optical highway, UK—Belgium no. 5 system', Proc. 'Subtopic '86', Versailles

28 MURPHY, R.H., *et al.*: 'Components: the key to optical submarine systems reliability', *ibid*

29 BAKER, N., GODDARD, I.J., and DYKE, P.J.: 'Wavelength multiplexing techniques applied to the subscriber access area'. Proc. 2nd IEE Nat. Conf. on Telecoms., York, 1989, IEE Conf. Pub. 300

Chapter 11

Optical fibre transmission systems

L. Bickers

BT Laboratories

11.1 Introduction

It is now some 30 years since the basic components for optical transmission systems began to emerge from the research laboratories. Early optical fibre had an attenuation >1000 dB/km, but by 1970 this had been reduced below 20 dB enabling the first system demonstrations by Bell Labs (at 45 Mbit/s) and British Telecom (at 140 Mbit/s) on installed cables. The success of these trials led to the first production systems which operated at a wavelength of 850 nm over multimode fibre. By 1982 the lower-loss windows at 1300 nm and 1500 nm were being exploited both in the laboratory and on field trials. Subsequently, singlemode fibre has been installed throughout most modern telecommunication networks and is now being considered for application in the local loop to serve domestic subscribers.

In this chapter we first consider each system element (transmitter, receiver and channel) of an optical transmission system, before we concentrate on system design, showing how the system elements interact to achieve a given design requirement.

11.2 Optical transmitter options and design

Semiconductor diodes are generally the only sources used in optical communication systems because of their small size, ease of modulation, relatively high electron-to-photon conversion efficiencies [1] and reliability. There are only two broad categories of device: the laser diode (LD), and light-emitting diode (LED) but within these categories are hundreds of different variants in terms of their structure, material and operating parameters. Currently-available devices operate in the range 650–1550 nm, with the key fibre operating windows being centred on 850, 1300 and 1550 nm.

11.2.1 Device choice

A laser diode is characterised by threshold current for stimulated emission. Above this the carrier life time is very short, which allows high modulation rates, the spectral emission narrows to a few nm and the beam is highly directional. The LED generally has a broad spectrum (10s of nm), a broad-angle beam (>10°), and a modulation capability limited by the spontaneous carrier life time of about 1 ns under optimum conditions. Although LDs and LEDs can emit several mW of total power, their applications differ. The LED is generally only useful for limited bandwidth and short to moderate transmission distances, whereas the LD is useful for more-demanding, high-data-rate and long-distance applications. The LD employs higher current densities, is highly temperature dependent, and requires more complex drive hardware to accommodate the lasing threshold. It is therefore approximately four times as expensive as LED variants [1, 2].

There are currently two types of LED available for optical fibre applications: edge-emitting diodes (ELEDs) and surface-emitting diodes (SLEDs). The latter have low coupled power, typically 1.5 μW (-28 dBm) into single mode-fibre and 15 μW (-18 dBm) into multimode fibre. The former can be split into two sub groups: the low–current devices and the super-luminescent diodes (SLEDs). The high-power devices are currently capable of coupling 150 μW (-8 dBm) into single–mode fibres. They generally require drive currents in excess of 100 mA and some form of temperature control to stabilise their power output. As a result of this large drive current, these devices tend to have a shorter life time than would normally be expected for SLEDs. The lower-drive-current ELEDs typically couple 6 μW (-22 dBm) into single–mode fibre. Both low-current ELEDs and SLEDs can operate over the temperature range -40 to $+85C$ without cooling or power output stabilisation thus making the associated electronics both simple and inexpensive. Table 11.1 gives a brief listing of LED parameters for currently available devices employed in modern systems [3].

Two broad categories of LD are currently available, the Fabry–Perot and distributed feedback (DFB) device. The latter uses an enhanced structure of the former with transversal feedback to increase the Q of the cavity and hence reduce the number of possible oscillation modes (ideally, to just one). In contrast the Fabry–Perot laser can oscillate in many different modes, which can give rise to transmission performance limitations related to mode hopping under modulation. The advantage of the DFB is that it can be used at a wavelength of non-zero fibre dispersion due to its narrow linewidth ($\sim$0.2 nm) compared with its Fabry–Perot counterpart with a wide linewidth ($\sim$4 nm) [4].

Modern LDs can deliver 0.1–1.0 mW into a single or multimode fibre at modulation rates up to 10 Gbit/s. Table 11.2 gives a list of the various key parameters of LD devices currently available and employed in systems.

Table 11.1   Typical LED performance

Wavelength nm	Output power μw/mA	Spectral width nm	Fibre type
800—900	0·6—1·3	50	MM
1280—1300	0·1—0·15	80—140	MM
1280—1300	0·04—0·06	60—80	SM

Table 11.2 Typical laser performance

Wavelength nm	Output power mw/mA	Spectral width nm	Fibre type
820—850	0·06—0·15	1·5—3	MM
1280—1350	0·04—0·3	1·5—5	MM/SM
1500—1550	0·04—0·3	0·01—0·3	SM

11.2.2 Optical safety

Operational safety is presently focused on the possibility of human skin and eye damage by emission from any optical device. For communication systems, eye damage is the only cause of concern, as power levels are currently, and likely to continue to be, far below that necessary for skin tissue damage. Lasers are the only devices capable of producing a sufficient power density to transverse the BSI safety limits set at −6 dBm at 850 nm, −2.1 dBm at 1300 nm and currently undefined at 1550 nm [5].

Most system designers achieve safe operation by using launch powers at least 1 dB lower than the safety requirements, i.e. −3 dBm at 1300 nm. Despite the near impossibility of actually being able to induce eye damage at the stated safety levels, some operators demand automatic laser shut down under fibre break or system fault conditions. This introduces a considerable amount of additional hardware and software complexity for the detection and control process, and in truth it is virtually impossible to achieve 100% safe working. Fortunately, for system designers, most operators only require a warning label!

11.2.3 LED drive circuit

Transmission system designers are principally interested in five key LED parameters:

(i) *Optical output Power*: An ideal LED would exhibit a linear relationship between optical power and drive current. In practice, the devices are linear for lower powers, but tend to limit in a non-linear fashion at high power levels (Fig. 11.1). LEDs can thus be used in both an analogue and digital modulation mode [4].

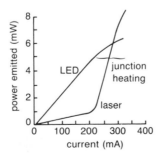

Fig. 11.1 LED and LD characteristics

(ii)   *Bandwidth*: This is determined by the doping level of the active layer, carrier life time and parasitic capacitances. For a small parasitic capacitance the power-frequency characteristic is given by [2]:

$$P(f) = \frac{P(0)}{\sqrt{1 + (2\pi f \tau)^2}}$$ (11.1)

and

$$\text{bandwidth } \Delta(f) = \frac{1}{2\pi\tau}$$ (11.2)

The value of τ varies with drive current; it can be >3 ns at 10 mA but reduces to >1 ns at 100 mA. Modulation bandwidths up to 1 Gbit/s have been recorded, but >200 Mbit/s are more typical.

(iii)   *Spectral width*: The spectral width of an LED is broad (Fig. 11.2): ~25—40 nm at 850 nm, ~50—100 nm at 1300—1550 nm. This gives rise to a significant amount of dispersion on both single-mode and multimode fibre. For a 50 nm width the bandwidth–length product for multimode is typically ~6 MHz/km, and for single mode ~500 MHz/km [4].

(iv)   *Temperature effects*: The optical output power for an LED varies exponentially with temperature and can be approximated by:

$$P_0(T) = P_0(T_0) \exp\{-T/T_0\}$$ (11.3)

where T is the operating temperature and T_0 is a device-related parameter.

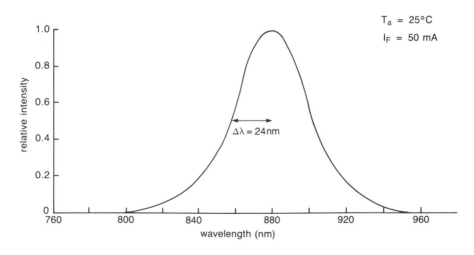

Fig. 11.2   Typical LED spectrum

T_0 varies from 80 to 145 K for 1.3 μm LEDs of different manufacture and reduces for longer wavelength, indicating a greater temperature sensitivity. Consequently, as temperature increases the light output falls and in order to utilise high power SLEDs and longer-wavelength devices at necessarily elevated temperatures, some form of cooling is required.

The peak wavelength of the spectral density of LEDs is determined by the band gap, but as temperature increases the band gap is reduced and the peak wavelength shifts towards longer wavelengths. The shift is small compared with the spectral width and is approximately 0.35—0.6 nm/C [1].

(*v*) *Reliability*: For commerically available devices that have survived the manufacturers' initial burn in, the optical power output degradation with time has been shown to follow the approximate relationship:

$$P_0(t) = P_0(0) \exp\{-\beta t\} \tag{11.4}$$

where β is the decay constant, which is dependent upon the device type. The half-power point is usually taken as the end of life condition,

$$\text{i.e.} \quad P_0(\tau) = \frac{P_0(0)}{2} \tag{11.5}$$

Values of $\tau \simeq 10^5$ hours (>10 years) are generally required by most designers as an engineering minimum [6].

The drive circuit arrangements for LEDs are essentially straightforward. The active device is normally fed from a voltage or current source depending upon the speed of operation (see Fig. 11.3). No attempt is normally made to monitor and control the optical power launched. The total device manufacturing tolerance, temperature and ageing uncertainty of drive-circuit current, LED electro-optic conversion and power coupled into the fibre thus have to be accounted for in the system design equation. The total launch power variation of ±2 dB is not insignificant in its impact on the receiver design. It has to accommodate this, as well as a further 3 dB ageing allowance, by a wide AGC and dynamic range over the system life [4].

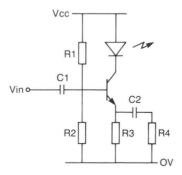

Fig. 11.3 LED drive circuit

11.2.4   Laser drive circuit

A laser diode is inherently a non-linear device with the onset of laser action defined at a distinct threshold current. It is thus necessary to bias such devices at a pre-defined operating point before applying modulation. When the lasing action has been stimulated the optical power-output current relationship is essentially linear, but it is extremely rapid and sensitive to temperature changes [2, 4]. The devices are thus generally used in a digital rather than linear mode; however, with the aid of control circuitry they can be modulated linearly.

For transmission system design, the principal LD parameters and characteristics to be taken into account are:

(i)   *Light/current characteristic*: This can be modelled (Fig. 11.4) by the approximation:

$$P(I) = \eta \frac{hc}{q\lambda} (I - I_{th}) \tag{11.6}$$

where

$$\eta = \text{quantum efficiency}$$

$$hc/\lambda = \text{lasing photon energy}$$

$$q = \text{electron charge}$$

Both the threshold current I_{th} and the quantum efficiency η are sensitive to temperature and vary in an exponential manner. This makes them

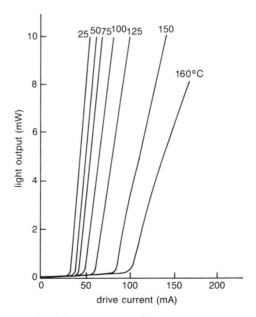

Fig. 11.4   LD output with bias current and temperature

extremely sensitive to temperature variations. Inadequate thermal control can lead to the constants of eqn. 11.6 being modulated indirectly by the signal, which in turn can lead to non-linear signal distortion.

For analogue modulation the operating point will normally be well into the lasing (50% max. power) region and no significant delay between electron input and photon output occurs. However, for digital modulation with binary signalling, it is necessary to bias the signal into the lasing region for the '0' state and well up the curve for '1'. If this is not done, a 'turn on' delay will result and pattern dependent jitter introduced.

Modern semiconductor lasers have threshold currents ranging from 15—50 mA and threshold temperature coefficients ~1 mA/C [3] (Fig. 11.4).

(ii)   *Bandwidth*: This is limited by the intrinsic laser action and parasitic capacitances of the package and drive circuit. The fundamental limitation, due to the intrinsic laser bandwidth, is defined by solving the small signal rate equations, which give the following power/frequency characteristic:

$$P(f) = \frac{P(0)f_0^2}{(f_0^2 - f^2) + j\alpha_f} \qquad (11.7)$$

where

α = device-related parameter

f_0 = frequency of natural relaxation oscillation

The useful laser bandwidth is taken as being $<f_0$. This can range from 2 to 20 GHz, and so intrinsic laser bandwidths from 1 to 10 GHz are generally possible. Packaging and drive electronic limitations tend, however, to reduce this figure considerably; thus values in the range 0.2—2 GHz are typical.

(iii)   *Spectral width*: This parameter has a strong bearing on the ultimate capacity of all optical transmission systems [2]. Depending on the type of device (Fabry–Perot index/gain guided, longitudinal/lateral moded, DFB, multi-moded or single-moded), spectral width can range from 10 MHz to 1000 GHz. What are of key importance, however, are the spectral properties under dynamic conditions.

Intuitively, it might be expected that if a laser is held at its threshold current and then intensity modulated by a bit stream, its spectrum will change with time. At the beginning of a pulse its spectrum will be broad, and as the photon density builds up the spectral width will be reduced and will eventually reach the steady-state value. For multimode lasers this is broadly true but the magnitude of each mode within the gain envelope is governed by specific features of the laser, which results in a broader spectrum under modulation than under CW conditions. The phenomenon is highly dependent upon the modulation format and results in a dynamic distribution of power between the individual modes, with the total sum of power remaining constant. Since the details of the spectrum depend upon a number of dynamic parameters, it is difficult to model or specify in advance. The spectrum under modulation must be measured and cannot be inferred from CW conditions (Fig. 11.5).

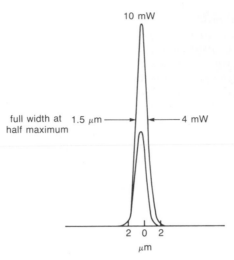

Fig. 11.5   Fabry–Perot LD power spectrum

(iv)   *Partition noise*: When light of variable spectrum content is launched into a fibre, each mode is subject to a different degree of chromatic or intermodal dispersion. The differential delay between modes gives rise to a shift in the pulse position at the receiver which leads to eye closure and consequently to a BER penalty. The magnitude of the penalty depends on the mode hopping distance (that is the spacing between modes) and the slope of the fibre dispersion curve. In most practical situations, the laser will have a large number of modes (>5) which leads to a randomisation of the effect; hence, it is termed partition noise [7].

Partition noise also introduces a further transmission limitation which is usually expressed as a bit rate × distance product (*BL*). The value of this product depends on the relative position of the laser spectrum and the zero-dispersion wavelength of the fibre. At the wavelength of minimum dispersion λ_0, the *BL* product is:

$$BL = \frac{1180}{\sigma^2} \text{ km Gbit/s} \qquad (11.8)$$

where σ = half power-point width of the laser spectrum
At some wavelength λ away from the zero dispersion wavelength:

$$BL = \frac{153.3}{(d\tau/d\lambda)\sigma} \text{ km Gbit/s} \qquad (11.9)$$

where $d\tau/d\lambda$ = slope of the dispersion curve.
The constants in eqns. 11.8 and 11.9 are based on an operating BER of 10^{-9}.

(v)   *Temperature effects*: Any variation in temperature >0.1°C can be guaranteed to affect laser operation in some way. Inadequate heat sinking and thermal control results in threshold current drift, centre-wavelength drift, changes in spectral width and mode hopping, all of which are governed by exponential relationships. Fig.11.6 gives an indication of the serious nature of this problem and the necessity for good temperature control [2].

(vi) *Reliability*: The greater current densities experienced by laser structures should, in theory, make them less reliable than LEDs by about an order of magnitude. In fact, they presently exhibit a similar MTTF figure of about 10^6 hours, with some manufacturers quoting 10^7 hours. In common with LEDs their reliability improves with increasing wavelength. This is a desirable and important feature, as it coincides with the complementary loss reduction in fibre [6].

In distinct contrast to the LED the drive-circuit arrangements for a laser are far from straightforward. It is necessary to provide a bias current that is able to track the peak of the current/optical-power curve, and to modulate up to a current giving the required peak optical power output. Until relatively recently, this latter objective was made functionally more complex by variations of current/power slope with bias current and laser lifetime. Fortunately, recent developments have seen lasers produced devoid of any significant slope variation [3].

A laser can thus be conveniently controlled and modulated by the incorporation of a monitor diode in the laser package. This is ususally a PIN photodiode placed close to the back facet of the laser chip. Including this in a feedback loop of the form shown in Fig. 11.7 provides all the necessary control to account for ageing and limited temperature variation. For a given drive-current swing and slope, it is only necessary to adjust the bias current to achieve a pre-defined mean power.

Whilst the laser is, in all respects, inherently more unstable than its LED counterpart, relatively precise control leads to a power output variation (<0.5 dB) that look far less formidable than that for the LED. Again, the end of life is taken to mean the point at which the optical power has reduced by 50% (−3 dBm).

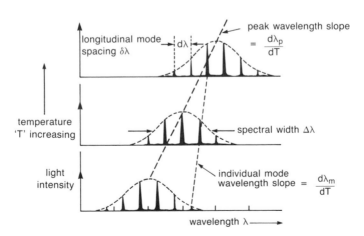

Fig. 11.6 LD spectrum with bias current and temperature

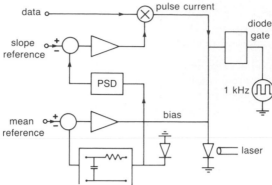

Fig. 11.7   LD drive circuit

11.2.5 Summary

At the time of writing, the majority of optical systems operate at 1.3 μm over single-mode fibres with Fabrey–Perot LDs at bit rates spanning 2—565 Mbit/s. Longer wavelength systems at 1.5 μm with DFB lasers are just being commercially introduced for the same data rates for long haul applications. Local-loop systems also employ single–mode fibre, whilst in the office environment, multimode is favoured. In both cases LED sources are being employed; however, LD solutions seem to be gaining in popularity as LED structures are proving to be equally complex and no more reliable [3].

11.3 Optical-fibre medium

The system design engineer has a choice of many different types of optical guiding media, ranging from the commonly-available plastic and silica glass to the more exotic research based materials such as fluoride. The same basic system design criteria are of interest for whichever type of guide is selected. These include attenuation per unit length (dB/km), microbending loss, dispersion, bandwidth, coupling efficiency, splice and connector loss, strength, cable type and structure, alignment with source and detector, splice strength, available connector and installation stress.

Before proceeding further to consider some of the above topics in detail, it is worth pausing to extend our understanding of how light propagates in the guiding fibre media [7].

11.3.1 Light propagation in fibres

The guidance of light rays inside thick glass rods, and around bends in such rods, is well known. With reference to Fig. 3.9, if a light ray R_1 inside a rod meets the surface of the rod at an angle θ_1, then total internal reflexion of the ray takes place provided that:

$$\cos \theta_1 < \cos \theta_c = n_2/n_1 \qquad (11.10)$$

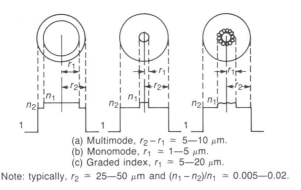

(a) Multimode, $r_2 - r_1 \simeq 5$—10 μm.
(b) Monomode, $r_1 \simeq 1$—5 μm.
(c) Graded index, $r_1 \simeq 5$—20 μm.
Note: typically, $r_2 \simeq 25$—50 μm and $(n_1 - n_2)/n_1 \simeq 0.005$—0.02.

Fig. 11.8   Cross-sections of fibre types

where

$n_1 = $ refractive index of the rod material,

$n_2 = $ refractive index of the surrounding medium
($\simeq 1$ of the medium is air)

$\theta_c = $ internal cut-off ray angle.

All the power in ray R_1 is then reflected to ray R_2. But if, for a given angle θ_1, the ratio n_2/n_1 is increased (for example by immersing the rod in a liquid having a suitably higher value of n_2) to a value greater than $\cos \theta_1$, there appears an external refracted ray R_3 at an angle θ_2 such that:

$$\cos \theta_2 / \cos \theta_1 = n_1/n_2 \qquad (11.11)$$

For the given higher value of n_2, total internal reflection can be restored by reducing the internal angle of incidence θ_1 until $\cos \theta_1$ again exceeds n_2/n_1. For a straight rod or fibre, (or, as in most practical guided-wave applications where the radius of curvature of the fibre is large compared with the fibre diameter) the angle θ_1 is equal, or almost equal, to the angle between the ray and the axis of the fibre (the off-axis angle of the ray).

The propagation time of a light pulse travelling along a ray of off-axis angle θ_1 exceeds that along a ray parallel to the axis by a factor $1/\cos \theta_1$. The intensity of any rays with off-axis angles greater than the cut-off angle θ_c will, however, decrease rapidly in the direction of propagation due to the leakage of a significant fraction of the power at every encounter with the fibre walls. Thus, only rays having $\theta_1 < \theta_c$ will survive any considerable distance of travel along the fibre. If an impulse of omnidirectional light is produced at the planar near-end face (normal to the axis) of a length L of fibre, a fraction of the light will arrive at the far end of the fibre as an extended pulse with an arrival time spread between Ln_1/c and Ln_1^2/n_2c, where c is the velocity of light in free space.

An additional requirement for usefulness in the present context is that the beam diameter should be small compared with the diameter of the optically important part of the fibre cross-section. It is, thus, just permissible to make use of the ray concept in considering light propagation in fibres of about 50—100 μm effective diameter, provided a ray is regarded as a beam of about 10 μm diameter. A brief consideration of the wave theory of light shows, however, that such a ray will not conserve its diameter over any great propagation distance. Any quantitatively accurate analysis of the propagation properties of these and smaller fibres at such wavelengths must make use of the wave theory.

The wave theory shows that significant optical-field penetration occurs to about a wavelength outside the surface of a simple rod or fibre under the conditions of total internal reflection noted earlier. In the absence of light-scattering or light-absorbing objects close enough to the surface, this phenomenon is of no consequence. But, in practice, the fibre must be mechanically supported and, necessarily, comes into contact with other materials at unpredictable points along its surface. The practical solution to this problem is to construct the fibre with a continuous cladding layer of sufficient thickness (at least several wavelengths) to ensure a negligible optical field at its own outer surface, and of sufficiently-low optical loss to make a negligible contribution to the loss of the fibre as a whole. The refractive index n_2 of the cladding material must ensure that the condition for total internal reflection is met, up to the desired cut–off angle θ_c [7].

11.3.2 Multimode fibres

Fig. 11.8a shows schematically the refractive index profile of a step index fibre, which is commonly referred to as the multimode type. The term arises from the wave-theory description of a fibre having a large enough core diameter $2r_1$ to permit, when proper account is taken of the ratio n_1/n_2, many independent wave modes [4] to propagate freely within the fibre. Although there is not a simple one-to-one relationship between wave modes and rays, a multimode fibre can be considered as capable of carrying rays over a range of off-axis angles up to θ_c. The ray description becomes more meaningful as the number of possible modes becomes higher.

An important attraction of multimode fibre is that it can capture a high proportion of radiation from an incoherent light source of given radiance B and optical power $P_1 = BA\Omega$, where A is the effective cross sectional area of the fibre, and Ω is its effective optical capture solid angle. For a multimode fibre:

$$A \simeq \pi r_1^2 \tag{11.12}$$

If the end of the fibre is immersed in a medium of refractive index n_1, Ω is approximately the solid angle of a cone of half-angle equal to θ_c,

$$\Omega \simeq 2\pi t(1 - \cos\theta_c) = 2\pi(n_1 - n_2)/n_1 \text{ sterad} \tag{11.13}$$

Thus, P_1 increases in proportion to: $r_1^2(n_1 - n_2)n_1$
A key disadvantage of multimode fibre is that a transmitted pulse is dispersed to an overall duration of:

$$\Delta t \simeq L(n_1 - n_2)n_1/n_2 c \tag{11.14}$$

after transmission over a fibre of length L. However, to maximise the economic benefit of a fibre transmission system, it is generally necessary to maximise the bandwidth/bit-rate-distance product. The resulting product may not be sufficient to render a multimode fibre system competitive with alternatives.

11.3.3 Graded index fibres
In a graded index fibre the sharp step in refractive index between core and cladding of the simple multimode fibre is replaced by a radial gradation of refractive index according to the parabolic law:

$$n_2 = n_1^2(1 - Kr^2/r_2^2) \qquad (11.15)$$

where K is a constant. This leads to a significant reduction in pulse spreading (dispersion).

Given this law, a ray launched into the fibre at not too large an off-axis angle oscillates sinusoidally in distance from the fibre axis as it propagates along the fibre, and the propagation time is almost independent of the angle.

It is physically almost impossible to reduce n below unity in a solid and it is technologically difficult to match the parabolic law closely over smaller, but still wide, ranges of n or r. Nevertheless, both theory and experiment have shown that a considerable reduction in the spread of propagation time can be achieved with rather poor approximations to the ideal parabolic law. It is not even necessary to have a maximum of n on the axis of the fibre. The technology for producing a suitable gradation of n over the radius may be more practicable if the maximum of n can be away from the axis (e.g. on a cylindrical surface centred on the axis). Fig. 11.8c shows a possible graded-index fibre of this type, in cross-section and in refractive index profile.

11.3.4 Monomode fibres
For single-mode propagation the core diameter $2r_1$ of the fibre is reduced until:

$$r_1 < 2.405\lambda_0/\{2\pi\sqrt{(n_1^2 - n_2^2)}\} \qquad (11.16)$$

where λ_0 is the wavelength of the light in free space and the fibre is of circular cross-section.

The ray description is hardly significant for such a fibre, except in the sense that a single ray is guided by the core/cladding boundary and no other rays can exist in a sustained form. Fig. 11.8b shows the cross-section and refractive index profile for such a fibre. A short pulse of light transmitted along this type of fibre suffers no dispersion due to multimode effects. Small amounts of dispersion remain, due to:

(i) The dependence of n_1 and n_2 on λ_0 (commonly known as optical dispersion, and generally different for different materials and different wavelength regions)

(ii) The dependence of the optical field distribution between the core and the cladding on λ_0, even if n_1 and n_2 are independent of λ_0 (commonly termed waveguide dispersion).

Single-mode fibre is inherently easier to manufacture and is therefore cheaper than multimode. It is currently being adopted by many PTTs as the prime or only fibre used in their telecom networks.

11.3.5 Loss mechanisms

Absorption loss: At present the major cause of loss in the low-melting-temperature glasses in the wavelength range, 800—900 nm, due to the presence of transition metal ions. Many of the transition metals in the first row of the periodic table of elements (titanium—copper) can exist in several ionic states in the glass, each ion contributing absorption bands of different intensity, shape and position. These depend on the melting conditions and the techniques used to make the glass. The presence of manganese results in a strong absorption band at a wavelength of 490 nm with a pronounced low-energy tail extending into the near-infra-red wavelength range. These ions are the dominant impurities in the spectral region of current interest, but other impurities such as chronium, cobalt and nickel are important contributors to the loss at shorter wavelengths. From the intensities of the absorption bands, the maximum acceptable concentration of such impurities, for an optical attenuation of 10 dB/km, was estimated to be in the range 1—100 parts in 10^9 [7]. Considerable progress has been made in reducing the loss attributable to this source.

Similar behaviour of impurities is found in silica and high-silica glasses. However, because of the completely different preparation techniques, which ensure a higher purity material and advantageous transition-metal attenuation coefficients, the transition-metal impurity problem is less severe.

Water is a common impurity in both glass and silica and is present as hydroxyl. Absorption bands occur between wavelengths of 2.7—4.2 μm, depending on the position of the hydroxyl in the glass network. These fundamental bands give rise to a series of overtone (harmonic) and combination bands extending into the visible-wavelength region. In most glasses and in silica, the water content is sufficient to result in substantial attenuation. Fortunately, the peaks are relatively sharp, and the loss attributable to the high water content in glass, in the region 800—900 nm, varies in the range 0.8 to 2.0 dB/km.

Colour centres can be produced in glass by exposure in ionising radiations. The induced absorptions are generally observed in the ultra-violet and visible regions and have low-energy tails extending into the near infra-red wavelength range. A typical background dose for a period of 20 years is 2 rd, and this would produce an incremental increase in the loss at 850 nm of about 0.03 dB/km and 0.15 dB/km for silica and sodium borosilicate glasses respectively. Therefore, it does not seem to be a serious problem.

Scatter loss: The most fundamental loss mechanism in glasses in the visible region is due to light scattering from inhomogeneities smaller than the wavelength of light. This is called Rayleigh scattering. The inhomogeneities arise from density and compositional fluctuations which are frozen in the glass structure on cooling. The density fluctuations increase with increase in freezing temperature; consequently, the scatter loss due to the density fluctuations is higher in fused silica than in the glass with a lower transition temperature. Multicomponent glasses have an additional scatter loss due to local fluctuations of composition. However, the total scatter loss of some glass compositions and fused silica are comparable, being about 0.3 dB/km at 1300 nm. More recent work has shown that some glasses have very low compositional fluctuations, and hence scatter losses which can be half that of fused silia. Rayleigh scatter loss is proportional to λ^{-4}, and consequently becomes unacceptably high for communication systems towards the blue end of the visible spectrum [1].

Mie scattering is caused by inhomogeneities comparable in size to the wavelength of light. This produces mainly forward scatter, and is avoidable in principle [2].

A further scatter loss mechanism, peculiar to fibre, arises from any small irregularity of the core-cladding interface. Such fluctuations arise from vibrations or other perturbations during the pulling process. This leads to two effects, one that is beneficial and one that is not. In a multimode fibre waveguide such undulations scatter energy from one guided mode to another. This mode mixing tends to average out the delay time for energy launched initially into different modes; hence, it helps to increase the usable bandwidth. Unfortunately, the same undulations scatter energy from the guided modes out of the fibre and thus increase its loss. In practice, striking a balance between these two mechanisms has proved to be a difficult problem.

Non-linear effects, such as stimulated Raman and Brillouin scattering, have little effect on the loss below a threshold power level ($\sim+14$ dBm). In fibres, the optical power is concentrated on a small cross-section; consequently, high power densities can be obtained at a modest total power levels. However, these threshold levels will not be reached in the systems envisaged at present [3, 7].

11.3.6 Coupling and splicing

As with other transmission lines, the optical fibre must be connected to its source and detector with the minimum amount of loss. The simplest and, in general, most-efficient way to couple to an LED is to butt the fibre directly against the diode. The radiation from the diode is nearly isotropic and the coupling efficiency is proportional to the solid angle of rays that can be accepted by the fibre, which is equal to the square of the numerical aperture. If the diameter of the source is smaller than that of the fibre core, as with a gallium arsenide laser, then the butting efficiency can be improved with a lens. Such lenses have been fabricated by a photographic resist method on the end of the fibre itself [1].

Good splices, with low transmission loss and a low reflection coefficient, are essential for a practical system. Both the source and the detector will probably be manufactured already coupled to fibre tails which will then be joined to the transmission line at each end. Repairs to a broken fibre cable are more difficult to deal with. It may be an advantage to arrange the fibres so that they can be easily identified and joined as a group; for example, by fixing them side by side along a flat tape. Several different types of joint have been demonstrated in the laboratory. The best have a loss as low as 0.1 dB. Single–mode fibres are the most difficult to splice because of their small cores. Graded-index fibres are marginally more difficult to joint than multimode fibre, but their alignment is not so critical as that of single–mode fibre [3].

At this point, it is worth considering the contributing factors which affect joint loss. These are:

Splicing technique: Whichever technique is adopted it is necessary to align accurately the fibre cores and butt the end faces. Where a machine is used in fusion splicing (which offers the best loss results in practice), machine intelligence can be traded against operator skill.

Cleaving: Correct preparation of the fibre ends (particularly in the case of single–mode fibre) is vital in achieving a low-loss joint. There are many cleaving tools commercially available, which are able to produce repeatable end angle (<1°), with a good tolerance to temperature and number of cleaves.

A typical splice-loss histogram for fusion splicing is shown in Fig. 11.9. The long tail of the distribution is a characteristic of the fusion splicing method. Over a long route length with several tens of splices this characteristic can introduce a loss variability of at least 5 dB.

11.3.7 Connectors

Optical-fibre connector technology has now matured to a state where many network operators have standardised on one particular connector type selected from the large number commercially available. The choice of

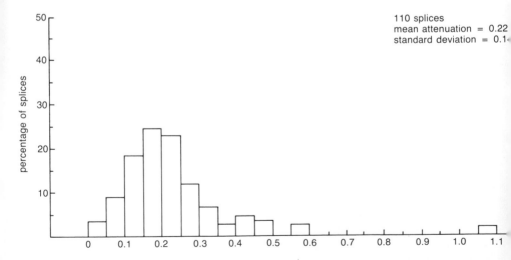

Fig. 11.9   Effective splice loss dB

connector is dependent on a number of inter-relating factors as indicated below:

- Insertion loss
- Repeatability of insertion loss
- Ease of connection (i.e. screw or push fit)
- Field or factory termination to optical fibre
- Equipment required to make connection
- Speed required
- Connector usage (connect/disconnect cycle)
- Effects of temperature, humidity, dust and dirt
- Overall cost, including equipment and labour

Current connector technology offers insertion losses of 0.2 dB for a high-quality to 2 dB for a low-quality connection. There is also a move towards receptacle connectors, where the optical device (transmitter/receiver) is mounted inside a connector housing. This obviates the need for pigtailed devices, but at the cost of a high coupling loss, which may exceed 10—30 dB depending on the cost and application.

11.3.8 Optical-fibre components

Optical fibre can also be used to manufacture components such as splitters, multiplexers and demultiplexers for use in the next generation of optical systems. Apart from the technical feasibility of bidirectional and wavelength-division multiplex (WDM) transmission there is a strong economic pressure from network operators to derive the maximum return from their large capital investment in optical plant. It should be expected that this trend will continue until the technical difficulties outweigh the economic benefits [8].

The combining of two optical signals can be performed by an optical-fibre device. This is constructed by bringing the core of two single–mode fibres into close proximity such that an interaction can occur between the optical fields being guided in the core. The power transfer characteristics of such a device are shown in Fig. 11.10. Over a range of wavelengths the 4–port device behaves as a 50% power splitter (or directional coupler), while at other wavelengths λ_1 and λ_2, it behaves as a wavelength division multiplexer. Table 11.3 shows the important parameters of duplex couplers.

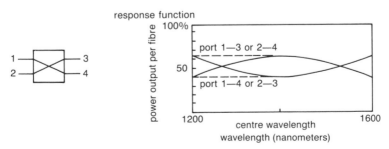

Fig. 11.10   Directional coupler

Table 11.3   Important parameters for duplex couplers

Parameter	Comment
Insertion loss	Manufacturers often give the excess loss (i.e. loss >3 dB) which is typically in the range 0·05—1 dB
Variation in split ratio over a given λ range	This indicates the 'flatness' of the coupler. The requirement will depend on the system application.
Return loss	The amount of power returned to a port with all other ports terminated. This is typically greater than 30 dB, which presents no problems to the transmitter.
Polarisation	It is important to ensure that the coupler has the same insertion loss for all input polarisation states. This enables its used with unspecified transmitter polarisation. Typical variations in insertion loss due to polarisation are less than 0·1 dB.
Stability of optical performance	Dependent on the sensitivity of the system design to changes in optical power (launch power, dynamic range, etc.)
Mechanical stability	Important for practical systems with a long life.

Directional couplers are produced commercially using two technologies, fibre-based types and bulk optic types. Each has its own parameters as detailed in Table 11.4.

It can be seen from Table 11.4 that the lowest near-end crosstalk is obtained with a fibre-based device. However, the basic 'unflattened' device has a narrow optical bandwidth which could be undesirable in some system applications.

Table 11.4   Typical coupler parameters (affecting system performance)

Parameter	Value		Units
	Device type		
	Fibre	Bulk optic	
Total insertion loss	3·05—3·5	3·5—4·5	dB
Optical bandwidth	40 (unflattened) 250 (flattened)	120	nm
Near end crosstalk (directivity)	50—60	30—50	dB
Return loss	>30	>30	dB

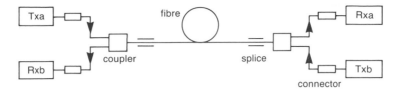

Fig. 11.11 Bidirectional system

Two different types of transmission system can be constructed using variants of this simple structure, as follows.

Bidirectional system: Signals of the same nominal wavelength are transmitted in opposite directions (Fig. 11.11). At each end of the system a transmitter and receiver are connected to the directional coupler (ports P_1 and P_2), while one of the output ports (P_4) is optically terminated by index matching, to avoid reflections. Power arriving from the optical line at P_3 is equally split between P_1 and P_2. Half the power is incident on the receiver at P_2, while the other half is directed at the transmitter P_1. However, since the line attenuation results in a low signal at P_3 (-10 dBm to -50 dBm), no transmitter interaction occurs. This type of system suffers a 3 dB insertion loss at each coupler, due to the 50% power split, while in practice devices have additional loss called excess loss usually in the range $0\cdot1$—1 dB [8].

WDM: Signals of different wavelengths (say λ_1 and λ_2) are transmitted in either the same direction (unidirectional system) or in opposite directions (bidirectional system), see Fig. 11.12. As an example, consider the unidirectional case; if λ_1 is input to port P_1 and λ_2 to P_2, power is transferred in both cases to the optical line at P_3 of the demultiplexer. Light of wavelength λ_1 will transfer to P_1 (and not P_2) while λ_2 will transfer to P_2 (and not P_1) [9].

First-generation WDM systems for long-haul transmission will concentrate on the two low-loss regions of 1300 nm and 1500 nm. These so-called 'one plus one' systems allow an addition to existing channels already operating at 1300 nm. There are two methods for configuring a WDM system, shown in Figs. 11.12 and 11.13.

(i)   Unidirectional transmission, where the data from the two transmitters travel in the same direction on the same fibre

(ii) Bidirectional transmission, where the data from the transmitters travel in opposite directions on the same fibre

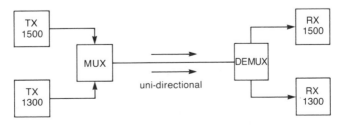

Fig. 11.12 Unidirectional WDM system

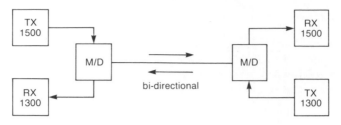

Fig. 11.13   Bidirectional WDM system

Table 11.5   WDM system penalty mechanisms and
parameter requirements

Transmission type	Penalty mechanism	Parameter requirement (isolation factor) dB
Unidirectional	Far end crosstalk	$\Delta P - 15$
Bidirectional	Near end crosstalk	$\Delta P - 15$

ΔP = difference between signal power and crosstalk power levels

It has been shown that the values of parameters required for the multiplexer and demultiplexer depend on the system configuration. The requirements are detailed in Table 11.5.

Note that if the signal and crosstalk powers are equal, an isolation of 15 dB is required. There are two types of devices suitable for two–channel WDM systems; these are detailed in Table 11.6, together with some of their important parameters.

It can be seen from Table 11.6 that the fused-fibre device can only achieve a large far-end isolation over a very narrow spectral width; therefore unidirectional systems enploying fibre-based devices require matching of the transmitter wavelength and WDM component passband. Bulk optic devices such as the dichroic filter have wide passbands and can be used in either uni- or bidirectional systems [10].

Table 11.6   WDM device parameters

Parameter	Device type		Units
	Fused fibre value	Bulk optic value	
Insertion loss	0·05–0·5	1—2	dB
Far end crosstalk isolation	30 (max)	25-60	dB
Polarisation coefficient	±1	±1	%

11.4 Optical-fibre losses and splicing—a practical view

Many texts cover the theoretical losses that optical fibre can achieve in ideal conditions. However, most system applications are in far from ideal environments. This section will give examples of the results obtained in a typical telecommunications environment [3].

11.4.1 Single–mode attenuation at 1300 nm

Since the first optical fibres were produced, a considerable research and development effort has reduced the overall attenuation by a factor of 1000. The primary effort was direct at the 1300 nm waveband, since this coincides with a minimum dispersion. Today, manufacturers produce fibre with losses in the range 0·3–0·6 dB/dm [7].

11.4.2 Single–mode attenuation at 1500 nm

It is well established that single–mode fibre has the potential for lower attenuation at 1500 nm (0·2 dB/km). However, it is also more sensitive to bending. In a practical environment each splice has several metres of fibre either side, since in most cases the jointing is done outside the manhole with the excess fibre being stored in loops inside the joint housing. Any small kinks or bends induced by packing the fibre cause additional loss at each joint. Changes in temperature or other mechanisms leading to strain will cause a variation in loss. Therefore, considerable care in the joint housing construction and fibre packing are necessary in order to obtain the full benefits of 1500 nm operation. Table 11.7 gives some sample results for jointed cable for three different cable manufacturers.

11.4.3 Single–mode attenuation at 850 nm

The cut-off wavelength for standard single–mode fibre is in the 1120–1250 nm region, below which the fibre becomes multimode and suffers from modal dispersion. Therefore the use of the 850 nm waveband is restricted to short links at low bit rates (e.g. 2 Mbit/s over 5 km). Attenuation is typically found to be in the range 2–2·5 dB/km, thus facilitating the use of 850 nm LEDs which launch around -30 dBm into the fibre. 850 nm lasers can also be used, but consideration has to be given to modal noise [11].

Table 11.7 Loss with wavelength comparison

Manufacturer	Loss (dB/km) at wavelength (nm)	
	1300	1500
1	0·51	0·32
2	0·67	0·4
3	0·53	0·48

11.4.4 Splice losses at 1300 nm

In addition to the basic fibre loss, there is an additional loss at each joint, which in the BT trunk network are around 2 km apart (the distance over which a cable can be pulled). If the cable can be directly buried or laid, as in a submarine system, the distance between joints can be increased, thus reducing the effect of joint loss.

As basic fibre attenuation has decreased, the overall attenuation of a link has become more sensitive to joint loss. If the link is short, its loss will be totally dominated by several poor splices. Consider a 1 km link with a basic loss of 0·35 dB; three splices with a loss of 0·35 dB each account for 75% of the link loss. Connectors and other passive optical components, such as splitters, will all require splicing to the fibre causing further erosion of the system power budget.

In a practical environment the joint loss depends on a number of factors [3] such as:

- Staff training and work practices
- Job completion timescales
- Availability of equipment
- Mismatch between fibres
- Reliability of equipment
- Cost

The results of an analysis will reveal a compromise between the joint quality (loss) and the elements mentioned above. It should be borne in mind that the jointing is usually a permanent feature which will be frozen into the network for its whole life (20 years), thus precluding possible network upgrades.

There are three fundamental factors which affect joint loss and can be addressed analytically. Table 11.8 summarises these to give an ideal splice loss figure, while Tables 11.9 and 11.10 give practical figures.

Table 11.8   Joint loss calculations (1300 nm)

Factor	Specification	Loss dB calculated mean	Comments
Field width mismatch	4·5—5·5 μm	0·02	Assumes Gaussian production statistics
Core concentricity	<0·7 μm	0·08 ·	Assumes 0·5 μm
Fusion joint	Many machines available	0·08 ·	Due to surface tension, fibre cleave angle and contamination
	Total 0·18		

Table 11.9   Practical joint losses (1300 nm)

Source	Mean Loss dB	No. of joints	Environment	Comments
	0·22	110	Lab.	Simulation of field conditions on land
BT	0·25	213	Field	Field conditions
	0·11	37	Lab.	Simulation of submarine cable
	0·1	72	Field	Submarine cable

11.5 Optical receivers

The engineer finds him or herself with a wide choice of commercial receivers with various characteristics. However, in practice, all receivers broadly fall into three categories using one of two types of device. Before exploring this further, it is worth reviewing noise sources and receiver front-end designs [1, 2, 7].

11.5.1 *Noise sources in optical receivers*
In any receiver the dominant noise sources occur in the early stages [6], as represented in the following equation:

$$F_1 + \frac{(F_2-1)}{G_1} + \frac{(F_3-1)}{G_1 G_2} + \cdots \qquad (11.17)$$

F = overall noise figure
F_1 = first stage noise figure
G_1 = first stage gain

Table 11.10   Joint loss comparison

Wavelength nm	Loss at joint number (dB)						
	1	2	3	4	5	6	7
1300	0·38	0·0	0·0	0·42	0·0	0·92	0·31
1550	0·49	0·15	0·1	0·44	0·17	0·94	0·12

The front-end noise figure is therefore critical as it dictates the ultimate noise performance of the optical receiver [2]. Sources of receiver noise are: shot, thermal, dark-current, signal-dependent and amplifier-induced. These are generally expressed in terms of the time-averaged mean-square spectral densities as follows:

Shot noise $\langle I_{sh}^2 \rangle = 2eM^2 FI$ $(A/Hz)^2$ (11.18)

where

e = electronic charge
F = noise figure of optical detector
M = multiplication factor of APD (=1 for a PIN)
I = mean signal and bias current

Thermal noise $\langle V_{th}^2 \rangle = 4KTR$ (V^2/Hz) (11.19)

$\langle I_{th}^2 \rangle = 4KT/R$ (A^2/Hz) (11.20)

where

K = Boltzman's constant
T = absolute temperature

Amplifier noise: Depends on the type of transistor used: FET or bipolar. For an FET the major noise source is the thermal noise associated with the channel resistance.

$$\langle V_A^2 \rangle = \xi 4KT/g_m \quad (V/Hz) \tag{11.21}$$

where

g_m = FET transconductance
ξ = material constant

The major noise source of a bipolar junction transistor is the shot noise associated with the base and collector bias currents, I_b and I_c.

$$\langle I_A^2 \rangle = 2eI_b \qquad\qquad (A^2/Hz)^2 \tag{11.22}$$

$$\langle V_A^2 \rangle = [2(KT)^2/eI_b\beta] \quad (V^2/Hz)^2 \tag{11.23}$$

where

β = current gain

11.5.2 *Receiver front end designs*

The three types of receiver front end are: high impedance, low impedance and transimpedance [4]. The first two are based upon the conventional voltage amplifier circuit, whilst the latter employs an overall feedback configuration.

Voltage Amplifier Circuit: A small-signal equivalent circuit representing a photodetector feeding a voltage amplifier is shown in Fig. 11.14.

$$V_I = RMI/(1 + jwfRC) \tag{11.24}$$

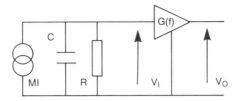

Fig. 11.14 Voltage amplifier

For the overall transfer function to remain frequency-independent, the following equation must be valid:

$$G(f) = G_0(1 + jwfRC) \qquad (11.25)$$

then

$$V_0 = RMIG_0 \qquad (11.26)$$

The noise equivalent circuit for the amplifier is shown in Fig. 11.15. A total RMS noise can be obtained by integrating the mean square noise amplitudes over the frequency range. The peak signal/noise ratio K can be expressed as:

$$K = I \left/ \left[\frac{V_A^2}{M^2}\left(\frac{1}{R^2} + \underbrace{\frac{4\pi^2}{3}(\Delta f)^2 C^2}\right) + 2eIF + \frac{4KT}{M^2 R} + \frac{I_A^2}{M^2} \right]^{1/2} (\Delta f)^{1/2} \right. \qquad (11.27)$$

$$\underset{(a)}{\uparrow} \quad \underset{(b)}{\uparrow} \quad \underset{(c)}{\uparrow} \quad \underset{(d)}{\uparrow} \ \underset{(e)}{\uparrow}$$

where Δf is the necessary bandwidth, typically $0.7 \times$ bit rate.
The signal/noise ratio can be improved in one of three ways [7]:

(i)   Increasing M until the shot-noise term (c) dominates.
(ii) Increasing front-end resistance R, provided terms (a) and (d) are significant.
(iii) Minimising front-end capacitance because of C^2 term of (b).

High impedance front end: If R is large, noise becomes dominated by the terms (b), (c) and (e). Therefore the peak signal/noise ratio is now given by:

$$K = I \left/ \left[\underbrace{\frac{V_A^2}{M^2}\frac{4}{3}\pi^2(\Delta f)^2 C^2} + 2eIF + \frac{I_A^2}{M^2} \right]^{1/2} (\Delta f)^{1/2} \right. \qquad (11.28)$$

$$\underset{(b)}{\uparrow} \quad \underset{(c)}{\uparrow} \ \underset{(e)}{\uparrow}$$

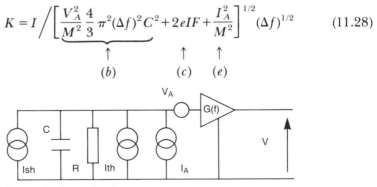

Fig. 11.15 Noise equivalent circuit

Equalisation is required for the high impedance design which must be tailor made for each individual circuit, because the values of R and C, in eqn. 11.27 will vary between individual devices and circuits. A high impedance design has a very good sensitivity but the dynamic range can be limited by low frequency saturation in the amplifier. The first-order pole of the amplifier is between 40 and 200 kHz and the gain rolls off at 20 dB/decade. Consequently, a distortion free signal may not be recoverable at the equaliser output for high level input signals.

Low impedance front end: If the front end impedance R is small and inequality 11.30 is met then no equalisation is required [7]:

$$R < 1/\pi C \Delta f \tag{11.29}$$

When this condition is met the shot noise limit can be reached using an APD. However, the use of a PIN, for which $M = 1$, results in a considerable sensitivity penalty.

Transimpedance front end: The advantage of a transimpedance amplifier is that equalisation is not necessary and the receiver performance can approach that of a high-impedance receiver. Consider Fig. 11.16. For

$$A \gg 1 + R_f/F \tag{11.30}$$

$$V_0 = \frac{-R_f MI}{1 + j2\pi f RC/A} \tag{11.31}$$

Then no equalisation is required providing condition 11.32 is valid [7]:

$$A > 2\pi CR_f \Delta f \tag{11.32}$$

The use of negative feedback reduces the effective impedance seen at the amplifier input and K, for the transimpedance design is given by eqn. 11.33, (see Fig. 11.17). (I_T includes shot noise, amplifier current noise, thermal noise of biasing and amplifier input impedance noise.)

$$K = I \left| \left[\frac{V_A^2}{M^2} \left\{ \left(\frac{1}{R} + \frac{1}{R_f} \right)^2 + \frac{4}{3}\pi^2 C^2 (\Delta f)^2 \right\} + 2eIF + \frac{4KT}{M^2} \left(\frac{1}{R} + \frac{1}{R_f} \right) + \frac{I_A^2}{M^2} \right] (\Delta f)^{1/2} \right. \tag{11.33}$$

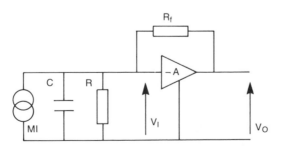

Fig. 11.16   Transimpendance amplifier

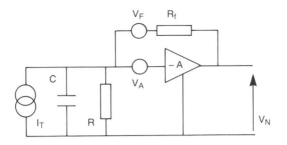

Fig. 11.17 Noise equivalent circuit

By comparing eqn. 11.33 with eqn. 11.28 it can be seen that the noise performance can approach that of a high-impedance design as R_f tends to infinity.

The main problem experienced with the transimpedance design is one of stability. Careful layout is necessary to avoid positive feedback via parasitic capacitance in the feedback path. Also, an amplifier with a small group delay must be chosen to maintain stability margins.

11.5.3 *APD characteristics*

The advantage of an APD is the greater-than-unity gain factor M, which is material-dependent. A material for which one carrier type dominates the impact ionisation process exhibits low noise and large gain–bandwidth product. The photocurrent generated in an APD is expressed in eqn. 11.34, where I_p is the photocurrent before avalanche multiplication occurs. Avalanche gain increases the diode current as follows:

$$I = I_p \times M \qquad (11.34)$$

Typical figures for the gain factor and required avalanche breakdown voltage for different materials are presented in Table 11.11.

The noise factor F associated with an APD is also material-dependent:

$$F = M^x \qquad (11.35)$$

where $x = 0.55$ for silicon and 0.85–1.0 for germanium.

The disadvantages of an APD are: its complex structure, a relatively high cost, high reverse bias voltage required and a gain mechanism that is temperature-dependent, requiring stabilisation circuitry to guarantee performance.

Table 11.11   Material properties

Material	Gain factor M	Break-down voltage V_{Br}/volts	Wavelength range g(nm)
Silicon	<150	<200	850
Germanium	<40	<40	850—1300
III–V*	<30	30—40	1300—1550

* Devices not readily available

Table 11.12   Comparison of receiver front ends

Front end	Noise performance		Equalisation required	Typical dynamic range* (dB)
	PIN	APD		
High impedance	Very good	Very good	Yes	20
Low impedance	Poor	Good	No	20
Transimpedance	Good	Very good	No	>20

* Dynamic range figures are dependent on characteristics of individual receivers and the type of line code used

11.5.4 PIN characteristics

PIN photodetectors have no internal gain; hence $M = F = 1$ and the photo-current generated in a PIN structure is given by the following equation [8]:

$$I_p = \frac{\eta e P}{hf} = RP \qquad (11.36)$$

where

η = quantum efficiency
e = electronic charge
f = frequency of incident photon
h = Planck's constant
R = responsivity, A/W
P = incident optical power

The advantages of using the PIN structure are: a simple structure, low cost, low reverse bias voltages ($<10V$) and insensitivity to temperature variations.

Table 11.13   Comparison of PINs and APDs

	APD	PIN
Associated circuitry	Requires large reverse bias voltage to achieve avalanche breakdown, typically >40 V. Hence the need for DC–DC convertors. Temperature stabilisation circuitry required.	Single power supply. Hybrid modules are available that contain front-end amplifiers. No temperature control circuitry necessary.
Noise performance	Good with all types of front end.	Good except with a low-impedance front end.

11.5.5 Receiver designs summary

The front end of the receiver can be configured in one of three ways: high impedance, low impedance and transimpedance. Parameters to be considered for all three designs are their relative noise performance (or sensitivity), whether equalisation circuitry is required and the dynamic range. A summary of the three designs is shown in Table 11.12 and the relative merits of PINs and APDs are summarised in Table 11.13.

The performance spread of the various receiver circuit options depends on the quality of the components used and the production techniques employed. At the top of the quality range a sensitivity standard deviation ~0·5 dB might be expected, whilst at the bottom end a figure ~2 dB (or worse) is common.

11.6 System design model

To design an optical transmission system, it is necessary to have an understanding of the relationships and interactions between the various system elements (transmitter, transmission medium, receiver). Fig. 11.18 shows a block diagram linking the system elements via the interactions which occur and can be broken down into two categories [7]:

(i)   Time variables (upper half)
(ii) Amplitude variables (lower half)

Amplitude variables, such as attenuation, lead to receiver power penalties (i.e. a receiver requires more power to obtain a certain BER) and do not affect the shape of the BER curve. Time variables modify the shape of the BER curve, causing error floors which are independent of optical power [4, 13].

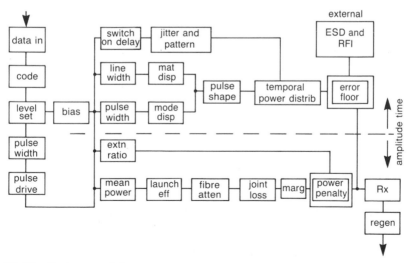

Fig. 11.18   System model

System margin is often used to define an effective BER by stating the excess received power, over a given BER (often 10^{-9}). This is convenient owing to the time it would take to measure very low BERs. The implied assumption is that the BER curve follows a gaussian plot. If an error floor exists, this definition of margin no longer holds, since the margin is only the difference between the received-power BER and the error-floor [13, 14].

In Fig. 11.18 the system margin is shown in the amplitude path to allow for connector loss variations, laser output-power reductions etc. Although the system designed may have 10 dB of margin to allow for loss variables, the system error floor may be only 1 dB below the operating power, thereby offering no advantage in long-term BER.

11.7 A 565 Mbit/s system design example

The optical power budget for $1 \cdot 3$ μm operation is given in Table 11.14. A mean launch power of -3 dB (NRZ pulses) is used. This is readily available from modern lasers and under normal operation does not exceed the safe working limits of a Class 1 laser as derived from British standards. The receiver sensitivity, which is referred to the line rate of 647 MBaud, is achieved using a front-end stage based on a hermetically-sealed 100 mm germanium avalanche photodiode (APD). These devices suffer from a high bulk leakage current at elevated temperatures and empirically have been found to incur a sensitivity penalty $\sim 0 \cdot 07$ dB/deg beyond 25°C. As the temperature within an equipment rack can, exceptionally, be as high as 60°C, this would result in a penalty of approximately $2 \cdot 5$ dB. To avoid this problem, a Peltier cooler has been included within the receiver and restricts the APD maximum temperature to 25°C over the full range of operating conditions [15]. The receiver sensitivity quoted in Table 11.14 refers to the

Table 11.14   Optical power budget

Mean launch power (NRZ)	$-3 \cdot 0$ dBm
Receiver sensitivity (10^{-10} EBER)	$-36 \cdot 0$ dBm
Repeater gain	$33 \cdot 0$ dB
Fibre loss	$23 \cdot 5$ dB
Cable repair margin	$2 \cdot 0$ dB
Cable allocation	$25 \cdot 5$ dB
Design penalty	$3 \cdot 0$ dB
Ageing and temperature margin	$2 \cdot 5$ dB
Connector loss (two connectors)	$2 \cdot 0$ dB
Optical receiver dynamic range	$> 30 \cdot 0$ dB

expected measured performance at an ambient temperature of 60°C using an optical attenuator instead of fibre.

In practice, additional impairments, referred to as the design penalty, arise owing to fibre dispersion, reflection noise, intersymbol interference, jitter and component tolerances. Although the bandwidth of single-mode fibres is large at 1·3 μm, chromatic dispersion and the related effects of pulse broadening, mode-partition noise and reflection noise are potentially serious problems for 565 Mbit/s and higher-bit-rate systems. To contain this within acceptable limits, it is important to specify tightly the centre wavelength and spectral width of the optical source and to provide temperature stabilisation. The latter normally has the added bonus of improving the laser reliability and preventing thermal runaway. In view of the extremely rapid onset of dispersion-related penalties and the resulting increase in the error floor, it is prudent to constrain the penalty to less than 2 dB.

Although the dispersion penalty is dependent on the particular receiver configuration and laser characteristics, for 565 Mbit/s transmission with a 7B8B line code [16] the dispersion should be constrained to approximately <500 ps. Single-mode fibre currently installed by BT has a dispersion zero at approximately 1·31 μm and dispersion <3·5 ps/nm km over the window 1·3—1·325 μm. With the laser wavelength specified within this window, and with a spectral width of <4 nm, repeater spans of 40 km can be realised. In practice, installed fibres often achieve even lower dispersion and even longer repeater spans are possible.

In general, installed fibre loss can be expected to be <0·6 dB/km. For the fibre allocation quoted in Table 11.14, this also allows a repeater span of greater than 40 km. Thus, repeater sections can either be loss or dispersion limited, depending upon the precise characteristics of installed fibres. This generally allows 40 km spacing with adequate systems margins [5, 15].

11.8 References

1 WILSON, J., and HAWKES, J.F.B.: 'Optoelectronics—an introduction', Prentice Hall, 1983
2 HOWES, M.J., and MORGAN, D.V.: 'Optical fibre communication', John Wiley, 1980
3 Special Issue on 'Fibre optic systems for terrestrial applications', *IEEE J.*, 1986 **SAC-4**, (9)
4 SENIOR, J.: 'Optical fiber communication', Prentice Hall, 1985
5 HP Fiber Optics Handbook, Hewlett Packard, 1989
6 GOWER, J.: 'Optical communication systems', Prentice Hall, 1984
7 MIDWINTER, J.E.: 'Optical fibre for transmission', John Wiley, 1979
8 HUNWICKS, A.R., and ROSHER, P.A.: 'Duplex — diplex optical transmission in the 1300 and 1500 nm fiber windows – an installed system', *IEE Elect Lett*, **23**, no 10, 1987, pp. 542–544
9 COCHRANE, P.: 'The evolution of optical fibre transmission in the British Telecom Network', Telematica 86, Stuttgart, Tiel 1, pp. 215–231

10 BROOKS, R.M. *et al*: 'A highly integrated 565 optical fibre system', *BTTJ*, **4**, No 4, 1986, pp. 28–40
11 BETTS, R.A., MOSS, J.P., and HALL, R.D.: 'A low cost 2 Mbit/s optical local line transmission system using single mode fibre', *IEE Elect Lett*, 1986, **22**, No 3, pp. 143–144
12 BYLANSKI, P., and INGRAM, D.G.W.: 'Digital transmission systems', Peter Peregrinus, 1980
13 BAILEY, A.E.: 'Microwave Measurements', Peter Peregrinus, 1985
14 CATTERMOLE, K.W., and O'REILLY, J.J.: 'Mathematical topics in telecommunications, Vols 1 & 2', Pentech Press, 1984
15 COCHRANE, P., BROOKS, R.M., and DAWES, R.: 'A high reliability 565 MBit/s trunk transmission system', *IEEE J.*, 1986, **SAC-4**, (9), pp. 1424–1431
16 Special Issue on: 'Coding for digital transmission systems', *Int. J. Electronics*, 1983, **55**, (1)

Chapter 12
Microwave radio links
M.J. de Belin

GEC Plessey Telecommunications Ltd, Coventry

12.1 Introduction

The foundations of microwave radio were laid in the massive effort devoted to radar research during the 1939–45 war. The application which gave terrestrial line-of-site radio systems impetus was the transmission of the television signal, a broadband baseband signal of many octaves which cable systems had difficulty in carrying. The first microwave link in Europe to go into public service was established between London and Birmingham in 1949 to extend television from London to the Midlands [1]. The link operated at 900 MHz and the transmitter delivered 7.5 W to the antenna using disc-sealed triodes for the RF stages. In 1954, the Eurovision Link was established with equipment in the 2 GHz band. Towards the end of the 1950s, systems for telephony transmission were being installed with a capacity of 240 telephone channels. During the 1960s, the capacity of each bearer advanced, to 300, 600, 960 and 1800 channels as the design capability for broadband circuits improved, and the carrier frequency moved up through 4 GHz to 6 GHz. Today, systems with 2700 and 3600 channels per bearer are in operation.

Also during the 1960s, there was a major technology change from thermionic devices to all-semiconductor equipment, based on transistor oscillators and amplifiers and diode multipliers. Initially, the output power of the transmitter had to be generated at VHF and multiplied to the required output frequency before the modulation was applied in the final output stage. Gradually, improved devices allowed amplification of the modulated signal at the output frequency, and only in the high-capacity high-frequency systems, e.g. 1800 channels at 6 GHz, did the last thermionic device remain in the form of the travelling-wave-tube (TWT) amplifier.

Until the 1970s, frequency modulation (FM) had been universally adopted, because the necessary linearity could be achieved in the modulation and demodulation process, and the modulated signal was not distorted by the limiting amplifiers in the output of the transmitter. In the late 1970s, pressure for increased capacity in the available frequency bands resulted in development of single-sideband (SSB) modulation for 6000 telephone channels on a single bearer [2].

The late 1970s saw another major change to the industry as the requirement for digital transmission grew, stemming mainly from the proposed introduction of digital exchanges [3]. In the UK, the decision to convert the whole of the trunk network to 140 Mbit/s transmission by 1990, gave clear direction to industry.

In the UK, the 11 GHz frequency band was available and it has been used to provide a digital network overlaying the existing analogue network. Next, the analogue equipment in the lower 4 GHz and lower 6 GHz bands is being replaced with digital equipment. The introduction of a second public telecommunications operator, Mercury Communications, has greatly increased the pressure on the available frequency bands for efficient use of the spectrum. The higher frequency bands of 19, 23, and 29 GHz are being used for short links for various purposes, and use of the millimetric frequency bands above 30 GHz is currently being planned. Introduction of the synchronous digital hierarchy (SDH) into the networks may result in new requirements for trunk radio transmission.

The history of microwave radio has seen a continual development of equipment to meet the users' changing requirements, with improved performance and better utilisation of the available frequency spectrum. Each achievement has been made possible by new technology, new devices, better test equipment, and engineering effort in research, development, production and operation.

A microwave link is an alternative to the line transmission systems described in Chapter 10. Its relative advantages and disadvantages may be summarised as follows:

(a) Capital cost is usually lower.
(b) Installation is easier and quicker, since no work is required along the route between repeaters.
(c) As a result of (b), problems are not caused by the route crossing difficult country between repeaters.
(d) The system is immune to interruptions caused by events, such as road works, which damage cables.

However,

(e) Operation is restricted to line-of-sight distances.
(f) Repeater stations may need to be sited in remote locations, where there are problems in obtaining access by road.
(g) Provision of adequate power supplies can be a problem in remote locations.
(h) Adverse weather conditions can cause fading of the signals. Measures must therefore be taken to combat this.

The choice between microwave and cable systems thus depends on several economic and practical factors. Microwave-radio systems predominate in the trunk networks of countries where the terrain is difficult. Other countries use cable systems for some routes and microwave radio for others, as appropriate. In some countries, such as the UK, both types of system may sometimes be used on the same route, in order to obtain security of operation in the event of breakdown of either system.

12.2 Transmission medium

Radio waves propagate from transmitter to receiver through the air; hence the transmission medium is unbounded, except for the earth's surface. Unlike coaxial cable or optical fibre, the transmitted signal propagates in all directions. We focus the energy in the desired direction by using narrow-beam antennas, but the receiver will take in any signal which falls within its bandwidth. To carry more than one channel on a given path, we need frequency separation. The bandwidth of each channel depends on the amount of information it carries and it controls the spacing between channels. Available frequency spectrum is therefore the limiting factor on the amount of information which can be carried on a given route and has to be treated as a scarce resource. Channel spacings and frequency assignments for the various frequency bands are agreed internationally within the forum of the CCIR [4]. Fig. 12.1 shows the frequency plan for the two 6 GHz bands. The equipment designer has to apportion his filtering to minimise the adjacent-channel interference, whilst maintaining an acceptable in-channel performance. The system planner has to choose his stations and his antennas to restrict the interference from other paths into each receiver to an acceptable level.

Not only is the transmission medium unbounded, it is also uncontrolled, in that propagation between two stations may be affected by climatic conditions. We choose the path; nature creates the conditions. Engineers endeavour to understand, live with and overcome the effects of these conditions. A simple example is the attenuation of the signal due to rain, sleet or snow. This has little affect on frequencies below 10 GHz, becomes

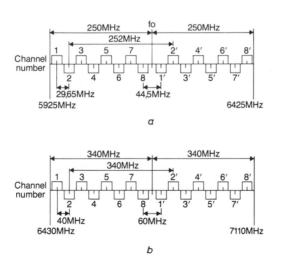

Fig. 12.1   CCIR frequency plans

a Lower 6 GHz band (Recommendation 383)
b Upper 6 GHz band (Recommendation 384)

a limitation on path length between 10 and 20 GHz and becomes the limiting factor above 20 GHz.

More significant are the effects which arise when signals reach the receiver by more than one path, i.e. multipath propagation effects [5]. Fig. 12.2*a* shows a main path with two possible secondary paths. The upper one arises from layering of the atmosphere, causing bending of the rays due to different refractive indices. The level of this signal can be close to that of the main signal with a very short additional delay. The lower path arises from reflections from the earth's surface, particularly reflections from water. Again, the level can be close to that of the main signal, but the delay can be much larger depending on the clearance of the path. Fig. 12.2*b* shows the variation of amplitude against frequency of the sum of two signals, one delayed in time due to the extra path length. When the signals are in phase, the received signal can be up to 6 dB above that received by the direct path only. However, when added in antiphase, the received signal may be reduced by 30, 40, 50 dB or more, depending on the relative levels of the two signals.

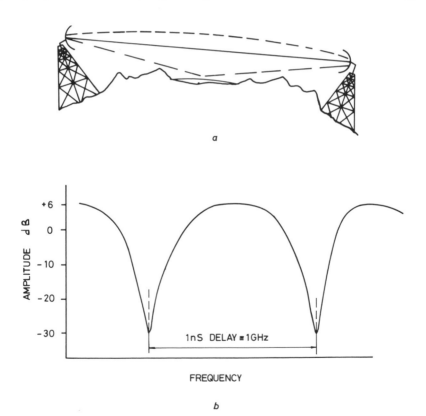

Fig. 12.2 Multipath propagation

a Main and secondary paths
b Variation with frequency of amplitude of received signal

The spacing in frequency between successive minima depends on the delay between the signals, e.g. 1 ns equates to 1 GHz. Therefore, in addition to the change in the receive level, the signal also suffers distortion both in amplitude and phase; the higher the capacity of the system the wider the channel bandwidth and the greater the distortion.

To summarise, the equipment designer has to provide accurate filtering in the transmitter and receiver to prevent adjacent-channel interference. Also, he must provide variable gain in the receiver to combat fading and rain attenuation. His signal will suffer distortion during multipath fading which he may be able to equalise adaptively. The system planning engineer has to choose his paths to minimise the interference from other systems and keep the multipath effects to an acceptable level. Where excessive multipath effects cannot be avoided, they can be significantly reduced by the addition of a second antenna at the receiver to provide height-diversity reception [6]. These considerations apply equally to the transmission of information in analogue or digital form. There are, however, quite basic differences between the two types of system which will be reviewed in the next section.

12.3 Information to be carried

12.3.1 Analogue transmission
The baseband signal normally consists of a number of telephone channels multiplexed in FDM or, alternatively, a TV signal with or without a number of sound channels. There are serious problems in trying to mix different services on the same bearer, e.g. TV and 240 telephone channels, owing to the need for sophisticated filters and the problems of distortion arising from intermodulation. In practice, it is rarely done. The system will be designed to carry a specific number of channels per RF bearer. However, for much of the time the loading is light; the system then works quite happily, carrying any number of circuits up to and sometimes beyond its design capacity.

The aim of the system is to carry the information from A to B with minimum addition of noise, which occurs in two forms:

(a)   White noise (i.e. thermal noise) is added in the various circuits of the equipment, particularly the front end of the receiver.
(b)   Intermodulation noise arises owing to imperfections in the attenuation and phase characteristics of the transmission path.

In an analogue system, every picowatt of noise is added cumulatively. During 'normal' propagation, there will be a reasonable balance between thermal and intermodulation noise. However, during periods of fading, the noise arises overwhelmingly in the front end of the receiver, and is directly proportional to the signal level. To meet the CCIR requirement for the hypothetical reference circuit [7] during non-fading periods, a relatively-high signal level is necessary to reduce the thermal noise from the receiver to a few picowatts. This then results in a fade margin on high-capacity systems well in excess of 50 dB.

Table 12.1 Digital transmission hierarchical levels

Hierarchical level	Europe (Mbit/s)	North America (Mbit/s)	Japan (Mbit/s)
1	2·048	1·544	1·544
2	8·448	6·312	6·312
3	34·368	44·736	32·064
4	139·264	274·176	97·728
5	(564·992)		397·200

12.3.2 Digital transmission

The baseband signal for a digital system is always at a fixed rate, usually one of the hierarchical levels: 2, 8, 34, 140 Mbit/s. The transmission system will only run at this rate (allowing very little tolerance on frequency) and it is not possible to under- or overload it. Since all information is in digital form, there is no problem for the radio channel in carrying mixed information. Telephony, TV and data can all be carried, provided they are multiplexed together (outside the radio equipment) to give a standard signal at the input.

Information in the digital network is normally interfaced at one of the hierarchical levels. Table 12.1 shows the internationally-agreed levels for three geographical areas. It should be noted here that the proposed synchronous digital hierarchy (SDH) based on 155.52 Mbit/s and multiples thereof may significantly alter the trunk transmission rates in the future.

Where digital radio links operate to the same frequency plan as existing analogue systems, the aim has been to use the most-effective package of hierarchical levels. In the UK, 140 Mbit/s is the trunk transmission rate, with 34 Mbit/s for small networks. In North America, $1\times, 2\times,$ or 3×45 Mbit/s is used, and in Europe 140 Mbit/s or $1\times, 2\times 34$ Mbit/s are common.

The aim of the system is to carry the information from A to B with the minimum number of errors. These will arise owing to noise, distortion or interference. In addition, the jitter performance must meet the CCITT requirements [8]. Because the signal is regenerated at each repeater, noise is not added cumulatively and the performance of a link will be dominated by the performance of the most-difficult hops.

12.4 Carrier/noise calculation

In planning a radio system, the engineer has to calculate the percentage of time a given noise allocation or a given error rate is exceeded. To do this, he uses propagation statistics to arrive at a *fade margin* for each hop, i.e. the reduction in received signal level below the clear-sky-propagation level which can be permitted before the threshold condition is exceeded. The relation of the fade margin to equipment parameters can be seen in the calculation of carrier/noise (C/N) ratios; i.e. we can establish for a given receiver noise figure how much power output is necessary to achieve the required fade margin. It will be appropriate here to show, by means of an example, how carrier/noise ratios are calculated for practical microwave radio systems. We can then establish, for a given transmitter power and receiver noise factor, the fade margin that is available before a given error rate is exceeded. Conversely, we can establish the transmitter power needed to provide a required fade margin. Let us suppose it is the latter we need to know.

We will take for our example a 140 Mbit/s 4-phase modulation 11 GHz digital radio system, with a repeater section length of 48 km, with a feeder length of 60 m at each end and 3 m-diameter paraboloid antennas.

To establish the normal (i.e. unfaded) carrier/noise ratio we need to calculate the received signal level and the receiver noise power.

If we take

    $P =$ Transmitter power (dB relative to 1 watt)
    $L =$ path propagation loss (dB)
    $X =$ feeder losses plus incidental losses associated with combining
        or separating several radio channels on one antenna (dB)
    $G =$ Antenna gain (dB) relative to an isotropic radiator,
        (i.e. one which radiates in all directions)

The received signal level (RSL) will be:.

$$\text{RSL} = (P - L - X + 2G)\,\text{dBW} \tag{12.1}$$

where P is at present unknown.

The signal power is proportional to the square of the distance (inverse square law) and the square of the operating frequency, as shown in Section 3.3.2. The path loss in decibels may be written

$$L = 10 \log\left(\frac{4\pi d}{\lambda}\right)^2 \text{dB} \tag{12.2}$$

where d is the distance and λ the wavelength (both measured in the same units).

Therefore

$$L = 32.5 + 20 \log d + 20 \log f \,\text{dB} \tag{12.3}$$

where d is in km and f is in MHz.

If $d = 48$ km and $f = 11$ GHz, this works out to 147 dB. (Note that this figure is for isotropic transmit and receive antennas.)

The feeder and other losses, X, amount typically to about 7 dB at each end of the repeater section, i.e. 14 dB total. (For example, circular waveguide has a loss of about 4.5 dB per 100 m, but some higher-loss waveguide has to be used for entries into buildings).

The power gain G of a paraboloid antenna relative to that of an isotropic radiator is given by

$$G = E\left(\frac{\pi D}{\lambda}\right)^2 \tag{12.4}$$

where D is the diameter, λ is the wavelength, and E is a factor representing the antenna efficiency.

E normally lies in the range 50—70%. Thus, the power gain of a paraboloid antenna increases as the square of the diameter and as the square of the frequency. (This tends to offset the increase of path attenuation with frequency.)

In decibels, the gain may be written

$$G = 10 \log E \left(\frac{\pi D}{\lambda} \right)^2 \, \mathrm{dB} \qquad (12.5)$$

For a 3 m-diameter antenna at 11 GHz with an efficiency of 60%, this works out to 48.5 dB. Thus, for our example, the received signal level (RSL) is

$$\mathrm{RSL} = P - 147 - 14 + (2 \times 48.5)$$

$$= (P - 64) \, \mathrm{dBW}$$

Next, we evaluate the receiver noise power (RNP). As shown in Section 2.6, this is given by:

$$\mathrm{RNP} = (10 \log kTW) + F \, \mathrm{dBW} \qquad (12.6)$$

where

k = Boltzmann's constant,
T = temperature (K),
W = receiver bandwidth (Hz)
F = receiver noise figure (dB).

In the present example of a 140 Mbit/s 4-phase system, the required receiver bandwidth corresponds to half the bit rate, i.e. 70 MHz. Typically, the noise figure will be 4 dB. Thus, at normal ambient temperature of 20°C, the receiver noise power is

$$\mathrm{RNP} = -121 \, \mathrm{dBW}$$

Now let us suppose our objective is to maintain an error rate of not worse than 1 in 10^5 in a fade depth of 40 dB. An error rate of 1 in 10^5 theoretically corresponds to a C/N ratio of about 14 dB for 4-level PSK; however, it is necessary to allow for other degradations, interference effects etc. Let us assume that the allowance due to thermal noise only is set at a C/N of 17 dB. Then, to meet our 40 dB fade margin, the free-space (unfaded) C/N must be 57 dB.

The free-space C/N is simply the difference between the RSL and the receiver noise power. We can therefore write

$$C/N = (P - 64) - (-121) \, \mathrm{dB}$$

$$= P + 57 \, \mathrm{dB}$$

Thus, to obtain a C/N of 57 dB, we need

$$P = 0 \, \mathrm{dBW}, \quad \text{i.e. 1 watt.}$$

Repeater sections will vary in regard to their path length and feeder lengths, so some flexibility in output power is desirable. An amplifier of output 1 W which can be reduced over a 10 dB range, plus a 2 W option for longer hops would meet this case.

From the C/N calculation, we note that the following parameters of the equipment affect the fading performance:

(*a*) Transmitter output power
(*b*) RF multiplexing losses
(*c*) Feeder losses
(*d*) Antenna gain
(*e*) Receiver noise figure
(*f*) Receiver bandwidth

12.5 Channel frequency allocations

Several interference paths can exist in a microwave link, as illustrated in Fig. 12.3. This shows two terminal stations and two intermediate repeaters. The two most serious interference paths are those labelled 1 and 2 in this Figure. The signal received from path 2 is reduced by the high back-to-back ratio of the two directional antennas. However, the signal received from path 1 is only attenuated by the side-to-side loss between the antennas, which is much lower. If a common antenna is used for transmitting and receiving, protection is further reduced to only the isolation of the antenna circulator.

A very large difference exists between the signal levels at the input and output of a repeater. In spite of the highly-directional antennas used, it would be difficult to obtain sufficient attenuation between input and output to achieve stability. Consequently, a frequency change is made in the repeater so that filtering can ensure sufficient attenuation under all conditions.

The 500 MHz of the lower 6 GHz band and the 680 MHz of the upper 6 GHz band are each divided into two blocks of eight channels, as shown in Fig. 12.1. At any station, all the transmit channels are in one block and all the receive channels are in the other block. At an intermediate repeater, a frequency shift is made to the received signals before they are retransmitted. Thus, if the frequency shift is 252 MHz for the L 6 GHz band or

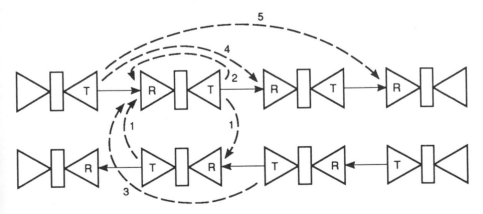

Fig. 12.3 Interference paths in a microwave link

340 MHz for the U 6 GHz band, channels 1 to 8 are retransmitted as channels 1′ to 8′ (or vice versa). This eliminates interference over path 1.

If such a two-frequency plan is used, signals are received from the opposite directions at the same frequency and at about the same field strength. Interference over path 3 is reduced by the front-to-back ratio of one antenna, which is about 70 dB. If this is inadequate, a four-frequency plan may be used by splitting the channels and using half of each for each direction of transmission.

Interference over path 4 in Fig. 12.3 is not a problem because the adjacent hops use different frequencies. The 'over-reach' interference over path 5 should have a high attenuation because of the long distance involved (i.e. 3 repeater sections). It can be avoided by ensuring that at least every third station is not situated in a direct straight line with the others.

When a system uses several channels, these must be separated by filters at each station. In the L 6 GHz band, adjacent carriers are spaced at 29.65 MHz and in the U 6 GHz band they are spaced at 40 MHz intervals. The bandwidth of each channel must be slightly less to allow for a guard band between them. However, the requirements on the filters are reduced because adjacent channels are orthogonally polarised, as shown in Fig. 12.1.

Finally, interference between adjacent transmitters and receivers in the centre of the band requires very sharp filters to provide sufficient isolation.

12.6 Equipment design

12.6.1 General

Some examples of equipment design will be found in References 9—11. Fig. 12.4 shows a basic block diagram of the transmitter and receiver at a

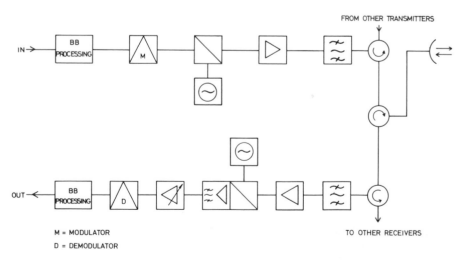

M = MODULATOR
D = DEMODULATOR

Fig. 12.4   Block diagram of transmitter and receiver at terminal station

terminal station. The transmitter and receiver can each be divided into three different parts, each of which provide a particular function. The functions performed in the transmitter are as follows and the receiver performs the inverse functions:

(a)   Baseband processing; which converts the input signal to one which is suitable for the modulator

(b)   *Modulator*; which converts the information into a modulated carrier suitable for transmission over the radio path

(c)   *Radio-frequency (RF) equipment*; which provides the means of transmitting the modulated signal.

We shall consider each of these in turn.

12.6.2 Baseband processing

In analogue systems, the baseband processing at the transmitter provides amplification for the baseband input signal. A continuity pilot is added, together with any sub-baseband signals for supervision and engineers' communication channels. Some frequency shaping, i.e. emphasis, is generally applied.

For digital transmission, the first task of the baseband processing is to establish a clock frequency. This may be the frequency of the incoming traffic stream or, where the input consists of two or more plesiochronous streams, it may be internally generated. In all cases, the line code is removed. The data stream is then divided into the appropriate number of streams for the modulation process to be employed, i.e. 2 for QPSK, 4 for 16 QAM etc. These streams are then scrambled to remove strong spectral lines and to ensure adequate timing information for clock recovery in the demodulator. They are then retimed and the frame is established by inserting additional bits. Thus, each frame consists of a frameword, a block of data bits and a number of spare bits. Some of these are used for inserted supervisory or wayside traffic channels and the remainder for coding. Parity coding is normally employed to allow error monitoring along the route. In addition, some form of coding may be applied to permit the correction of errors by a technique called *forward error correction* (FEC) [12]. The streams then pass to the modulator.

In the receiver, the baseband processing performs functions which are the inverse of those at the transmitter. In digital systems, an important additional function that is performed is regeneration of the demodulated signal.

Most of the circuits employed in baseband signal processing lend themselves to integration. CMOS VLSI is widely used to reduce size and cost and to improve reliability.

12.6.3 Modulator and demodulator

In the transmitter, the modulator converts the baseband information into a modulated carrier suitable for transmission over the radio bearer. This may be done either at the output frequency of the transmitter, or at a lower intermediate frequency (IF) usually 70 MHz, which is then converted to the transmitter frequency. In the receiver, the demodulator recovers the data from the modulated signal. The modem is one of the most critical parts of the system and a very high degree of accuracy is essential to avoid performance degradations.

Initially, frequency modulation was universally adopted for analogue microwave links. Only by this means was it possible to obtain economically the very-high degree of input/output linearity essential for multi-channel telephony or colour TV transmission. Since the modulated carrier is a signal of constant amplitude, it is unaffected by limiting amplifiers such as the TWT amplifier and the output stage of an IF amplifier. There is an additional advantage in improved signal/noise ratio, depending on the baseband-frequency/deviation-frequency ratio. The demodulator is normally a simple frequency-discriminator circuit, which is preceded by a limiter to remove any unwanted amplitude modulation picked up from the transmission characteristics of the channel. Later, owing to pressure for more capacity in a given frequency band, single-sideband (SSB) amplitude modulation has been employed for very-high-capacity transmission.

For digital transmission, a single-frequency carrier can be modulated by amplitude, phase, or frequency. Phase-shift keying (PSK) or a combination of phase and amplitude, i.e. quadrature amplitude modulation (QAM), are prefered for high-capacity systems [13]. Constellations and eyes for different modulation schemes are shown in Figs. 12.5 and 12.6. The points in the constellation represent the amplitude and phase of the transmitted signal. All PSK-modulated signals have a constant amplitude (except during the transition from state to state), and the information is carried in the phase of the signal. The eye diagram shows alternative tracks that the signal can follow as it changes from state to state. The height of the eye opening relates to the spacing between the states; as they get closer together, the eye height reduces and the system becomes more sensitive to noise. The next level of PSK beyond 8 would be 16, but this would halve the eye height. A better spacing can be achieved by using a 4×4 matrix. This is 16 QAM. The 16 states produce 3 different amplitudes and 12 different phases of the transmitted signal. The signal for each state is obtained by adding together two signals in phase-quadrature. A 4-level in-phase (I) signal added to a 4-level quadrature-phase (Q) signal produces the 16 modulation states. Similarly, for 64 QAM, the I and Q signals have 8 levels.

Fig. 12.7a shows the block diagram of a digital modulator. The carrier to be modulated is generated by a frequency-stable oscillator, e.g. a crystal-controlled 70 MHz oscillator. It is fed to two double-balanced mixers. A 90° phase shift is added to the signal to one of the mixers to put the two modulated signals into phase quadrature. In a 4-state modulator, the two data inputs are each a single data stream. The D/A convertor followed by the low-pass filter applies the modulating signal to the mixer. The upper channel produces a modulated signal which is in phase with the driving voltage (the I channel) and the other in quadrature (the Q channel). Adding the two modulated signals produces the four states of 4 PSK. If two streams are fed to each input, each D/A convertor produces a 4-level signal, and the addition of the two modulated signals produces 16 QAM. The low-pass

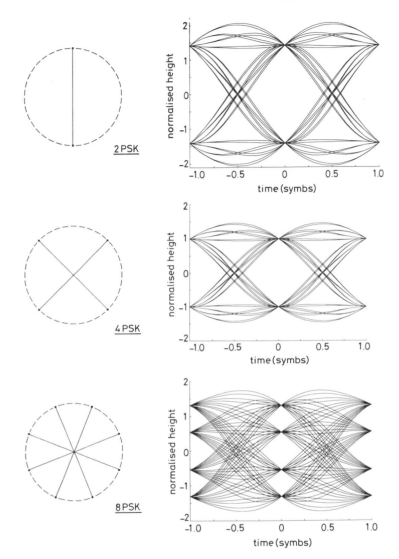

Fig. 12.5 Constellation and eye diagrams for PSK modulation

a 2 PSK
b 4 PSK
c 8 PSK

filters constitute the major part of the transmitter spectrum shaping. The band-pass filter at the output removes harmonics of the modulation process.

The demodulator recovers the data streams from the modulated IF signal. Fig.12.7 *b* shows the block diagram. The IF signal is fed to two double-balanced mixers together with a locally-generated reference signal, one in phase and one in quadrature. The reference signal must be coherent in

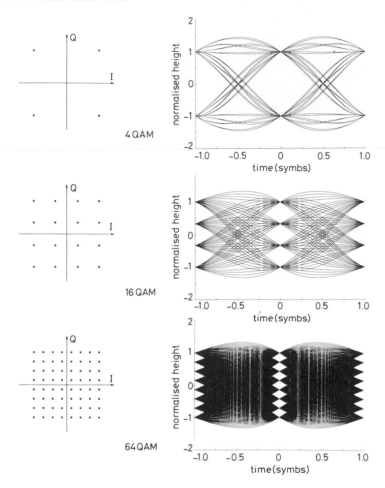

Fig. 12.6   Constellation and eye diagrams for QAM modulation

a 4 QAM
b 16 QAM
c 64 QAM

frequency with the modulated signal and maintained in the correct phase relationship. Various types of circuit, some based on correlation of the output data streams, are used to control the frequency and hence the phase of the reference signal.

The second process is regeneration, i.e. converting the analogue signals to the original digital streams. This requires a clock frequency, which is extracted from the analogue signals by squaring or slicing. The phase of the clock is crucial in sampling the analogue signals at the instant of maximum eye opening. Carrier recovery (CR) and timing recovery (TR) become more difficult as the received signal is distorted by multipath activity.

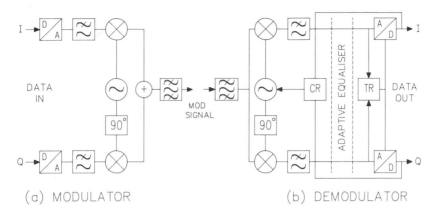

(a) MODULATOR   (b) DEMODULATOR

Fig. 12.7   Block diagram of QAM modem

a Modulator
b Demodulator

An adaptive equaliser is therefore often used to reduce the signal distortion. It operates on the analogue signals before regeneration.

The demodulator has no way of knowing which of the phases of the incoming signal is the correct reference phase. It therefore has four alignment positions which are equally likely; however, only one produces the same data streams as the modulator input. To overcome this difficulty, the data streams are differentially encoded in the modulator. The data are then carried as changes in modulation state rather than the absolute states and these are the same for any demodulator reference phase.

The band-pass filter at the input protects the demodulator from harmonics of the IF signal. The low-pass filters remove any IF signals from the mixers and form part of the overall receive selectivity.

12.6.4 RF units

The microwave source must provide sufficient output power, e.g. 10 to 50 mW. It must also maintain an accurate frequency, i.e. within 30 parts/million (PPM), over the working temperature range of the equipment, e.g. 0° to 50°C. Transistor, Impatt or Gunn-diode oscillators can provide the required power level. The necessary frequency stability can be achieved either by coupling the oscillator to a high-Q resonator (a cavity or a puck of high-dielectric-constant material), or by providing electronic frequency control of the oscillator and coupling this via a phase-locked loop (PLL) to a reference frequency derived from a crystal oscillator. The resonator has the advantage of simplicity and cheapness, but it has difficulty in achieving the required temperature stability and the ability to tune to different frequencies. The PLL oscillator overcomes these problems, but with greater technical complexity and higher cost. An ideal source would be a single active oscillator device coupled to an inexpensive passive stabilising device, such that either part could be changed without expensive test equipment to realign the combination.

The output power amplifier must convert milliwatts to watts over a broad frequency range with good DC–RF efficiency. For many years, the TWT amplifier has most adequately fulfilled these requirements. A modern packaged amplifier provides 40 dB gain and 10 or 20 watts of output power over several hundred MHz of bandwidth with an efficiency of about 25%. The major disadvantage is that, since it uses a thermionic emitter, it has an average life of 2 to 3 years and then has to be replaced. It is this replacement cost, in labour and materials, which encourages the change to all-semiconductor amplifiers.

Over the years, GaAs FET transistors have been developed to give more output power at higher and higher frequencies. 20 W at 6 GHz and 1 W at 19 GHz are currently available. The overall efficiency of a 4 W 6 GHz amplifier is about 13% for a saturated amplifier, but only about 5% for a linear amplifier.

Both TWT and GaAs FET amplifiers give the best efficiency and lowest AM/PM conversion when operating in a limiting condition. SSB and QAM systems both require a high degree of linearity in the transmission path. Consequently, the amplifiers have to be operated in a non-saturated condition, and the performance may be further improved by a linearising circuit.

The receiver downconverts the input signal to a lower intermediate frequency, usually 70 MHz, at which it is easier and cheaper to provide the necessary gain and apply automatic gain control (AGC) to hold the input level to the demodulator constant over the full fading range of the receiver.

The noise figure of the receiver directly affects the C/N ratio, and hence the threshold of the receiver. In modern receiver design, a low-noise amplifier (LNA) forms the input stages of the receiver with sufficient gain to make the noise added by the succeeding downconverter and IF amplifier a very small contribution to the overall noise figure. Since the LNA amplifies the noise on the image frequency as it does the wanted signal, an image-rejecting mixer is required as the downconvertor. Because of the large dynamic range of the input signal, particularly where enhanced input levels of 15 dB above normal are possible (up fading), AGC is normally employed as part of the LNA. An IF bandpass filter is included as part of the overall receiver selectivity, particularly where additional selectivity is required to prevent interference between transmitter and receiver.

The transmitters and receivers of several channels at different frequencies can share a common antenna, as shown in Fig. 12.4. Their signals are multiplexed by making use of the directional properties of ferrite circulators together with the filters connected to their ports. Usually, one channel is reserved as a protection channel. It is automatically switched in to replace an operational channel if it fails. For an analogue bearer, the switching (which is carried out at IF or baseband) occurs if the level of the pilot signal of a channel falls by 6 dB or the noise at the pilot frequency increases beyond a predetermined threshold. For a digital bearer, it is carried out if the bit error ratio exceeds a preset threshold.

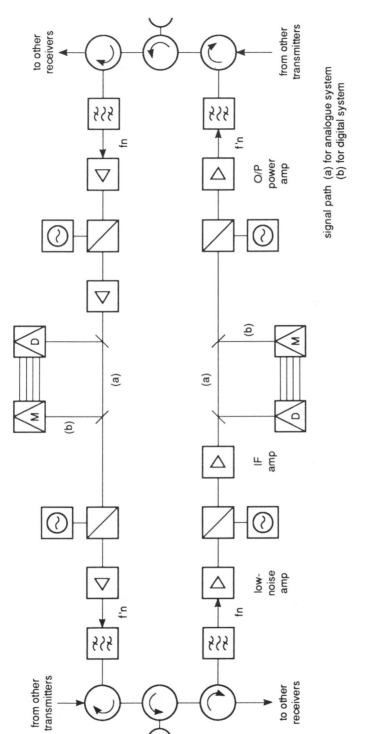

Fig. 12.8 Block diagram of repeater. (*a*) for analogue system (*b*) for digital system.

12.6.5 Repeaters

The block diagram of a repeater is shown in Fig. 12.8. The input signal of a particular channel is separated from the others by a circulator and a band-pass filter. It is then downconverted to the intermediate frequency where it is amplified. The IF amplifier has automatic gain control in order to cope with fading. In an analogue system this IF signal then passes to the upconverter of the transmitter (connection (a) in Fig. 12.8), to shift the signal to the frequency to be transmitted. For the L 6 GHz band, this is 252 MHz above or below the input frequency and for the U 6 GHz band the difference is 340 MHz. Finally, the signal is amplified by the output power amplifier, combined with the signals of other channels and retransmitted.

In a digital system, the signal is demodulated so that noise can be removed by regeneration (connection (b) in Fig. 12.8). The regenerated outputs from the demodulator are connected directly to the modulator inputs at the signalling rate. It is not necessary to provide any digital processing to interface the demodulator and modulator at the data rate. For example, for a 140 Mbit/s system employing 16 QAM, the interface will be four streams of approximately 37 Mbit/s. The modulator output at the intermediate frequency is then upconverted to the frequency to be transmitted.

12.7 Performance assessment of radio systems

12.7.1 Analogue systems

Broadband analogue radio systems, whether their application is multichannel telephony or colour TV plus sound channels, must satisfy certain requirements in respect of the following transmission characteristics:

(a)   Gain/frequency response
(b)   Group-delay/frequency response
(c)   Linearity of modulator and demodulator transfer characteristics (derivative).

Departures are cumulative on multi-hop systems. If these are not controlled sufficiently, there will be consequential distortion in the demodulated baseband signal.

In the case of telephony, the distortions cause intermodulation products which can fall into other channels [14]. In the case of colour TV, distortion results in colour subcarrier amplitude and phase errors as the luminance level (on which the subcarrier is imposed) varies between black and white.

Examination of transmission characteristics provides useful diagnostic information in the case of performance deficiency and it is carried out using specialised test equipment. This produces a low-frequency sweep voltage on which is superimposed a low-level 'search' frequency of around 500 kHz. Detection of the amplitude and time-delay variation of the search frequency as the sweep signal causes excursions either side of the centre frequency of the IF passband is used to display the transmission characteristics. Fig. 12.9a shows typical characteristics for an 1800-channel telephony system.

Another diagnostic technique which is also used is to measure the high-frequency derivative response. As the frequency of the search signal is

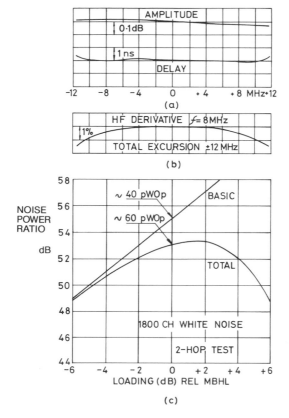

Fig. 12.9 Typical performance characteristics of an 1800-circuit telephony system

a gain and group-delay responses
b high-frequency derivative response
c overload response

increased, the sensitivity to amplitude departures increases very rapidly and this is a useful technique to search out hidden amplitude responses e.g. in limiters [15]. Fig. 12.9*b* shows a typical characteristic.

Baseband performance assessment on analogue radio systems for telephony invariably makes use of white-noise test signals [16] of the same bandwidth as the FDM baseband signal (e.g. 300 kHz-8.2 MHz for 1800 channels). A facility is provided for suppressing the noise at the transmit end in narrow 'slots' at a selection of test frequencies over the baseband range. At the receiving end, highly-selective filters permit the measurement of the noise level falling into corresponding slots. Three measurements are normally made:

(*a*) Reference level with the white-noise signal applied and transmit slot suppression removed.

(b) Total noise (thermal plus intermodulation) with the white-noise signal applied and transmit slot suppression invoked.
(c) Basic (thermal) noise only with the white-noise signal removed at the transmit terminal.

The difference between noise levels (b) or (c) and the reference level (a) is called the *noise power ratio* (NPR).

By increasing the level of the white noise relative to the normal signal power, termed the mean busy-hour loading (MBHL), it is possible to assess the overload characteristic of the system while observing total noise (b). By attenuating the receiver input level while observing basic noise (c), it is possible to determine the fade margin of any particular hop. Fig. 12.9c shows a typical non-fading NPR performance for a two-hop 1800-channel system. Under free-space conditions, the objective is to meet the CCIR requirement of 3 pW per km and fade margins of 40 to 50 dB are usually required to meet CCIR requirements under fading conditions [7].

Baseband performance assessment for colour TV is normally done by waveform analysis of CCIR-defined test signals, including 50 Hz and 15 kHz square waves, sine-squared pulse, staircase with superimposed colour subcarrier and luminance/chrominance inequalities. Random and periodic noise are also measured. The CCIR gives performance guidelines [17, 18].

12.7.2 Digital systems

The performance of digital radio transmission is measured by the bit error ratio (BER). This can be measured, using external test equipment, by sending a pseudo-random sequence down the link and counting the errors at the receive terminal. Alternatively, the violations of the parity coding applied in the transmitter digital processing equipment can be counted.

The CCIR recommends a number of error requirements which should be met for a link of length L between 280 and 2500 km for any month [19], as follows:

(a) Severely errored seconds; $BER \geqslant 10^{-3}$ (Integration time, 1 sec)   not more than $\dfrac{0 \cdot 054 \times L}{2500}$ %

(b) Degraded minutes; $BER \geqslant 10^{-6}$ (Integration time, 1 min)   not more than $\dfrac{0 \cdot 4 \times L}{2500}$ %

(c) Errored seconds;   not more than $\dfrac{0 \cdot 32 \times L}{2500}$ %

(d) Residual BER (RBER):   not more than $\dfrac{L \times 5 \times 10^{-9}}{2500}$

Apart from equipment faults, errors occur on radio links either when flat fading reduces the signal to a point where thermal noise plus any interference causes the regenerator in the demodulator to make an incorrect decision, or when selective fading distorts the pulses and again causes errors.

There may also be a combination of flat and selective fading. Each of these can be assessed by different tests.

Performance in the presence of thermal noise can be measured by reducing the signal level to the receiver, thereby degrading the C/N ratio. The range of levels of interest are those where the error rate increases rapidly over the range 10^{-8} to 10^{-2}. Fig. 12.10 shows the characteristic of a 140 Mbit/s QPSK system. The difference between the measured performance and the theoretical line is a measure of the imperfections in the equipment. From this measurement, we determine the signal level at which the error rate will be 10^{-6} or 10^{-3}, and hence the fade margin. At higher signal levels, the characteristic will reach a 'floor' which should be very low, e.g. 10^{-13}. This cannot be measured directly, but can be calculated from error rates measured with an interfering carrier [20].

Whilst the test of bit error ratio (BER) versus signal level gives the basic fade margin of the equipment, it is also desirable to know how the system will behave under multipath propagation conditions. The distortions arising from multipath transmission can add to the effects of signal attenuation to produce an effective fade margin which may be considerably reduced from the basic fade margin. The sensitivity of the equipment to such distortions is known as the *system signature* [21]. The signature is the locus of points of equal error ratio, say 10^{-3}, produced by the distortion arising from two-ray transmission. An example is shown in Fig. 12.11c. The plane of the signature represents all combinations of relative amplitude and phase of the two signals. It is normally calibrated in terms of the depth of the amplitude notch and its offset in frequency from the centre of the channel. It forms

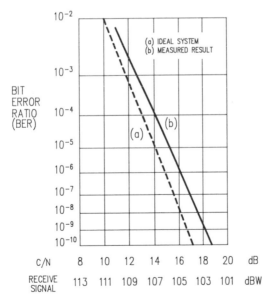

Fig. 12.10   Variation of bit error ratio with thermal noise for a 140 Mbit/s QPSK system

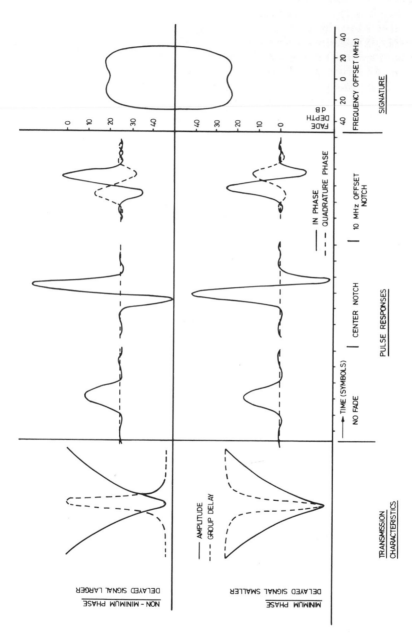

Fig. 12.11 Effect of two-ray propagation on a 140 Mbit/s QPSK system

a Transmission characteristics
b Pulse responses
c Signature

a very sharp boundary between the notches that fall outside, which have very little effect on transmission performance, and those falling inside, which cause complete loss of transmission. There are two halves to the plane; that where the delayed signal is the smaller, termed 'minimum-phase' fades, and that where the delayed signal is the larger, termed 'non-minimum-phase' fades. Although the amplitude response of the transmission path is the same for any given notch depth, for non-minimum phase fades the group delay is inverted and the distortion to the pulse response occurs before the peak amplitude (a precursor) rather than after.

The height and width of the signature vary with the delay between the two signals relative to the bit period. Hence: the longer the delay, the larger the signature; the longer the bit period (lower signalling speed), the smaller the signature for a given delay.

12.8 Spectrum efficiency

In any digital radio transmission, there is a three-way balance between the impairments due to thermal noise, intersymbol interference and adjacent-channel interference. For any given transmission rate, say 140 Mbit/s, these factors control the channel bandwidth and channel spacing, and hence the number of bearers which can be accommodated in the available frequency band, i.e. the *spectrum efficiency*. A simple measure of spectrum efficiency is the traffic bit-rate divided by the channel spacing, e.g.

$$\frac{140 \text{ Mbit/s}}{40 \text{ MHz}} = 3 \cdot 5 \text{ bit/s/Hz}$$

To improve spectrum efficiency, we need to reduce the required channel bandwidth so as to reduce the channel spacing. This can be achieved by one of two methods:

(*a*) We can increase the number of modulation states, e.g. by using 8PSK, or using 16 or 64 QAM. The channel bandwidth is thereby reduced to $\frac{2}{3}$, $\frac{1}{2}$ or $\frac{1}{3}$ of that required for 4 PSK.

(*b*) We can deliberately reduce the bandwidth of the channel by narrow filters. This generates intersymbol interference, spreading each bit into the successive bit or bits, and rapidly closes the eye. Because the intersymbol interference is fixed by the filters, it is possible to cancel the interference by feeding back some part of the previous bit, or bits. This process, called decision feedback, is described in Section 7.3. When applied to 4 PSK modulation, it produces a modulation system termed reduced-bandwidth quaternary phase-shift keying (RBQPSK) [22].

Improved spectrum efficiency is not achieved without penalty. In all cases, the eye aperture at the regenerator is reduced and the C/N required for a given bit error ratio (BER) is increased. This makes the system more sensitive to low signal level, multipath propagation distortion and external or internal interference. Table 12.2 illustrates this by comparing the parameters of RBQPSK, 16 QAM and 64 QAM systems with QPSK.

Table 12.2   Comparison of spectrum efficient modulation schemes with QPSK

Modulation	Effective eye opening	Performance at 10^{-5} BER		Co-polar channel spacing for 140 Mbit/s MHz	Spectrum efficiency bit/s/Hz
		Ideal systems CNR dB	Practical systems CNR dB		
QPSK	0·88	13·8	15·5	134	2·2*
RBQPSK	0·62	17·0	18·4	90	3·1†
16QAM	0·32	22·7	25·1	80	3·5*
64 QAM	0·14	29·5	30·5	59·3	4·7*

* Systems using interleaved cross-polarised channels
† Systems using co-frequency cross-polarised channels

Early systems established in new frequency bands tended to choose spacings to suit the modulation process, e.g. 11 GHz QPSK on a 67 MHz cross-polar spacing and RBQPSK on a 90 MHz co-polar spacing. More recently, there has been a move towards using the same frequency plan for digital transmission as originally set by analogue transmission, e.g. those shown in Fig. 12.1 for the 6 GHz bands. As a result, the question becomes; what modulation is required to obtain the best spectrum efficiency in a 40 MHz or 30 MHz channel spacing? On this basis, 16 QAM can easily fit 140 Mbit/s within a 40 MHz plan, but not into 30 MHz where 64 QAM is required. These systems have the benefit of filtering plus cross-polar discrimination (XPD) to protect the channel from interference from adjacent channels. The capacity of the band can be doubled if channels can be added to operate co-frequency cross-polar. The channel now has increased interference from adjacent channels on the same polarisation, requiring better filter protection, and from the co-frequency channel, protected only by the XPD. Tighter filtering is achieved by reducing the roll off, α, of the channel filtering from 0·5 to 0·19, and by improving XPD by the use of better antennas perhaps supported by some form of cross-polar canceller (AXPIC) unit [23].

256 QAM is the next target in the quest for higher spectrum efficiency. This would permit 140 Mbit/s in 20 MHz channel spacing or 2×140 Mbit/s in 40 MHz.

The advantages and disadvantages of increased spectrum efficiency are summarised in the following statements:

(*a*)   Increased spectrum efficiency reduces the cost per channel, either by increasing the number of channels per bearer, or by increasing the capacity of each frequency band. Hence, it postpones the requirement for the new antennas, feeders etc. required to move into a new frequency band.
(*b*)   Increasing the number of modulation states reduces the signalling frequency, and hence reduces the size of the signature, as shown in

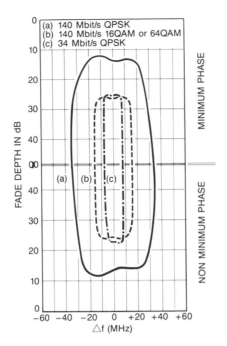

Fig. 12.12   Two-path signatures (Delay = 2·5 ns, BER = 10^{-3})

Fig. 12.12. The effect of selective fading is consequently reduced for a given data rate.

(*c*)   Multilevel modulation schemes reduce the eye height, and hence require a higher C/N for a given error rate. Some of this is recovered, because the smaller bandwidth of the receiver reduces the thermal noise and therefore improves the C/N. However, overall, the fade margin is reduced for a given output power.

(*d*)   The sensitivity of the modulated signal to equipment imperfections increases rapidly with the number of modulation states. For example, the need for better linearity results in a larger back-off in the output amplifier and this either reduces the fade margins still further or increases cost by requiring an amplifier with a higher saturated output power.

(*e*)   Because of the small eye, the system is very sensitive to all sources of interference. This makes it difficult or impossible to plan the system into a dense network.

12.9 Adaptive techniques

12.9.1 General
Apart from equipment failure, all outages on radio systems arise from anomalous propagation or rain attenuation. Little can be done to overcome ray bending or ducting, which takes the energy away from the receiving antenna. This results in flat fading; i.e. a uniform reduction of level at all

frequencies. More common is multipath propagation, where a number of
signals from the transmitter arrive at the receiver with varying levels and
varying delays relative to the main (unfaded) signal. High-capacity digital
transmission is extremely sensitive to delayed signals. They cause distortion
of the pulses, which closes the eye and upsets the operation of the receiver
in extracting its carrier and clock frequencies. We will therefore review the
adaptive techniques which have been employed to-date on digital systems
[24].

12.9.2 Height diversity

Height diversity has been used for many years to avoid the loss of signal to
the receiver which occurs when equal signals arriving at the antenna cancel
completely. Two antennas are used; ideally they are spaced so that when
signals cancel at one antenna they add at the other. This is relatively easy
when the reflection point is known, as in over-water paths; however, it is
less easy when the delayed signal arises from layering of the atmosphere.
Initially, two receivers were provided and the output switched to the one
with the highest signal level. More recently, techniques have been developed
to combine the signals from the two antennas, either at RF or at IF.

The block schematic of an RF-combiner diversity receiver is shown in
Fig. 12.13. The variable phase shifter is used to adjust the phase of the
signal from the diversity antenna, so that it adds to that from the main
antenna. This is termed 'maximum power' combining. However, for a
two-path situation, each antenna receives a main signal and a delayed signal;
thus, the phase shifter can be adjusted to give maximum cancellation of the
delayed signal in the combined output, i.e. 'minimum-distortion' combining.
In either case, the drop in signal level is drastically reduced and the amount
of distortion is also reduced. For this reason, space diversity makes a major
improvement in outage on high-capacity digital systems. Improvement factors reported from field evaluation range from 20 to over 50 [25].

One other advantage is gained by using space diversity. Most systems use
two polarisations, horizontal and vertical, to achieve additional discrimination between adjacent channels. Although high values of discrimination can
be obtained from well-designed antennas, i.e. over 40 dB, this is reduced
dB for dB as the signal fades. Space diversity therefore not only preserves
the receive signal level; it also maintains the cross-polar discrimination, and
thereby adjacent-channel interference protection.

12.9.3 Frequency domain equalisers

The earliest adaptive equalisers were designed to reduce the effect of
multipath propagation on the amplitude response of the received signal.
In their simplest form, a variable amplitude tilt is applied. The magnitude
and polarity of the tilt are controlled by two narrow-band detectors which
monitor the level of the received signal at each end of the spectrum.
More-sophisticated equalisers use more detectors to control higher-order
amplitude responses, i.e. 'bumpers and tilters'. Yet another approach uses
a single tuned circuit which is variable in frequency and amplitude, and is
automatically adjusted to minimise the distortion to the spectrum of the
received signal.

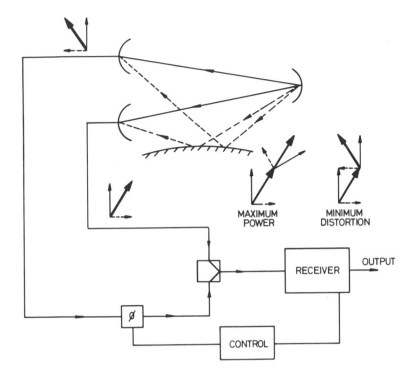

Fig. 12.13 Height diversity using RF combining

The improvement provided by this class of equaliser is partly due to the reduction of intersymbol interference. However, they also assist the carrier-recovery and clock-recovery circuits to hold in longer and recover more quickly in severe multipath conditions. Their limitation is the inability to detect and equalise the phase characteristic of the distorted signal when an in-band notch flips from minimum to non-minimum phase, thus causing the group-delay characteristic to invert repeatedly.

12.9.4 Time domain equalisers

The time-domain equaliser is designed to reduce, at the sampling instant, the offsets in the pulse response which arise from multipath distortion. It does this by adding to the signal, before regeneration, variable amounts of the received signal which are advanced and delayed in time, usually by an integral number of bit periods.

In a transversal equaliser (TVE), equalisation may be applied before or after demodulation (i.e. at IF or baseband). Fig. 12.14*a* shows a 5-tap baseband transversal equaliser with two forward and two backward taps. The amplitude of the signal added by each of the taps is controlled by an attenuator. The control of these tap weights is derived from the correlation of an error signal with the appropriate advanced or delayed signal. Determining the algorithm for the control signals is a major part of the design.

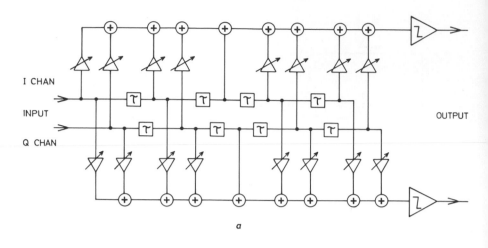

a

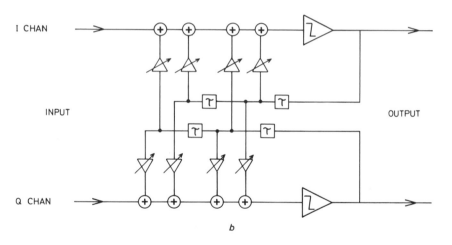

b

Fig. 12.14 Time-domain equalisers
a 5-tap transversal equaliser
b 2-tap decision-feedback equaliser

The simplest is termed 'zero forcing', which sets the correlated error to zero [26].

In a decision-feedback equaliser (DFE), the added signals are taken after the regenerator. The summing takes place after demodulation to baseband and the technique only works for backward taps. Fig. 12.14*b* shows the schematic.

Comparing these alternatives, the TVE has the disadvantages that it adds noise to the signal and it may modify the selectivity of the receiver. The DFE cannot correct for precursor distortion; consequently, it only improves the minimum-phase signature. A combination of TVE for the forward taps and DFE for the backward taps gives the best technical improvement at the expense of increased complexity [27].

12.10 Performance of radio links in the network

For analogue transmission, the estimation of performance for a given route or network has become a well-established practice based on Rayleigh fading.

In the case of digital transmission, the sensitivity to multipath distortion has made the task more difficult [28]. It was recognised that there was no longer a simple relationship between depth of fade and error rate probability. Measured results for a given equipment were used to determine the *effective fade margin*. This is the depth of fade which has a 50% probability of producing an outage. Performance is then calculated using this value of fade margin in the usual way. More-recent studies relate the performance to the measured signature characteristic of the equipment [29, 30]. This, in turn, requires an estimation of the probability function of the delay on the particular path.

Fig. 12.15 (taken from Reference 28) shows the results of field measurements for a number of different equipments, using different modulation schemes, in different frequency bands, from different parts of the world. It can be seen that the outage increases rapidly with path length, so the performance of a route is dominated by the performance of its longest hops. The improvement gained from the use of diversity and adaptive equalisers is equally clear.

12.11 Current trends

Digital microwave radio systems have operated in most countries for many years. Trunk transmission at 140 Mbit/s has been provided using QPSK, 16 QAM and 64 QAM modulation. Higher-capacity systems based on 155·52 Mbit/s and multiples thereof can be expected, using multicarrier 256 QAM or possibly 512 TCM (Trellis Code Modulation) [31]. Much effort has been devoted to improving adaptive-equaliser techniques to make the spectrum-efficient systems more robust. As the frequency bands in the 2–12 GHz range become saturated, greater use will be made of the higher microwave frequencies. Equipment 2, 8 and 34 bit/s is available in the 13, 15 and 19 GHz bands. Millimetric frequency bands such as 38 GHz and 60 GHz can be expected to provide short-haul systems in the near future.

Radio links will continue to play their part in transmission networks, particularly where service is required quickly, or the terrain is difficult for putting in the necessary ducts for cable or fibre. Generally, the world is changing from analogue to digital transmission and these trends will continue. However, some countries may have many years of frequency-band sharing for analogue and digital transmission. Advances in equipment

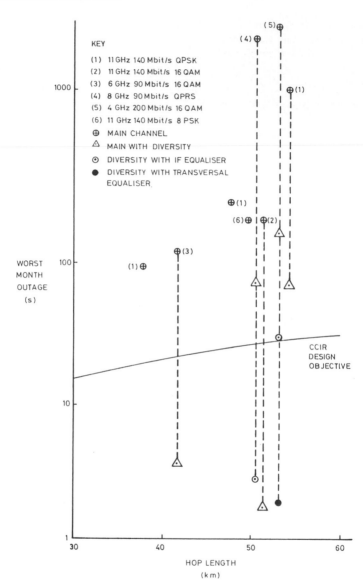

Fig. 12.15   Worst-month outage for high-capacity digital radio

design to make cheaper more-reliable and more-available systems will continue, stimulated by advances in technology and, of course, engineering effort [32].

12.12 References

1 CLAYTON, R.J., ESPLEY, D.C., GRIFFITHS, G.W.S. and PINKHAM, J.M.C.: 'The London-Birmingham television radio-relay link', *Proc. IEE*, 1951, **98**, Part 1, pp. 204–226

2 CCIR: XVIth Plenary Assembly, Dubrovnik, 1986, Volume IX, Report 781
3 HARTMAN, P.R. and TAYLOR, D.P.: 'Telecommunications by microwave digital radio', *IEEE Comms. Mag.*, 1986, **24**, pp. 11–16
4 CCIR: XVIth Plenary Assembly, Dubrovnik, 1986, Section 9B
5 RUMMLER, W.D., COUTTS, R.P. and LINIGER, M.: 'Multipath fading channel models for microwave digital radio', *IEEE Comms. Mag.*, 1986, **24**, pp. 30–42.
6 CCIR: XVIth Plenary Assembly, Dubrovnik, 1986, Report 376
7 CCIR: XVIth Plenary Assembly, Dubrovnik, 1986, Rec. 393
8 CCITT: VIIIth Plenary Assembly, Malaga–Torremolinos, 1984, Rec. G823
9 HYAMSON, H.D., MUIR, A.W., and ROBINSON, J.M.: 'An 11 GHz high capacity digital radio system for overlaying existing microwave routes', *IEEE Trans.* 1979, **COMM–27**, pp. 1928–1937.
10 MATSUMOTO, S., SANGO, J. and SEGAWA, J.: '200 Mbit/s 16 QAM digital radio relay system operating in 4 and 5 GHz bands', *Japan Telecom. Review*, Jan 1982. pp. 65–73 (see also July 1988, pp. 34–41)
11 BATES, C.P., ROBINSON, W.G. and SKINNER, M.A.: 'Very high capacity digital radio systems using 64-state quadrature amplitude modulation', IEE 3rd Int. Conf. on Trans., London, 1985, pp. 48–51
12 MICHELSON, M.A. and LEVESQUE, A.H.: 'Error-control techniques for digital communication', J. Wiley, 1985)
13 NOGUCHI, T., DAIDO, Y. and NOSSEK, J.A.: 'Modulation techniques for microwave digital radio', *IEEE Comms. Mag.*, 1986, **24**, pp. 21–30
14 'Transmission systems for communications', Bell Telephone Laboratories, 1971 4th edn., chap. 21
15 'Differential gain and phase at work' Hewlett-Packard Application Note 175–1
16 'Multichannel communications systems and white noise testing'. Marconi Instruments Library Series
17 'Television video transmission measurements'. Marconi Instruments Library Series
18 CCIR: XVIth Plenary Assembly, Dubrovnik, 1986, Volume XII, Rec 567
19 CCIR: XVIth Plenary Assembly, Dubrovnik, 1986, Volume IX, Rec 634
20 'Determination of residual bit error rates in digital microwave systems'. Hewlett Packard Product Note 3708–3
21 GIGER, A.J. and BARNETT, W.T.: 'Effects of multipath propagation on digital radio', *IEEE Trans.*, 1981, **COMM–29**, pp. 1345–1351
22 DUDEK, M.T., ROBINSON, J.M. and CHAMBERLAIN, J.K.: 'Design and performance of a 4·5 bit/s/Hz digital radio using reduced bandwidth QPSK' *IEEE Globecom '82*, D3·4, pp. 784–788
23 LANKL, B.: 'Cross-polarisation interference canceller for QAM digital systems with asynchronous clock and carrier signals', *IEEE Globecom '86*, 15.3.1 pp. 523–529
24 CHAMBERLAIN, J.K., CLAYTON, F.M., SARI, H. and VANDAMME, P.: 'Receiver techniques for microwave digital radio' *IEEE Comms. Mag.* 1986, **24**, pp. 43–54
25 CCIR: XVIth Plenary Assembly, Dubrovnik, 1986, Volume IX, Report 784
26 FENDERSON, G.L., PARKER, J.W., QUIGLEY, P.D., SHEPARD, S.P. and SILLER, C.A.: 'Adaptive transversal equalisation of multipath propagation for 16 QAM, 90 Mbit/s digital radio', *AT&T Bell Lab. Tech. J.*, 1984, **63**, pp. 1447–1463

27 DUDEK, M.T., CAMILLERI-FERRANTI, M., PRICE, A.J. and ROBINSON, J.M.: 'The performance of adaptive equalisers in an experimental 16 QAM digital radio relay system', IEE 3rd Int. Conf. on Trans., London, 1985, pp. 52–56

28 HART, G.: 'Performance of digital radio-relay systems', *Brit. Telecom. Eng.*, 1984, **3**, pp. 201–209

29 CAMPBELL, J.C.: 'Outage prediction for the route design of digital radio systems', *Austral. Telecom. Res.*, 1984, **18**, pp. 37–48

30 GREENSTEIN, L.J. and SHAFI, M.: 'Outage calculations for microwave digital radio', *IEEE Comms. Mag.* 1987, **25**, pp. 30–39

31 CHOULY, A. and SARI, H.: 'Application of trellis coding to digital microwave radio.', Conf. Rec. ICC '88, PA, USA, paper 15.1.1 pp. 468–472

32 YAMAMOTO, H. *et al.*: 'Future trends in microwave digital radio', *IEEE Comms. Mag.* 1987, **25**, pp. 41–52

Chapter 13

Satellite communication

M. Nouri

Microwave Communications Division, Marconi Communication Systems

13.1 Introduction

The launch of the Early Bird satellite in 1965 heralded the commercial exploitation of the geostationary orbit (GSO) [1]. This circular orbit, approximately 36 000 km above the equator, offers a number of advantages compared to non-geostationary circular or highly-elliptical orbits. A satellite repeater positioned in the GSO appears stationary to users on earth, which results in constant and predictable transmission parameters and simpler earth stations. As shown in Fig. 13.1, only three satellites are required to provide a global coverage of the earth's surface (except for polar regions). The Early Bird (Intelsat I) satellite transmission medium terminated at each end in gateway earth stations having large-diameter (32 m) antennas and located in selected sites away from centres of population. The services were trunk telephony using FDM/FM carriers and television using TV/FM carriers, with all tranmissions in the C band (i.e. 4 and 6 GHz).

It was recognised from the early days that satellite links not only offer a service which is independent of distance, and thus economical for long-distance communication, but are a 'natural' medium for mobile communication and for point to multi-point applications such as TV broadcasting. Other advantages were identified, such as easy application to disaster-relief communication and communication from areas which are difficult to access.

The decade following the launch of the Early Bird witnessed an increasing deployment of Intelsat gateway earth stations around the world. Further expansion of the Intelsat network followed the introduction of smaller earth stations (13 m) and thin-route digital carriers. The emergence of domestic networks from 1973 (Canadian Telesat, etc.) using dedicated satellite or leased satellite capacity resulted in a rapid worldwide growth of satellite networks. Furthermore, the first mobile satellite service was established in 1976 via the Marisat satellite, offering L band (i.e. 1.5 and 1.6 GHz) channels to maritime users and hence diversifying the use of the satellite transmission medium.

During the 1980s the satellite medium was made more and more accessible to individual users by a gradual reduction of earth-station size and cost. This was made possible by use of the Ku band (11 and 14 GHz), spot beams and improvements in performance of low-noise amplifiers. A link between 3 m earth stations using digital carriers for data communication, video

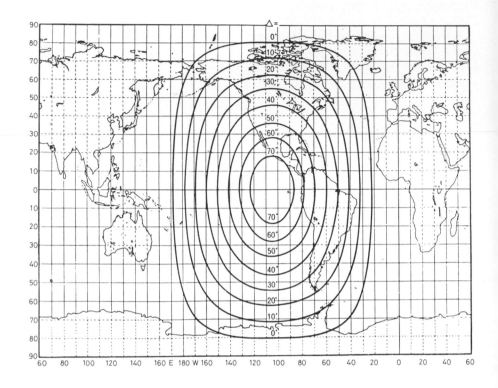

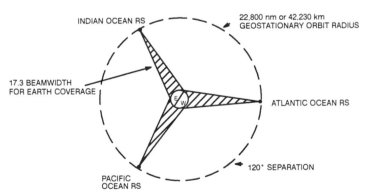

Fig. 13.1 Coverage of earth by geostationary satellites

a Typical coverage area for one satellite (with elevation angle as a parameter)
[*Courtesy: CCIR, volume IV, Geneva, 1982*]
b System with three satellites
[*Courtesy: SPILKER, Jr., J.J.: 'Digital Communication by Satellite, Prentice Hall, N.J.*
1977]

conferencing and a host of other business services was a typical scenario, almost completely eliminating the use of terrestrial tail-ends. An important development in Europe in this decade was the emergence of the Eutelsat regional (Ku band) network providing digital (TDMA) trunk communication between major cities, as well as digital business services and TV distribution via small terminals.

The trend towards smaller and more-economical user terminals has continued in the last few years with the advent of microterminals with a minimum antenna diameter of 60 cm for receive-only and 1·2 m for two-way communication. TV direct broadcasting satellite (DBS) services require only a 30 cm dish (or 20 cm square antenna) with a terminal costing no more than £250.

The major application areas and frequency bands are summarised in Table 13.1. This chapter gives a general introduction to satellite communication systems. Fuller information is given in References 2 to 7.

13.2 Satellite access methods

13.2.1 General
One of the key features of the satellite medium is its accessibility from anywhere within its *footprint*, i.e. the coverage area of the satellite antenna beam. Multiple-access methods enable various users within the footprint to share a satellite repeater (transponder). The following multiple-access methods are currently employed in satellite communication systems:

(*a*)   Frequency-division multiple access (FDMA)
(*b*)   Time-division multiple access (TDMA)
(*c*)   Code-division multiple access (CDMA)

13.2.2 Frequency-division multiple access
Frequency-division multiple access (FDMA) was the first access method employed in satellite links. It is based on dividing the transponder bandwidth between various carriers, with the provision of guard bands between adjacent

Table 13.1   Major application areas and frequency bands

Types of service	Frequency band (GHz) uplink/ downlink	Band designation	Allocated bandwidth (GHz)
Fixed satellite service (FSS)	6/4	C band	1·0
	14/12 (or 11)	Ku band	1·1
	30/20	K band	3·5
Broadcasting satellite	14/12	DBS band	0·8
service (BSS)	18/12	DBS band	0·8
Mobile satellite service (MSS)	1·6/1·5	L band	34/29 MHz

carriers to prevent interference. The individual carriers originate from any earth station and can be modulated by a multiplexed analogue signal (e.g. FDM/FM) or a digital signal (TDM) or a non-multiplexed analogue signal (single FM voice channel) or a digital signal.

The major international satellite systems used for telephony, e.g. Intelsat, transmit FDM groups similar to those used in terrestrial systems, with large blocks of channels allocated semi-permanently to each country involved [8]. Usually, a block of 972 channels (12 to 4028 kHz) occupies a transponder of 36 MHz bandwidth, using frequency modulation.

For the lower-traffic requirements of domestic satellite systems, a large number of carriers is amplified by each transponder with each carrier conveying an individual channel. This is termed *single channel per carrier* (SCPC) multiple access. Each carrier may transmit a companded analogue voice channel (using FM) or a 64 kbit/s digital channel (using PSK). The flexibility of SCPC is advantageous when small numbers of channels are transmitted from each of a large number of different earth stations in a satellite system. The satellite multiservice system (SMS), provided by Eutelsat for business applications, has up to 1600 carriers in its 72 MHz bandwidth [9]. These can be reallocated extremely quickly in response to changes in demand.

FDMA carriers are normally assigned to various earth stations according to a fixed frequency plan (i.e. pre-assigned). However, in applications where limited space-segment resources are to be shared by a relatively-large number of users (such as some SCPC systems), the satellite channels are accessed on demand (i.e. demand assigned). The control of the demand-assigned access can be carried out from a central earth station or can be distributed amongst the user earth stations.

13.2.3 Time-division multiple access

Time-division multiple access (TDMA) is based on time-sharing a transponder between different earth stations. Each earth station transmits a burst (a short-duration SHF signal) according to a *burst time plan*. The time plan is designed such that different bursts reach the transponder in the assigned positions within a TDMA frame without any overlap, as shown in Fig. 13.2. Guard times are configured in the frame to minimise interference. Fig. 13.3 shows a typical TDMA frame, corresponding to the Intelsat or Eutelsat international trunk network [10]. The link transmission rate in these sytems is 120 Mbit/s and the frame period is 2 ms. This leads to a capacity of about 1800 one-way channels per transponder.

Each earth station in the network is assigned one or more traffic bursts depending on the required capacity. The TDMA frame also contains reference bursts, RB1 and RB2, transmitted from different reference stations to provide frame positional information or frame reconfiguration data to traffic earth stations. The frame is received by all earth stations in the same beam, and each earth station extracts the data destined to it in accordance with the burst time plan.

In TDMA operation, since a transponder handles one burst at a time, it can be operated near saturation; this increases the link power efficiency

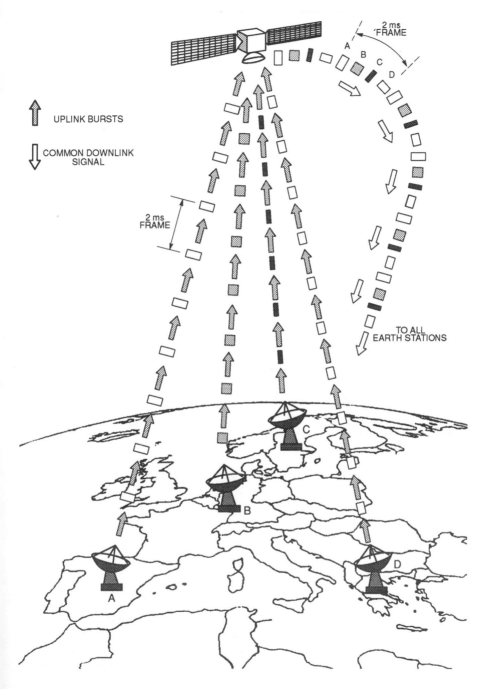

Fig. 13.2 Typical TDMA system

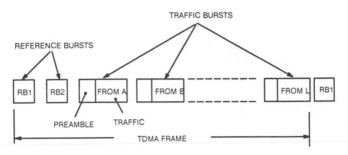

Fig. 13.3   TDMA frame format [*Courtesy: Intelset*]

compared to multicarrier operation as in the FDMA. Another advantage of the TDMA method is its flexibility of application to varying traffic situations by altering the time plan. For example, the individual earth station's capacity can be altered easily to suit a new requirement.

A TDMA earth station includes the following two major subsystems:

(*a*)   A TDMA control subsystem
(*b*)   A terrestrial interface subsystem

The TDMA control subsystem is responsible for frame acquisition and synchronisation, time-plan processing, preamble generation, burst generation and control, scrambling, error correction, modulation etc.

The terrestrial interface subsystem provides a direct or buffered interface to the terrestrial network and performs data formatting, synchronisation, multiplexing, echo suppression etc. In addition, terrestrial interface modules implement digital speech interpolation (DSI) on groups of incoming PCM voice signals [12] to provide a 2-to-1 capacity improvement (as described in Section 2.9). Facilities are available in this subsystem to insert order-wire channels and to satisfy the signalling requirements of the TDMA system.

13.2.4   Code-division multiple access

In code-division multiple access (CDMA), the user's signal is spread or hopped over a significant portion (or all of) the transponder bandwidth using spread-spectrum techniques [13]. No frequency or time co-ordination is necessary between the users, and the transponder access can be at random. The spreading is carried out by a family of orthogonal pseudo-noise (PN) codes, each member assigned to a user signal. An exact replica of the PN-code must be generated and synchronously maintained at the receiver for de-spreading the signal. Since PN-codes used by other signals in the same band are uncorrelated to the wanted signal, they will be rejected in the receiver and their effect significantly reduced. The main application of CDMA is in military systems. However, in recent years it has been used in commercial microterminals, primarily to combat interference from adjacent satellites and other sources.

13.3 Digital tranmission and business services

The bulk of present and near-future traffic on satellite systems consists of voice. In recent years, the traditional FDM/FM systems, widely used in international links, have adopted companding techniques to increase the number of voice channels carried by a given bandwidth. However, it is the emerging digital transmission methods which promise to achieve higher bandwidth efficiencies. The digital carrier types currently used in routes with a heavy telephony traffic, such as Intelsat major routes, are the TDMA and Intermediate Data Rate (IDR) carriers. The IDR carriers are generally TDM carriers with data rates up to 45 Mbit/s. The FDMA access method is used for narrower IDR carriers which share a transponder. The IDR carriers are being increasingly deployed in preference to FDM/FM carriers. They are simpler to implement than the TDMA carriers, but they are not as flexible in traffic reconfiguration.

Both TDMA and IDR carrier systems benefit from using DSI. The DSI process makes use of the inactivity of one way voice traffic during a conversation (listening to the other speaker, pauses, hesitations) to reduce the output bandwidth requirement by up to (and even exceeding) a factor of 2. Low-rate encoding (LRE) of speech is another technique recently adopted to increase the voice-traffic capacity of satellite links. The bit-rate requirement for toll-quality speech has been reduced from 64 kbit/s PCM to 32 kbit/s adaptive differential PCM (ADPCM), as described in Section 6.8. Moves are underway to introduce 16 kbit/s speech coding in the near future. Thus, using DSI plus LRE can increase the capacity by a factor of more than 8.

In recent years, digital satellite links using small terminals located near or at users' premises have been used increasingly for business services. The satellite link suffers in general from the following principal limitations:

(a) Additive wideband white Gaussian noise
(b) A transmission delay of 280 ms (one way)

A further limitation in a satellite link using small earth stations, as in business-services applications, is the power limitation. These features have prompted link designers to employ efficient and rugged modulation/coding techniques. Most satellite links have opted for BPSK or QPSK modulation, coupled with forward error correction (FEC). FEC is used to avoid data re-transmission. Techniques such as ARQ (Automatic Repeat Request), which are widely used in terrestrial data networks, are inefficient in satellite links because of the large two-way delay. The most common FEC method is convolutional coding and Viterbi (maximum likelihood) decoding [14]. Using a half-rate coding, a coding gain of 5·5 dB is achievable at the expense of doubling the bandwidth.

Business-services networks, such as the Intelsat Business Services (IBS) operated by Intelsat and the Satellite Multiservice System (SMS) operated by Eutelsat, use earth stations as small as 2·5 m in open network applications. An open network is carefully structured such that different corporate users complying with network specifications can communicate with each other. In contrast, a closed network could be a private network designed to cater

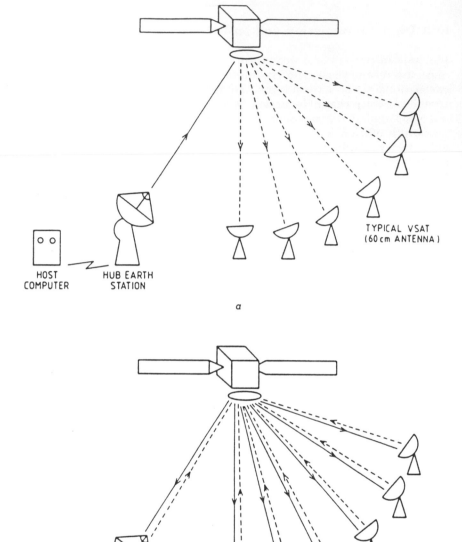

Fig. 13.4 Typical VSAT network configurations

a One-way VSAT network (data broadcasting)
b Two-way VSAT network (interactive data)

for the communication requirements of one business entity. In this case, the network designer is free to select the satellite link parameters freely, subject to meeting certain interference constraints.

An example of closed-network application of satellite links is the *very-small-aperture terminal* (VSAT) network [15] shown in Fig. 13.4. VSAT networks have had a rapid growth in the USA in recent years. These networks are characterised by a star topology with a large central (hub) earth station communicating with small (0·6 m to 1·8 m) remote earth stations. The networks are primarily used for data communication with computer-based transactional traffic. A typical VSAT link utilises a random-access TDMA method with carriers at 56 kbit/s from remote to hub and a TDM carrier at 256 kbit/s from hub to remote. However, the bit rates and access methods vary considerably from network to network.

The steady expansion of data networks such as LANs and WANs, the emergence of ISDN and broadband ISDN and the communication between data terminals and workstations are presenting the greatest challenge to satellite links. The interface speeds, codes and protocols vary greatly with user terminal types. The packet-switched satellite links already in operation in VSAT applications rely on protocol conversion to match the satellite-link features to existing terrestrial protocols, such as X.25, SDLC etc. Further work is in progress to define, the role of satellite links within future networks such as the broadband ISDN [16].

13.4 Earth station technology

Since the introduction of Intelsat's international gateway (standard A) earth stations over 20 years ago, there has been a dramatic rise in the type and number of earth stations worldwide. With the steady increase in the complexity of communication satellites there has been a general trend towards smaller and simpler earth stations. The range extends from multi-million pound 32 m Intelsat standard A earth stations to 30 cm DBS receivers costing about £250.

The functional block diagram of an earth station is shown in Fig. 13.5. The complexity of an earth station depends on whether it is used for both transmitting and receiving (TX/RX), for transmitting only (TX) or for receiving only (RX). It also depends on factors such as redundancy, single or dual-polarisation operation and, most importantly, on carrier types and volume of traffic.

Large antennas have very narrow beams, e.g. 0.3° for an 11 m antenna at 6 GHz. Thus, automatic satellite tracking is an essential feature of an earth station, except in very small earth stations. A satellite in the GSO undergoes small diurnal movements around its nominal position. This is kept to a minimum by satellite thrusters. The residual movement is called satellite station keeping, which is 0·1° in N/S and E/W directions in typical satellites. In large earth stations, the narrow antenna beam can easily miss the satellite because of the satellite station keeping and the wind loading of

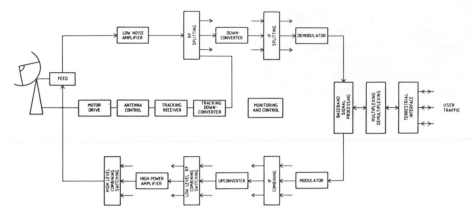

Fig. 13.5   Block diagram of a typical earth station

the earth-station antenna. If no provision is made for satellite tracking, this results in a considerable gain reduction and variation. Fig. 13.5 shows a subsystem for autotracking (called step tracking) currently used in many earth stations. The step-tracking technique is based on the reception of a satellite beacon signal via the low-noise amplifier (LNA), beacon down converter and beacon receiver. The antenna control unit is provided with an algorithm to move the antenna in small steps in order to maximise the received level of the beacon signal.

In almost all earth stations, the antenna employs a parabolic reflector, traditionally an axi-symmetric front-fed or Cassegrain design. More recently, offset-fed antennas (front-fed or Gregorian) have been introduced increasingly in small and some medium-size earth stations. The offset-fed antennas exhibit a lower sidelobe pattern, which results in lower interference into adjacent satellites. This is becoming an important issue, since a more efficient use of the GSO requires a closer spacing of satellites; 2° is the current minimum. Another advantage of a lower sidelobe envelope is the ability to transmit a higher *effective isotropic radiated power* (EIRP) compared to a symmetric antenna for the same level of unwanted off-axis radiation.

On the uplink, the traffic signal modulates an intermediate-frequency (IF) carrier, which is usually at 70 or 140 MHz. FM is used for analogue signals and PSK for digital signals. The modulated carrier is then frequency-converted, in a single converter or a series of converters, to the required transmitter frequency. This will be in the range 5·925—6·425 GHz for the C band or between 14·0 and 14·5 GHz for the Ku band. The high-power amplifier (HPA) and the antenna gain determine the power radiated towards the satellite. At C band, an antenna of 32 m diameter has a gain of 64 dBi and one of 11 m diameter has a gain of 54·5 dBi. Thus, an HPA of 3 kW output with an 11 m antenna gives an effective isotropic radiated power (EIRP) of about 89 dBW (over 800 MW). This power must be closely controlled to maintain the performance of the satellite repeater.

On the downlink, the extremely-weak signal received by the antenna is amplified by the low-noise amplifier (LNA). This signal is in the range 3·7—4·2 GHz (C band) or 10·95—12·75 GHz (Ku band). Its frequency is changed to the IF (70 or 140 MHz) in the down converter and then the demodulator recovers the traffic signal.

The RF transmission and reception capabilities of an earth station are specified by the EIRP and the G/T factor. The G/T factor is the ratio between the gain G of the antenna at the receive frequency and the noise temperature T of the antenna and the receiver referred to its input. The G/T factor is thus a measure of the sensitivity of the earth station to the very-weak signals received from a satellite. The minimum requirement specified by INTELSAT for its current largest C band earth stations (standard A) is 35·0 dB/K. The G/T factor is mainly determined by the antenna receive gain and the noise contributed by the antenna and the LNA at the front end of the receiver. The EIRP is determined by the gain of the antenna at the transmit frequency and the output power from the HPA of the transmitter. In earth-station design, a judicial choice of antenna size, LNA noise temperature and HPA power rating is sought to achieve a minimum-cost solution.

The most-commonly-used LNAs in current earth stations are the GaAs FET amplifier and, to a lesser degree, the parametric amplifier. Parametric amplifiers are more costly and more difficult to maintain, but they exhibit a lower noise temperature. However, the gap between the two noise temperatures has narrowed significantly in recent years. In medium-size large earth stations, cooled versions of the above amplifiers are often used to reduce the noise temperature. In early Intelsat standard-A earth stations, cryogenically-cooled parametric amplifiers were used with a cooling temperature of 20 K. However, because of advances in low-noise device technology, almost all current cooled LNAs use thermoelectric cooling, typically to −25°C. This, uses no liquid coolant and is less bulky and more economical.

Microwave tubes are extensively used as HPAs, except in very-small earth stations, such as microterminals used in business services and mobile earth stations, which use solid-state amplifiers. The RF power range required in earth stations is from around 1 W to 3 kW. The travelling-wave-tube (TWT) amplifier is the most-commonly used HPA; it covers a range of 20 W to 3 kW and provides a wide operational bandwidth (500 MHz). Klystron amplifiers are the other commonly-used type of microwave tube. Klystron HPAs are generally used in high-power (>600 W) applications in which the RF power falls within a transponder bandwidth. In comparison to tube amplifiers, solid-state amplifiers (single output stage, GaAs FET design) provide about 8 W at Ku band and 30 W at C band.

The combiners and splitters shown in Fig. 13.5 at IF (normally 70 or 140 MHz) and RF are used to add more carriers to the earth station. For example, SCPC carriers sharing the same transponder are frequently multiplexed (and demultiplexed) at the IF, whilst carriers destined for different transponders are combined at RF. The RF combining is carried out either at low level (pre-HPA) or at high level (post-HPA), depending on the carrier-power requirement and the type of HPA used.

13.5 Satellite repeaters

A satellite usually carries a number of repeaters, for example up to 50 on Intelsat 5. However, some of the these will be reserved for standby use. The repeaters in a satellite perform a similar transponder function to those in a terrestrial microwave link, as described in Section 12.6.5. However, the size of antennas that can be used is limited by the dimensions of the spacecraft. Also, the primary power source for the satellite is an array of solar cells; thus, the power available for the equipment is strictly limited. The severe restrictions on frequency spectrum, together with the large bandwidth of each channel (typically 36 or 72 MHz) require economical use of bandwidth. This may be used twice by means of dual polarisation; this can be orthogonal linear polarisation or right-hand and left-hand circular polarisation.

The block diagram of a typical repeater is shown in Fig. 13.6. As in a terrestrial microwave repeater, a frequency change is made so that filtering can ensure that the loop gain is well below zero at all frequencies in order to guarantee stability. Fig. 13.6 shows a transparent repeater. In a regenerative repeater, the incoming signal is demodulated to baseband and regenerated. The output from the regenerator then modulates the transmitter.

Modern communication satellites transmit multiple spot beams to maximise the power transmitted to the several earth stations in the system. In a TDMA system, the beams used must be changed frequently, since incoming signal bursts from each uplink beam must be steered to the correct downlink beam. This necessitates incorporating a switching matrix in the repeater. In a transparent repeater, the incoming signal is down-converted to IF in order to perform the switching. In a regenerative repeater, the

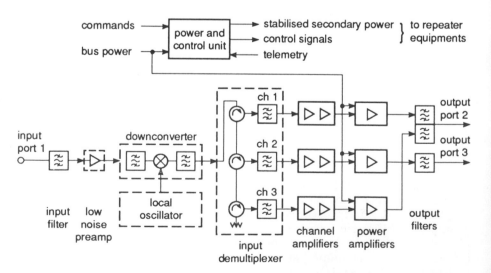

Fig. 13.6 Block diagram of satellite repeater

switching is performed at baseband, since it is necessary to demodulate the signals in order to regenerate them.

Because of the very low level of the incoming signal from its antenna, the receiver of the transponder must have a low-noise amplifier and field-effect transistors (FETs) are usually employed. The transmitters in most spacecraft employ travelling-wave-tube (TWT) power amplifiers. In some ways, the TWT appears most unsuitable for satellite applications. It has a limited life and it requires a high-voltage supply. Nevertheless, its wide bandwidth and high efficiency make the TWT the best choice. (High efficiency is essential because of the limited power supply.) The need to obtain high efficiency means operating the TWT near to saturation and non-linearity in this region causes intermodulation. It is therefore necessary to control signal levels carefully when FDMA or SCPC signals are to be amplified.

13.6 Link budgets

The parameters of the uplink and the downlink must be chosen to obtain the required carrier-to-noise ratio (C/N) at the satellite and the earth station respectively. These C/N ratios are rarely the same. The performances of the downlink signal source and the uplink receiver are determined by the designer of the satellite; the performances of the uplink signal source and the downlink receiver are determined by the designer of the earth station. At the earth station, the only common feature of the two links is a high-gain antenna, which will be optimised for one link or the other.

For the uplink, the earth-station antenna must be large and the RF power high in order to produce sufficient signal strength at the satellite to overcome the noise due to the 'hot' earth at which its antenna is pointing. For the downlink, the power transmitted by the satellite is normally much lower. Fortunately, the earth station's antenna is looking at the coldness of space, so much less noise is received.

For each link, the received carrier-to-noise ratio is given by

$$\frac{C}{N} = E - L + G - N - M \text{ dB}$$

where

$E = \text{EIRP}$
$L = \text{path loss}$
$G = \text{gain of the receiving antenna}$
$N = \text{thermal noise power}$
$M = \text{required operating margin}$

The EIRP is determined by the gain of the transmit antenna and the output of the HPA. The gain of the antenna (given by eqn. 3.9) depends both on its diameter and the frequency employed. The output power of the HPA will be less than its full output rating because of the need to 'back off', i.e.

reduce the output level, in order to obtain an acceptable intermodulation performance. The reduction made is usually between 1 and 6 dB.

The path loss (given by eqn. 3.10) depends on frequency and path length. Although the satellite is stationary, the slant path length is not the same for every earth-station location. The path loss needs to be calculated for each case. At 6 GHz, it is of the order of 200 dB. The operating margin is required to accommodate variations in path length and transmitter power and additional attenuation due to rain. A typical margin is 4 dB.

The thermal noise power N at the receiver is given by eqn. 2.13*b*. This may be rewritten as

$$N = -228 \cdot 6 + 10 \log_{10} T + 10 \log_{10} W + 60 \qquad \text{dBW}$$

where

T = noise temperature, K
W = bandwidth, MHz

Since the antenna gain increases the signal power, this is equivalent to reducing the noise power. Thus, the equivalent noise power is

$$N_e = -228 \cdot 6 - (G/T) + \text{bandwidth factor} \qquad \text{dBW}$$

where G/T is in dB/K and the bandwidth factor is

$$10 \log_{10} W + 60 \qquad \text{dBW}$$

Because of the limitation on satellite EIRP, the downlink is usually more critical in terms of noise than the uplink. As an example, the budget will be calculated for an 11 GHz downlink to an earth station with a receiver bandwidth of 36 MHz.

Satellite HPA output rating (20W)	+13	dBW
Back off	−2	dB
Waveguide losses	−1	dB
Antenna gain	30	dB
Satellite EIRP	40	dBW
Path loss at 11 GHz	205·5	dBW
Received power	−165·5	dBW
Boltzmann's constant	−228·6	dBW/Hz/K
Bandwidth factor (for 36 MHz)	75·6	dB Hz
Earth-station G/T factor	34	dB/K
Equivalent noise power	−187	dBW
$C/N = -165 \cdot 5 - (-187)$	= 21·5	dB

Allowing a 4 dB operating margin reduces this to 17·5 dB.

In a transparent-repeater system, however, noise on the uplink and intermodulation noise in the satellite repeater are also present. If these contribute one third of the total system noise, the downlink noise power is increased by 50% (i.e. 1·8 dB). The resulting C/N ratio is therefore reduced to 15·7 dB.

13.7 Satellite system capabilities

A steady growth in traffic handled by international, regional and domestic satellite networks has resulted in more than 100 operating satellites with an increasing complexity in satellite design. For example, the new generation of Intelsat satellites (Intelsat VI) provides nearly 200 times the capacity of Intelsat I, as shown in Fig. 13.7.

The technologies being developed for satellite links are heavily influenced by the need to make efficient use of the two scarce resources: i.e. available

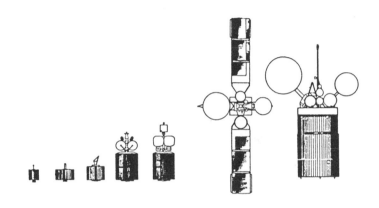

designation: Intelsat	I	II	III	IV	IV A	V	V A/V B	VI
year of first launch	1965	1966	1968	1971	1975	1980	1984/85	1990
prime contractor	Hughes	Hughes	TRW	Hughes	Hughes	Ford Aerospace	Ford Aerospace	Hughes
width (m)	0.7	1.4	1.4	2.4	2.4	2.0	2.0	3.6
height (m)	0.6	0.7	1.0	5.3	6.8	6.4	6.4	6.4
launch vehicles		Thor Delta		Atlas-Centaur		Atlas-Centaur and Ariane	Atlas-Centaur and Ariane	STS and Ariane
spacecraft mass in transfer orbit (kg)	68	182	293	1 385	1 489	1 946	2 140	12 100/3 720
communications payload mass (kg)	13	36	56	185	190	235	280	800
end-of-life (EOL) power at equinox (W)	40	75	134	480	800	1 270	1 270	2 200
design lifetime (years)	1.5	3	5	7	7	7	7	10
capacity (number of voice channels)	480	480	2 400	8 000	12 000	25 000	30 000	80 000
bandwidth (MHz)	50	130	300	500	800	2 137	2 480	3 520

Fig. 13.7 Evolution of INTELSAT satellites [*Courtesy: Intelsat*]

bandwidth and the geostationary arc. The re-use of the spectrum by the same satellite has been a pivotal development in order to increase capacity. This has been achieved by the introduction of spot beams covering a smaller portion of earth's visible disc from the GSO (Fig. 13.8) and, to a lesser degree, by employing dual-polarisation operation (Fig. 13.9). For example, as shown in Fig. 13.8, Intelsat VI satellites offer a sixfold frequency re-use in comparison with Intelsat I satellites. Moreover, narrower spot beams provide a higher EIRP, resulting in smaller and less-costly earth stations.

The current communication satellites can be termed 'low power' satellites; they use solid-state or TWT amplifiers with an output power up to 20 W. However, there is a trend to use higher-power transponders for television broadcasting to small domestic receivers and for data communication

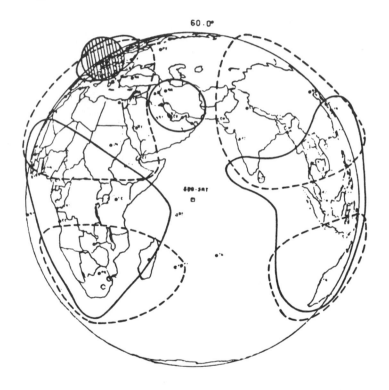

```
INTELSAT VI beam coverages for the
Indian Ocean Region - 60°E
------- Hemispheric beam
————— Zone      beam
  ⊘     Spot     beam
```

Fig. 13.8   INTELSAT VI beam coverage for the Indian Ocean region (60°E)
[*Courtesy: Intelsat*]

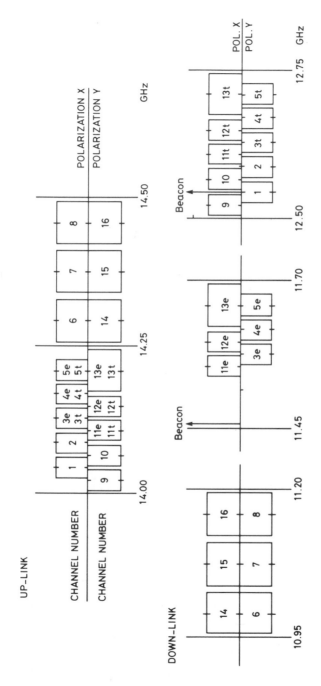

Fig. 13.9 Eutelsat II transponder plan [*Courtesy: Eutelsat*]

between business users via dedicated earth stations. This has led to the emergence of medium-power (40—100 W) and high-power (100—250 W) transponder satellites. Satellites such as ASTRA (medium power) and TDF1 (high power) have been launched since 1988 and others, such as Olympus and BSB (high power) and Eutelsat II (medium power), have followed recently.

Further increases in the capacity of a satellite require a higher degree of frequency re-use in the existing operational bands and the exploitation of new bands such as 30/20 GHz. This trend, plus the need for an efficient use of the GSO, requires deployment of larger satellites. The limit to the satellite size is imposed by the launch vehicle capability. With current launch vehicles, an increase in the satellite size by a factor of 2 is possible.

Feasibility studies have been carried out on the concept of large platforms in space, assembled at low earth orbit and transferred to the GSO. Such structures require a multiple launch to an orbiting space station, where assembling takes place. One concept pursued by Intelsat proposes a 245 × 236 m structure capable of generating 100 kW of prime power and providing five-million telephone channels and 200 TV programmes to earth users [17]. However, the implementation of such concepts are at least two decades away. Other developments which are more likely to be implemented in the 1990s to increase the capacity of the satellites are:

(*a*)   On-board processing
(*b*)   Narrow spot beams
(*c*)   Scanning spot beams
(*d*)   Widespread use of the 30/20 GHz band
(*e*)   Inter-satellite links

In parallel, improvements in interference-reduction techniques, better antenna design, a more-homogenous utilisation of the GSO and a tighter satellite station keeping should result in a closer spacing of satellites in geostationary orbit.

13.8 30/20 GHz operation

Most traffic growth forecasts predict a severe bandwidth congestion in C band and Ku band frequencies by the late 1990s. A natural solution will be to introduce the higher allocated band, i.e. 30/20 GHz, into service. This frequency band has the following advantages:

(*a*)   A large allocated bandwidth (3·5 GHz); this is wider than all lower satellite-communication bands combined
(*b*)   Narrower satellite spot beams; these allow increased scope for frequency re-use
(*c*)   Reduced frequency sharing with terrestrial networks
(*d*)   Smaller earth-station antennas

However, 30/20 GHz operation suffers from the following disadvantages:

(*a*)   High attenuation due to atmospheric precipitation

(*b*) High atmospheric depolarisation in frequency re-use applications; initially this is not a problem since the large available bandwidth makes frequency re-use unnecessary.

(*c*) Less mature technology; however, this is not a long-term problem.

A number of methods have been under investigation to combat the atmospheric effects [18]. Site diversity is one solution, but it is considered uneconomic because of the equipment duplication. Other methods rely on adaptive power control to compensate for rainfall losses. This can be carried out in the earth station for the uplink and on board the satellite for the downlink. Adaptive methods can also be used to reduce depolarisation [19].

The Japanese CS–2 satellites [20] are currently the only operational satellites using the 30/20 GHz band. A typical margin of 10 dB is used in critical links to allow for rainfall losses, which results in significantly-larger antennas. A number of European satellites, namely, Olympus, DFS Kopernikus and Italsat [21], have pre-operational 30/20 GHz payloads. NASA's ACTS satellite is another example [22]. This is an experimental satellite which exploits the 30/20 GHz band for commercial use and experiments. It has a highly-sophisticated onboard baseband processor along with beam-scanning antenna technology. The data collected from these satellites could pave the way for the most cost-effective methods of combating high rainfall losses in high-availability links.

13.9 TV distribution and DBS service

From the late 1970s there has been a large demand, initially in North America, for TV signal distribution via communication satellites to television receive-only (TVRO) terminals. Most early TVRO terminals were located at cable television (CATV) headends for subsequent redistribution by CATV networks. The satellites generally used 30 W TWTs at 4 GHz, requiring a TVRO antenna of at least 4·5 m diameter. This method of TV distribution was also adopted by domestic networks, mainly in developing countries using Intelsat leased transponders. In these networks, local radio broadcasting was used instead of CATV distribution networks.

The availability in the Ku band of narrow spot beams, such as those offered by ECS–F1 and Intelsat V satellites since the mid 1980s, has given rise to TV distribution services using TVRO antennas as small as 1·8 m. The use of smaller and less-expensive TVRO terminals has resulted in the adoption of this service in large numbers, including more individual users. Shortly after the launch of the ECS–F1 satellite, most of its transponders were leased for the TV distribution service. A major part of the Eutelsat system's revenue (approximately 80%) has come from this service.

The TV distribution service via satellite has received a major boost by the emerging medium-power satellites such as ASTRA, making TV signal reception possible by TVROs as small as 60 cm in diameter. The uplink signal for a TV distribution service is generally via a 5–8 m earth station.

Another TV service which is primarily aimed at individual receivers is the direct broadcast satellite (DBS) service [23]. The foundation for this service was laid down in the 1977 World Administrative Radio Conference (WARC–77). In this conference a downlink plan was agreed for ITU Regions 1 and 3 (i.e. all continents except America), which allocated to each country five TV channels, polarisation and an orbital position. A downlink frequency band of 11·7 to 12·5 GHz was chosen for the DBS service. This band is adjacent to the communication band which has been used for TVRO services. It was agreed in WARC–77 that a satellite TWT power in excess of 200 W is required to beam the TV signal with a high quality to individual DBS receivers using 90 cm dishes in the largest countries. The Regional Administrative Conference (RARC 83) produced the DBS reference parameters, including an uplink plan for the North American continent.

The use of high-power transponders is no longer essential for the DBS service because of a reduction in the noise figures of GaAs FETs from around 8 dB in 1977 to lower than 2 dB at present. In fact, the TVRO and DBS technologies are merging, although their frequency bands and polarisations are different. The former service uses linear polarisation, whilst the latter employs circular polarisation. The first true DBS satellites, i.e. BSB, TDF1, TV–SAT and Olympus have been launched for Europe. The US organisations have opted for medium-power satellites. The additional power of DBS satellites allows a 30 cm dish to be employed. However, the potential of DBS satellites lies more in future high-definition television (HDTV) broadcasting.

13.10 Mobile satellite services

Satellites provide an ideal transmission medium for mobile users, whether at sea, in the air or on land. At present, the Inmarsat system [24] is the only operating mobile satellite system. It provides a near-global coverage via three global beams to maritime users. The current maritime earth terminal, designated standard A, provides capabilities for a single voice channel (companded FM), telex, facsimile and data communications.

The data rates vary from 2·4 kbit/s over a normal voice channel to 56 kbit/s in the ship-to-shore direction, using a demand-assigned digital carrier. The demand assignment, also implemented for voice carriers, is necessary in order to alleviate the problem of shortage of L-band spectrum. All ship-to-satellite communication is at this frequency band, whilst satellite-to-shore links (which have an earth station on land) operate at C band. The Inmarsat service has proved specially valuable for ocean-going vessels. It provides a reliable high-quality service, whereas traditional HF radiocommunication suffers from interference and ionospheric effects. The number of standard-A ship terminals had risen from 1000 in early 1982 to 5000 at the end of 1986 and is expected to reach to about 17000 by 1995.

Inmarsat has introduced a digital low-bit-rate (600 bit/s) service using low-cost terminals, designated standard C. These terminals are compact and lightweight. They are designed for use in the smallest of vessels, including fishing boats, to provide telex and low-speed data services. This terminal can also be used for land mobile applications.

In the past few years, the Inmarsat network has been utilised for tests and demonstrations related to aeronautical satellite services. Two terminal types have been developed: one has a high-gain antenna providing voice and data services and the other has a low-gain antenna providing data-only services. The aeronautical terminal differs from its maritime counterpart mainly in its antenna design and in provisions to take into account the Doppler effect which is more significant in aeronautical satellite links.

Land mobile satellite services are planned by a number of organisations, mainly for operations within their national boundaries. Examples are the M-sat system in Canada and the Mobilesat in Australia. Fig. 13.10 shows M-sat's nine L band beams covering Canada and the USA for mobile-to-satellite links. Base-station-to-satellite links in the M-sat and Mobilesat systems are at Ku band, which allows the use of smaller earth stations compared to Inmarsat coast earth stations. Two types of land mobile terminals are emerging: a data-only terminal (similar to Inmarsat standard C) and a single-channel vocoded voice and data terminal. The vocoding is likely to be at 4·8 kbit/s with PSK and Amplitude-Companded Single-Side-Band (ACSSB) as possible modulation schemes. Inmarsat is also active in the land mobile area, offering space-segment capability for trials and early introduction of services by land mobile operators. Inmarsat has also developed a standard terminal specification (standard M) for land mobile telephone applications.

The mobile satellite service at L band suffers from a severe spectrum limitation. As shown in Fig. 13.11, only 29 MHz of downlink and 34 MHz of uplink spectrum are available. Hence, the link has to be designed for

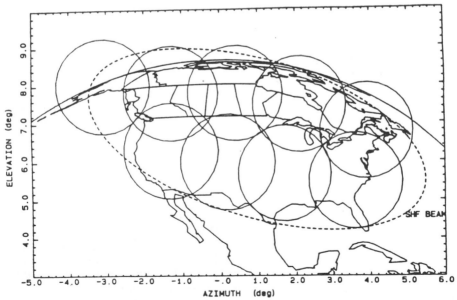

Fig. 13.10   Canadian M-sat beam coverages [*Courtesy: Canadian Telesat*]
　　　——— L band
　　　– – – – Ku band (SHF beam)

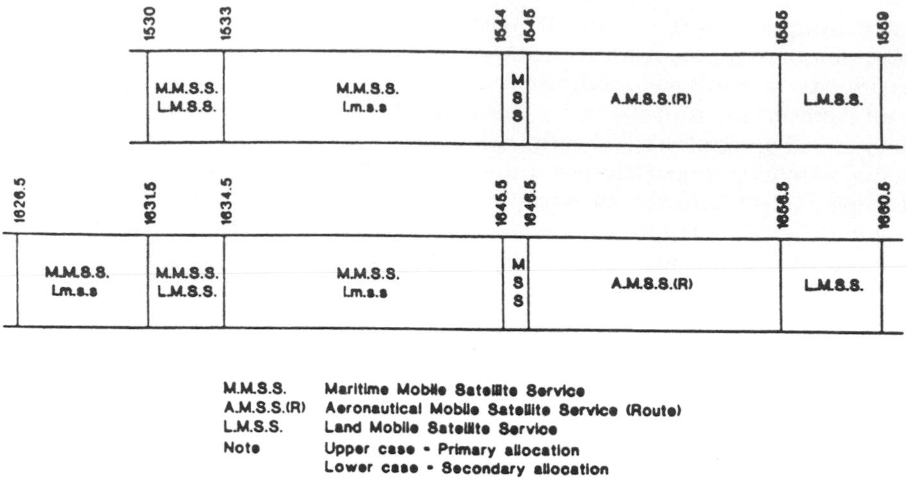

1530	1533		1544 1545	1555	1559
M.M.S.S. L.M.S.S.	M.M.S.S. Lm.s.s	M S S	A.M.S.S.(R)	L.M.S.S.	

1626.5	1631.5 1634.5		1645.5 1646.5	1656.5	1660.5
M.M.S.S. Lm.s.s	M.M.S.S. L.M.S.S.	M.M.S.S. Lm.s.s	M S S	A.M.S.S.(R)	L.M.S.S.

M.M.S.S.   Maritime Mobile Satellite Service
A.M.S.S.(R) Aeronautical Mobile Satellite Service (Route)
L.M.S.S.   Land Mobile Satellite Service
Note   Upper case = Primary allocation
             Lower case = Secondary allocation

Fig. 13.11   L-band frequency allocation for satellite mobile services
[*Courtesy: Inmarsat*]

maximum bandwidth efficiency. The long-term solution to the problem of spectrum congestion is frequency re-use via a large number of narrow spot beams. The Canadian M-sat and Inmarsat III will be the first operational satellites to use L-band spot beams. The Inmarsat second-generation satellites are not configured with spot beams. The additional capacity requirement in the Inmarsat system is achieved by using an increased bandwidth (current-generation satellites only use 8·5 MHz), together with voice activation and more-efficient digital carriers. The use of two satellites over the Atlantic ocean is also being introduced by Inmarsat. However, Inmarsat is considering 19 spot beams for its third-generation satellites. Narrow spot beams at L band require satellite antennas larger than the launch-vehicle dimensions. This has led to the development of unfurlable antennas which unfold in space after launch. 5 m antennas have already been tested and antennas as large as 55 m diameter are under consideration for future US mobile systems.

13.11 Future trends

The first satellite communication systems were introduced to provide large numbers of telephone circuits across the oceans. However, optical-fibre transmission systems are now sufficiently reliable for undersea use and are competitive in cost with both coaxial-cable systems and satellite systems. The first optical-fibre system across the Atlantic (TAT–8), with a single cable, nearly doubled the number of circuits on this route. It will be several years before optical-fibre systems exist worldwide. Nevertheless, in future, satellite systems are less likely to be introduced for long-distance high-capacity telephone routes and more likely to be used in those applications where the technology is uniquely suitable.

The trend towards smaller and more-user-specific earth stations and more-powerful and sophisticated satellites will continue. Applications will increase in areas of broadcasting and mobile and business services, including value-added services based on networks using ISDN concepts. The mobile application will include land, maritime and aeronautical users, with a general trend towards integrated mobile systems and personal communications. The satellites will employ regenerative transponders, scanning spot beams and a high degree of signal processing. This would require an extensive use of VLSI and MMIC technologies in powerful satellites and ultimately in space platforms [25]. The space platforms are large structures, requiring a multiple launch to the low earth orbit (e.g. to an orbiting space station), followed by assembly in space and transfer to the GSO.

Digital techniques will be used extensively, with efficient coding techniques to enhance system capacity. Satellite spacing in the GSO will be reduced by better antenna design and other improvements. Intersatellite links will enhance the general capacity of the GSO. An increase in satellite lifetime from the present 7–10 years to nearly 20 years will have a significant impact on overall circuit costs. The 30/20 GHz bands are likely to enter into service extensively for fixed services, whilst experiments using this band for mobile services are expected to start before the year 2000.

In some fixed-services applications, the satellite links will compete on a global scale with terrestrial digital fibre-optic networks, which are currently undergoing a rapid expansion. Although the terrestrial networks may prove more cost competitive in many medium and high-capacity routes, the satellite links could have the edge in private business networks, low-capacity long-distance routes and in remote-area and emergency applications.

13.12 References

1 WALKER, J.G.: 'A condensed oribital history of Intelsat satellites', *Int. J. Sat. Comm.*, 1984, **2**, pp. 23–28
2 MARAL, G., and BOUSQUET, M.: 'Satellite communication systems' (J. Wiley, 1986)
3 EVANS, B.G. (ed.): 'Satellite communication systems' (Peter Peregrinus, 1987)
4 DALGLEISH, D.I.: 'An introduction to satellite communications' (Peter Peregrinus, 1989)
5 SPILKER, J.J.: 'Digital communication by satellite' (Prentice Hall, 1977)
6 BHARGAVA, V.K. *et al.*: 'Digital communications by satellite' (J. Wiley, 1981)
7 FEHER, K.: 'Digital communications: satellite earth station engineering' (Prentice Hall, 1983)
8 KELLY, T.M.: 'Leased services on the Intelsat system', *Int. J. Sat. Comm.*, 1984, **2**, pp. 29–39
9 McGOVERN, D., and HODSON, K.: 'The European Communications Satellite multi-service transponder', *Brit. Telecom. Eng.*, 1983, **2**, pp. 32–36

10 PONTANO, B.: 'The Intelsat TDMA/DSI system', *Int. J. Sat. Comm.*, 1985, **3**, pp. 5–9
11 LEWIS, J.R., and ELLIOT, D.T.: 'Time-division multiple access for satellite communications', *Brit. Telecom. Eng.*, 1985, **4**, pp 85–98
12 CAMPANELLA, S.J.: 'Digital speech interpolation', *Comsat. Tech. Rev.*, 1976, **6**, pp. 127–159
13 SKAUG, R., and HJELMSTAD, J.F.: 'Spread spectrum communication' (Peter Peregrinus, 1985)
14 MICHELSON, A.M., and LEVESQUE, A.H.: 'Error control techniques for digital communication' (J. Wiley, 1985)
15 RAYCHAUDHURI, D.: 'Ku-band satellite data networks using very small aperture terminals', *Int. J. Sat. Comm.*, 1987, **5**, Part 1, pp. 195–212; Part 2, pp. 265–278
16 NOURI, M., *et al.*: 'Satellite links and integrated broadband communication networks' IEE. Conf. on Broadband Services and Networks, 1990, IEE Conf. Pub. 329, pp. 276–282
17 EDELSON, B.I., *et al.*: 'The evolution of the geostationary platform concept', *IEEE J.*, 1987, **SAC–5**, pp. 601–614
18 LOPRIONI, M., and MANONI, G.: 'The design of a 30/20 GHz regenerative payload' Int. Conf. on Digital Sat. Comm., 1981, pp. 477–482
19 NOURI, M., and BRAINE, M.R.: 'Methods for forward uplink depolarisation control in satellite link earth stations', *GEC J. Research*, 1983, **1**, pp. 59–71
20 MORI, T., and Iida, T.: 'Japan's space development programs for communications: an overview', *IEEE J.*, 1987, **SAC–5**, pp. 624–629
21 MORELLI, G., and MATITTI, T.: 'The Italsat satellite programme', Proc. AIAA 12th Int. Comm. Sat. Systems Conf., 1988
22 WRIGHT, D.L., *et al.*: 'Advanced communications technology satellite (ACTS) and potential systems applications', *Proc. IEEE*, 1990, **78**, pp. 1165–1175
23 PRITCHARD, W.L., and OGATA, M.: 'Satellite direct broadcast', *ibid.*, **78**, pp. 1116–1140
24 GHAIS, A., *et al.*: 'Inmarsat and the future of mobile satellite services' *IEEE J.* 1987, **SAC–5**, pp. 592–600
25 CLOPP, W., *et al.*: 'Geostationary communications platform payload concepts' 11th Communications Sat. Systems Conf., San Diego, 1986, pp. 577–585

Chapter 14

Signalling

S. Welch

Ministry of Defence, Telecommunications

14.1 Introduction

Signalling is the interchange of information between different functional parts of a telecommunication network. In order to establish a connection between a calling customer and a called customer, and to release it when a conversation is over, information must be exchanged between the customers and their exchanges and between these exchanges and any others required to complete the connection. These essential actions are performed by various types of signalling system [1].

Signalling systems link the switching machines in a network to enable the network to function as a whole. Signalling systems must be compatible both with the switching systems at the network nodes and with the transmission links between them. Consequently, transmission has considerable influence on signalling methods. Also, transmission systems must cater for the signalling requirements. This chapter will therefore discuss both the effects of transmission requirements on the design of signalling systems and the effects of signalling requirements on the design of transmission systems.

14.2 Basic signalling requirements

14.2.1 Loop-disconnect signalling

Basic signalling is provided by the 'on-hook' and 'off-hook' states of a customer's line. In the on-hook state, the line is disconnected and no direct current flows. In the off-hook state, the line is looped and DC flows. These states give two continuous signalling conditions in the forward direction (i.e. from the calling customer towards the called customer) and two in the backward direction (i.e. from the called customer towards the calling customer). Dialling consists of a sequence of on-hook and off-hook conditions produced by interruptions of the callers' loop at approximately 10 pulses per second.

The minimum signalling requirements in the telephony service over the switched network, as shown in Fig. 14.1, are thus:

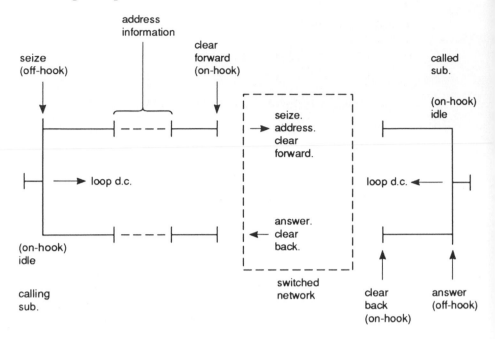

Fig. 14.1 Minimum basic signalling

Forward from caller
  Seizure (off-hook)
  Decadic address information
    (sequential on-hook, off-hook)
  Clear forward (on-hook)
Backward from called
  Answer (off-hook)
  Clear back (on-hook)

The clear forward signal usually serves to release the connection.
  This basic signalling comprises two functions:

(i)   Supervisory (seizure, answer, clear back, clear forward)
(ii) Selection (dialled address).

The signalling on the switched network should also include a further function, i.e. operational, which concerns network features (e.g. signals for subscriber and system facilities, error control, signalling control indications, etc.) and does not directly concern customers' on-hook and off-hook states.

14.2.2 Signalling from pushbutton telephone sets
Push-button telephone sets send digits by means of combinations of frequencies to speed up the selection signalling process over the subscriber line. A combination (compound) of two frequencies, one from each of two blocks of four frequencies, indicates the digit value (Table 14.1).

Table 14.1   Pushbutton telephone set signaling

Hz \ Hz	1209	1336	1477	1633
697	digit 1	2	3	spare
770	4	5	6	spare
852	7	8	9	spare
941	spare	0	spare	spare

This 2(1/4) MF type signal is sufficiently complex to minimise signal imitation. The telephone microphone, short circuited when a button is depressed, is restored to circuit between button depressions. It is thus liable to pick up background speech, music, room noise etc., with consequent danger of imitating the inband MF address signals.

14.3 Influence of switching on signalling

In direct control of switching (e.g. Strowger step-by-step), the dial operates the switches directly. The post-dialling delay is short, but the routing of the call connection is inflexible since the numbering scheme determines the routing. The facilities are limited owing to the lack of registers. The signalling system combines the supervisory (line) and selection (decadic address pulses) functions.

In common (indirect) control of switching (Fig. 14.2), a register-translator processes the address information to enable a connection to be switched.

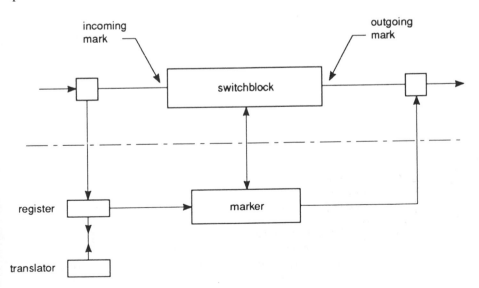

Fig. 14.2   Common control switching

This delays the start of switching and increases the post-dialling delay. However, routing flexibility and enhanced facilities are possible.
The interregister signalling may be:

(a)   Decadic loop-disconnect
(b)   Coded multifrequency (MF) signalling

Decadic loop-disconnect DC signalling is slow and not preferred. Should the line-signalling system be local DC, this system deals with the supervisory signals only. Line-signalling systems having both outgoing and incoming signalling terminals on the speech circuit (VF, outband, PCM, LDDC) transfer both the supervisory and selection (address) signals. Coded multi-frequency (MF) signalling is fast inter-register signalling and is separate from the line (supervisory) signalling systems. This fast address-information transfer between registers reduces the post-dialling delay.

Facilities are mainly concerned with the call-set-up process and require registers (or the equivalent), and thus common-control switching. As both the register and the inter-register signalling are concerned with connection set-up only, and then used to set up other connections, the signalling sophistication is given by the inter-register signalling system combined with the register. This permits per-speech-circuit line-signalling systems to be simple, dealing with the relatively-few supervisory signals in the main (Fig. 14.3). This simplicity is highly desirable in view of the large number of per-speech-path line-signalling systems in networks.

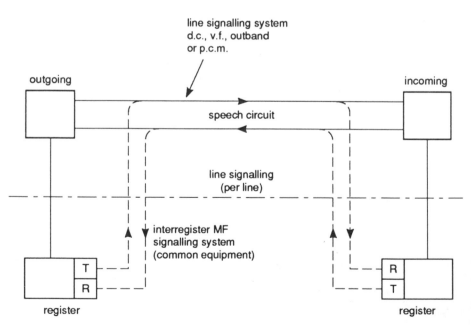

Fig. 14.3   Line and inter-register signalling

All the above comments apply to on-speech-path signalling in the analogue environment. Should the switching use stored-program control (as distinct from wired-logic control), as is sometimes the case in the analogue environment and always the case in digital, then common-channel signalling is preferred to on-speech-path signalling (see Section 14.12).

14.4 Influence of transmission on signalling

14.4.1 Transmission systems

Transmission plant varies in type (audio, FDM, digital). With on-speech-path signalling, the type of transmission has a profound influence on signalling. Particular transmission systems may permit, or preclude, particular signalling methods. On-speech-path signalling is relatively costly (some 25% of total switched network cost). Since the cost of the various signalling methods varies greatly, for economy the line signalling systems tend to be designed for different types of transmission media. This accounts for the variety of different line-signalling systems in most analogue networks.

Audio line plant (amplified and non-amplified) provides a metallic signalling path per speech circuit and DC signalling is usually applied, which is relatively simple, reliable and cheap. This DC signalling may be simple 'local', or long-distance DC (LDDC) when the local-signalling limit is exceeded. Low-frequency signalling (e.g. 50 Hz), which also usually requires a metallic signalling path, is sometimes applied instead of LDDC.

A DC signalling path is not available with FDM, so AC signalling (inband VF or outband) is usually applied for line signalling in analogue networks. Inband VF is the most-widely-applied line-signalling system in most long-haul analogue networks owing to its application flexibility. Outband signalling can only be used on FDM circuits and uses a signalling frequency (e.g. 3825 Hz) above the voice band (i.e. above 3400 Hz), but below the upper limit (4000 Hz) of the voice-band spacing. Various constraints preclude large-scale application of outband line signalling in most existing analogue networks.

Inband inter-register MF signalling is applied regardless of the type of transmission medium. Thus, the variation of signalling method as a consequence of the type of transmission system in analogue networks concerns the line-supervisory signalling systems; DC, VF, outband or PCM signalling is employed as appropriate when inter-register MF signalling is applied (Fig. 14.3).

14.4.2 2-wire and 4-wire circuits

Telephone signalling is duplex, meaning that simultaneous signalling in the two directions must occur without mutual interference. Loop DC signalling on 2-wire (and 4-wire amplified) audio circuits results in some complexity to achieve the duplex requirement. It should be noted that DC signalling uses the phantoms of 4-wire audio circuits, as shown in Fig. 2.3. Thus, the signalling is effectively 2-wire [1, 2].

Four-wire circuits give independent forward and backward speech trans-mission paths; advantage is taken of this by adopting independent forward and backward digital and AC signalling paths. Here, the duplex requirement is satisfied by the two simplex signalling paths, which greatly eases signalling-system design.

14.4.3 Echo suppressors
When fitted, echo suppressors [3] would normally disturb simultaneous forward and backward on-speech-path signalling unless precautions are taken, which may be:

(a) Locating the line-signalling terminals on the line side of echo-suppressors (Fig. 14.4).
(b) Disabling the action of echo suppressors, should these be located on the line side of the signalling terminals, by means of an appropriate condition extended from the signalling equipment to the echo sup-pressor while signalling is in progress.

Method (a) is preferred. Method (b) is sometimes necessary for inter-register MF signalling. Here, the echo suppressors are on the line side of the inter-register MF signalling terminals and disabling technique (b) is necessary for preferred forward and backward signalling (Fig. 14.4). An example of this is the CCITT R2 system. The disabling requirement does not arise should the inter-register MF signalling be forward only, as applies in the Bell R1 system (USA).

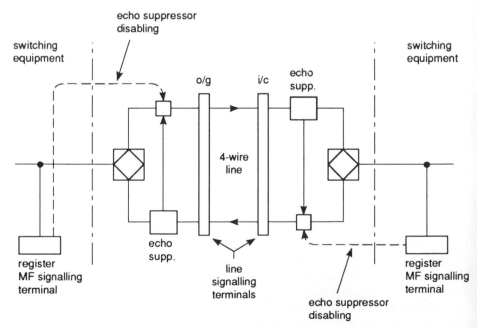

Fig. 14.4 Signalling in the face of echo suppressors

Echo suppressors are not fitted on common-channel signalling links and no problems arise.

14.4.4 Long-propagation time circuits

Long propagation times (e.g. from satellite links) are of concern with any form of acknowledged signalling [1]. The signalling is slowed down and the holding time of relevant equipment increased. Continuous compelled signalling is precluded for the transfer of address information, since the increase in post-dialling delay would be unacceptable.

14.4.5 Time-assignment speech interpolation equipment

When time-assignment speech interpolation (TASI) is employed, [4], time is required to associate a trunk with a channel, a signal being correspondingly clipped. Continuous signalling can tolerate such clips. However, as TASI is applied to relatively-long propagation-time circuits, such signalling is slow, the TASI signalling activity is increased, and the TASI speech activity and efficiency correspondingly reduced. This could imply preference for pulse signalling, but with the danger of unacceptable pulse clipping or excessive pulse lengths.

Continuous signalling is adopted for the line-supervisory signalling, and the TASI penalty accepted. This is inadmissible for address signalling, for post-dialling delay and TASI-efficiency reasons, so pulse inter-register signalling is adopted. This should be fast and arrangements should be adopted to maintain trunk-channel association during the inter-register signalling process to avoid signalling clipping. The arrangement may be:

(i)   Gaps between all pulse inter-register signals less than the TASI speech (signal) detector hangover time, thus maintaining trunk-channel association

(ii) Transmission of a TASI-detector-lock tone to maintain the trunk-channel association during the gaps between pulse signals.

Signalling system CCITT No. 5, used on transoceanic cables [1, 5] adopts method (i). The CCITT No. 5 bit system adopts method (ii). However, while this is a CCITT-specified signalling system it is not presently used, nor is it likely to be used.

14.4.6 Companders

Companders affect short-pulse compound signals (e.g. inter-register pulses) owing to distortion and the production of intermodulation frequencies. Detection of low-level signals is an onerous condition, and, as this arises with end-to-end inter-register signalling, link-by-link working eases the signalling problem in the presence of companders.

Usually, taking account of the permitted transmit-signal level, end-to-end signalling with a modest number of links is satisfactory in the presence of companders, provided the pulse length is adequate to take account of the pulse distortion.

14.4.7 Point-to-point PCM transmission

Some signalling over PCM systems could be achieved by sampling and coding AC analogue signals as for speech. However, the nature of PCM allows a convenient way of transmitting signal information as a built-in time-assigned function per system. Further, switching signals, owing to their relatively low requirement for signal distortion and speed, do not have to be sampled as frequently as speech.

Advantage is taken of this by combining the signals for a number of speech channels. In point-to-point PCM, the loop-disconnect DC signals are converted to bit-encoded signals in the PCM system and reconverted to loop-disconnect DC at the distant end (see Section 14.11).

14.4.8 Integrated digital networks

In integrated digital networks (IDN), common-channel signalling is preferred. The signalling is label-addressed as distinct from time-assigned and channel-associated. The digital transmission conveniently gives the common-channel-signalling bit stream.

14.5 Local junction network signalling

14.5.1 General

Local-area traffic is not so concentrated as in the long-distance network, and, as a consequence, local exchanges and junction-circuit groups are numerous. For economy, the numerous exchange, transmission and signalling equipments must be of simple type.

Junction line plant is usually simple 2-wire audio, and the signalling usually simple DC (loop-disconnect in the UK and 'loop reverse battery' in the USA). Loop DC with high/low resistance loops, or pulse DC signalling (loop and/or leg) are adopted by some administrations to increase the signal repertoire.

Point-to-point PCM transmission is being increasingly applied in analogue junction networks, which necessitates built-in channel-associated PCM signalling.

14.5.2 Supervisory units

Analogue space-division switching operates in the loop-disconnect DC mode for control purposes. Means must be provided on each switched connection at each exchange to detect the forward and backward supervisory signals. As supervisory equipment is only needed for call connections, which are much fewer than the number of subscribers on the exchange, it is accessed by switches to minimise the number and cost. A typical supervisory unit is shown in Fig. 14.5. It performs the following functions:

(*a*)   It receives and sends supervisory signals
(*b*)   When applicable, it repeats decadic address dial pulse signals
(*c*)   It incorporates a transmission bridge, which:
(i)   At the originating and terminal exchanges powers the respective telephones

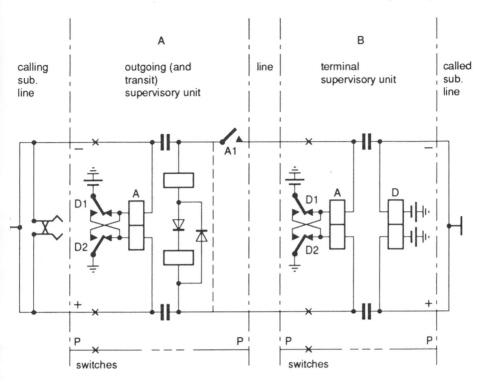

Fig. 14.5   Loop-disconnect DC signalling supervisory units

(ii)   Owing to the DC split, facilitates independent detection of the forward and backward supervisory DC signals

(iii)   Permits the through transmission of AC speech currents.

The form of supervisory unit varies with the transmission bridge (i.e. telephone DC feed arrangement), which depends upon the type of call, i.e. own-exchange, outgoing, incoming. In Fig. 14.5, unit A powers the calling telephone, and B the called. On an own-exchange call, one unit powers both the calling and called telephones, and could be of type B. Some administrations adopt a particular design of supervisory unit, with no current reversal, for own-exchange calls.

Both the supervisory and selection signals at least must be transferred to other exchanges on junction calls. In common-control switching, the selection signals are transferred between registers, the supervisory unit attending to the supervisory signals only. In direct-control switching, the supervisory and selection functions are not separated; the line-signalling system transfers both and the supervisory unit deals with both.

14.5.3 Loop-disconnect DC signalling

The signalling is performed between the supervisory units, which accounts for the relatively low cost of the method. Fig. 14.5 is typical. After being switched through on the initial digit(s), the supervisory unit A is connected to, and powers, the caller's telephone. Relay A operates to the caller's line loop and contact A1 repeats the seizure over the junction to operate relay A at unit B.

The called party's off-hook answer operates relay D at unit B. The answer signal is repeated by backward loop DC reversal (at D1 and D2 operated) to operate the rectified relay D at unit A. (The rectifiers polarise D to avoid its operation to the forward seizure loop-current polarity). Operation of D at unit A starts charging for the call.

Relay D operated at unit A reverses the loop DC backward to repeat the answer signal backward should there be an incoming junction to A. However, should A be the originating exchange (as in Fig. 14.15), the reversal simply reverses the feed current to the caller's telephone, which is of no consequence.

Restoration of the backward junction loop-current polarity to the preanswer condition on the called party's on-hook condition signals the clear back to release D at unit A. Relay A at unit A releases to the caller's on-hook clear forward. This releases the equipment at A. Contact A1 released disconnects the junction loop DC to (normally) release relay A at unit B and the equipment at B.

Similar arrangements apply on multilink connections. A supervisory unit is equipped on the connection at each exchange, the forward and backward signals being repeated at each exchange.

When the supervisory line signalling incorporates the address signalling, relay A at unit A responds to the dialled address and contact A1 repeats the signals to B by successive interruptions of the junction loop DC. Timing discriminates between dial-pulse breaks and the clear forward, as both are on-hook disconnections of loop DC. The transmission bridge is short-circuited during address-pulsing repetition to minimise pulse distortion, the short circuit being removed during interdigital pauses and at the end of address signalling.

The basic two-state continuous signalling of on-hook, off-hook, forward and backward is reproduced on the switched network with loop-disconnect DC signalling. The duplex condition on 2-wire circuits is achieved by the one loop current, operated by loop and disconnect forward, and by change of the loop-current polarity backward.

14.5.4 Limitations of loop-disconnect DC signalling

Limitations of loop-disconnect signalling concern the signalling range and the pulse distortion on address repetition [6]. Repetition distortion arises when an output pulse (break or make) is of different duration from the input pulse, the distortion being influenced by the different shapes of the arrival and decay wavefronts. This is of main concern in direct-control switching. In any transit-switched connection, the objective is to operate a terminal switch, the pulsing requirements of which determine the number and lengths of the individual junctions in a multilink connection. Some networks incorporate pulse correctors or pulse regenerators to ease this problem.

The resistance, leakance and capacitance of lines vary the wavefronts to result in various degrees of pulse distortion. The effect of line capacitance on long audio lines is a main problem. Increased line time constant degrades the arrival wavefront to 'make', tending to reduce the output make period. On 'break', the sending end is on open circuit (see Fig. 14.7a) and the line charges to the potential of the battery behind the receive relay, which delays relay release. On long high-capacitance lines, the decay wavefront is extremely gradual, resulting in significant decrease in the output break pulse, as shown in Fig. 14.6. The effect depends upon the relative values of line resistance and capacitance. Resistance tends to increase, and capacitance to decrease, the output break.

The high capacitance of long 2-wire circuits, and of the phantom (the DC signalling path) of 4-wire audio circuits generally causes a progressive decrease in the output break as the line length increases, despite the increasing line resistance. Owing to the slow waveforms, relay performance and battery voltage variations have a significant effect on the receive-relay output.

Registers perform address-pulse regeneration. Thus, cumulative distortion on multilink connections, which occurs with direct control, does not occur with common-control switching, each link being self-contained from the address-pulsing aspect. Relative to direct control, the permissible common-control junction limits are significantly greater.

14.6 Long distance DC signalling

14.6.1 General

With loop-disconnect DC, the send-end impedance varies during address pulsing, being zero on loop (make) and infinite on disconnect (break), as shown in Fig. 14.7a. This accounts for the non-symmetrical waveforms. The

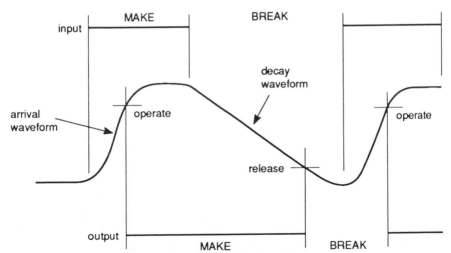

Fig. 14.6   Effect of line capacitance on pulse distortion

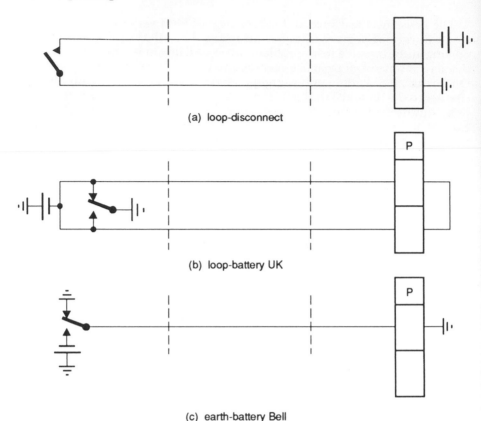

(a) loop-disconnect

(b) loop-battery UK

(c) earth-battery Bell

Fig. 14.7   DC signalling modes

a Loop-disconnect
b loop-battery (e.g. BT)
c Earth-battery (e.g. Bell)

method is unsatisfactory on long audio circuits, owing to the slow decay on high-capacitance lines giving high pulse distortion. Further, the single-current receive relay limits the signalling range.

For these reasons, long-distance DC signalling (LDCC) is usually based on the symmetrical waveform principle. An input pulse time T is reproduced betweeen crossover points on a zero datum. With a polarised device having operate values (in each direction) set equally about the zero datum. T is also reproduced between these points at all receive-signal levels. The polarised device, usually a polarised relay, also ensures signalling sensitivity.

Symmetrical waveforms are achieved by double-current working (or the equivalent). This requires:

(*a*)   the send-end impedance to be substantially equal during the pulse make and break periods; this is ensured by loop–battery in the UK (Fig. 14.7*b*), or by earth–battery in the USA (Fig. 14.7*c*), (*b*) the pulsing battery to be at the send end (it is at the receive end with loop-disconnect), and

(*c*) a polarised receiving device with bidirectional operation.

The above necessitates both outgoing and incoming signalling terminals on the circuit, the input to, and output from, the signalling system being loop-disconnect DC. The loop-disconnect supervisory function may be combined with the LDDC terminals.

There are a number of LDDC systems (for example, the UK SCDC systems and the Bell SX, CX, DX systems, etc.) [1, 7].

14.6.2 UK SCDC system

The British LDDC system (Fig. 14.8) adopts loop-battery signalling (Fig. 14.7*b*). A single send-end battery E is commutated by a single changeover contact A1; hence 'Single Commutation DC', more usually known as SCDC. The polarised relay AP responds to the symmetrical loop-current reversals. Contact AP1 repeats the signals forward, but in the loop-disconnect mode.

Polarised relay DP, with windings 1 and 2 in series opposition, responds to earth-return current, but not to loop current. Relay AP, with windings 1 and 2 in series aiding, is responsive to loop currents and, when operated to a received loop current of either polarity, is non-responsive to backward earth-return currents in either direction.

Duplex signalling applies; in general, loop-current signals apply in the forward direction (A1 commutating E) and earth return currents in the backward direction.

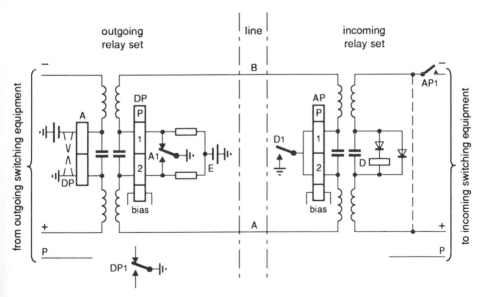

Fig. 14.8 Single-commutation DC long-distance signalling system (UK)

The following is a brief explanation of the principle, noting that signals into, and out of, the SCDC system are in the loop-disconnect DC mode:

(*a*) *Seizure*: A1 is operated. The line loop current operates AP. DP does not respond. AP1 (operated) repeats the seizure forward.

(*b*) *Address pulsing*: A1 commutates E. The line loop-current reversals successively release and operate AP. DP does not respond. AP1 repeats the pulsing forward.

(*c*) *Answer*: D is operated. D1 applies earth to the centre point of AP. The earth current on the line B wire maintains AP operated on winding 1, and operates DP on winding 1. DP1 repeats the answer signal.

(*d*) *Clear back*: D is released. D1 applies loop at AP. AP maintains operated to the loop current. DP releases to the loop current. DP1 repeats the clear back.

(*e*) *Clear forward*: A1 is released and reverses the loop current to release AP. AP1 released repeats the clear-forward signal forward. DP maintains released to the loop current.

Other signals are included, but the above is sufficient to explain the principle.

The signalling limit is about 8000 Ω loop resistance. The address pulse distortion is some 2–3 ms on the maximum circuit.

14.7 AC signalling codes

14.7.1 2-state continuous signalling
ON–OFF continuous tone is used in each direction, with tone-on idle (off during speech) or tone-off idle (on during speech) indicating the signal conditions. Tone-on during speech is inadmissible with VF signalling. It is admissible with outband signalling, but it tends to overload the FDM transmission systems.

Two-state continuous signalling simulates the customer's on-hook and off-hook conditions. It has potential for simple signalling, which can be realised for line signalling when inter-register MF signalling applies, but not for VF signalling. The signals are limited, are more subject to transient interruptions than pulse, and must be low level to avoid transmission-system overload.

14.7.2 Pulse signalling
Pulses of tone change the signal conditions (Fig. 14.14*b*). Relative to continuous signalling, pulse codes allow more signals and higher signal level. However, the signalling systems tend to be more complex owing to timing and signal-memory logic.

14.7.3 Acknowledged signalling
An acknowledgement (which may carry other signal information) confirms the receipt of a functional signal. Both signals may be continuous or pulse. When pulse signalling is used, then:

(*a*) Each functional signal may be acknowledged

(*b*)   One signal may acknowledge a group of functional signals

(*c*)   The functional signal may be repeated until acknowledged

With continuous compelled signalling (Fig. 14.14*a*), the functional signal persists until an acknowledgment is received, and the acknowledgment persists until it causes the cessation of the functional signal which gave rise to it. Cessation of the functional signal ceases the acknowledgment. Group acknowledgment is not possible, as each functional signal must be acknowledged.

In MF continuous compelled signalling, the signal meaning is given by the frequency coding of the signal, and not by ON–OFF of the signal tone. The signalling is slow, owing to the four propagation and four recognition times (two on, two off). In semi-compelled signalling, a pulse acknowledges a continuous signal. This is faster than fully-compelled signalling.

14.8 Inband voice frequency signalling

14.8.1 General

Long-haul circuits are invariably 4-wire, with separate forward and backward signalling paths. The VF frequency(ies) is within the voice band (inband); thus, the systems operate over any channel affording speech transmission.

2VF and 1VF systems are known, 1VF being preferred for simplicity and economy. Voice-frequency systems must be protected against false operation by speech. Thus, signalling during the speech period is precluded.

A buffer amplifier (unity gain forward, 60 dB loss reverse) at the receive end of each channel protects the signal receiver against interference from near-end switching surges. With link-by-link signalling (preferred), a receive-end line split prevents transmission of a signal over subsequent link(s) (Fig. 14.9).

14.8.2 Voice immunity

Protection against signal imitation by speech exploits the following differences between speech and signal currents [1, 8]:

(*a*)   Speech currents containing a signal frequency usually have other frequencies also present.

(*b*)   A signal frequency is used for which the energy in speech is low. (This implies a signal frequency in the range 2000—3000 Hz).

(*c*)   Signals are made longer than the normal persistence of the signal frequency in speech.

(*d*)   Signals of two frequencies compounded are less liable to occur, and persist, in speech than one frequency.

Protection (*a*) is a powerful factor, and VF signal receivers incorporate a signal-guard circuit. In concept, the receiver has an element accepting the signal-frequency band and a guard element accepting other frequencies (Fig. 14.10). The signal and guard outputs are compared. The signal output tends to operate, and the guard output to inhibit, operation of the receiver.

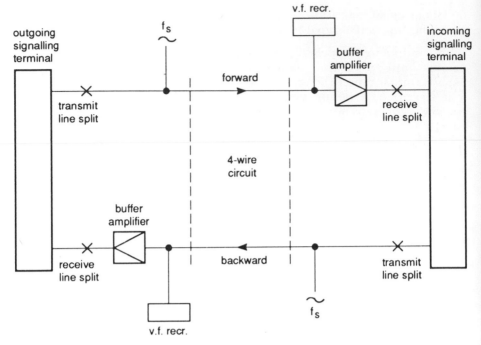

Fig. 14.9   VF line signalling: general arrangement

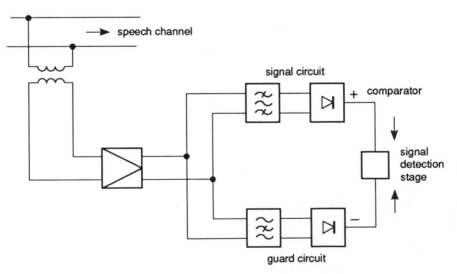

Fig. 14.10   VF receiver signal guard

High signal output and low guard output operate the receiver. Low signal output and high guard output inhibits receiver operation. Receiver operation is allowed with low-level guard output; otherwise, circuit noise could prevent receiver operation to a genuine signal.

Fig. 14.10 shows the principle of the signal-guard feature. The design detail of realisation varies depending upon particular design preferences.

14.8.3 Line splits

A receive line split in each direction, initiated by receiver operation and persisting with the signal, confines link-by-link signals to the link intended (Fig. 14.9). Spill over the signal tone to the next link(s) occurs during receiver operation, but its duration (typically 20 ms) is less than the signal-recognition time and false signalling does not occur.

A transmit line split, initiated immediately prior to, and preferably persisting with, VF signal transmission, minimises interference to the signal from the transmit end.

14.8.4 N. America Bell SF system

The Bell SF system [9] is a tone-on-idle link-by-link 1VF system of the continuous type (Fig. 14.11). It uses a signal frequency of 2600 Hz in each direction on 4-wire circuits. A 'forward transfer' signal is required to be

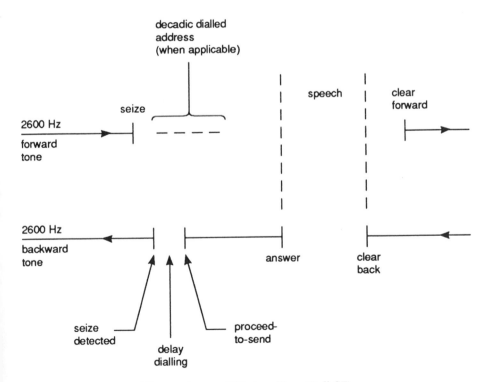

Fig. 14.11   Tone-on idle continuous VF signalling (Bell SF)

pulse. Blocking (backward busying) is by cessation of the backward tone. The signal repertoire is clearly limited.

The system aims to realise the simplicity of on-hook/off-hook 2-state continuous signalling. However, owing to the different operating conditions (VF and not outband or DC) the equivalent is not realised, as follows:

(*a*)   To avoid overloading transmission systems, the signals are required to be switched from an initial high level to operate a receiver to a low level for holding it.

(*b*)   The receiver must be safeguarded against false release owing to line interference during idle (to avoid false seizures), and against false operation during speech (to avoid false clears). These conditions conflict in regard to the receiver guard circuit. Therefore, the selectivity and sensitivity are required to be switched, low during idle and high during speech.

(*c*)   The signal-frequency band must be suppressed from the speech to permit backward speech on calls which are not answered by an electrical signal.

(*d*)   The line splits are somewhat complicated due to factor (*c*). The splits vary in character depending upon the signalling and circuit conditions.

(*e*)   Complications arise when ON–OFF signal conditions conflict (e.g. tone-off 'delay dialling' conflicts with tone-off 'incoming seizure' on bothway working).

14.8.5   UK AC9 system

The AC9 1VF system [1, 15, 16] uses 2280 Hz in each direction, with pulse link-by-link signalling (Fig. 14.12). It is typical of many pulse types. While the UK signal requirements for this system are modest, it is obvious that additional signals may be included by varying pulse timings. Pulse systems do not attempt to simulate the DC 2-state continuous signalling condition;

signal	forward pulse (ms)	backward pulse (ms)
seize	50 – 95	
decadic address (break) (when applicable)	55 – 65	
clear forward	700 – 1000	
answer		200 – 300
clear back		200 – 300
release		650 min.
blocking		continuous

Fig. 14.12   Pulse VF signalling (UK)

signal-memory logic is required. However, in the 1VF form, the systems are acceptably simple and are adopted by most administrations.

The release guard signal unbusies the outgoing end and indicates that the incoming equipment has released to the clear forward signal. This signal is not included in the Bell SF system. The release guard and blocking signals are typical of the 'operational' type.

14.9 Outband signalling

The general arrangement of an outband signalling system [7, 17, 18] is shown in Fig. 14.13. The signal frequency, 3700 Hz (CCITT Recomm. 3825) or 3850 Hz, in each direction is outside the speech band (300—3400 Hz). Thus, speech transmission is not degraded. As signalling and speech are separated by filtration, outband signalling has potential for simple arrangements as no speech interference problems arise. Moreover, signalling during speech is admissible.

The system can only be applied to 4-wire carrier circuits, being integrated with the FDM transmission systems (Fig. 14.13). Link-by-link working is implicit, as the speech band only is extended from the transmission terminals to the switching equipments. The signalling to and from the switching equipment is DC on E and M leads, which may be separate from the speech leads, or the phantoms of the speech tie cables.

Outband signalling simulates the basic on-hook/off-hook, signalling conditions with relative ease by 2-state continuous (low-level) AC signals, either tone-on idle or tone-off idle. Tone-on idle is preferred, similar to Fig. 14.11. The arrangements are less simple when more signals are required; pulse signalling is then necessary (which permits a higher signal level).

Application constraints to outband signalling arise when:

(a) A section (e.g. audio) of a traffic circuit is not capable of transmitting outband.
(b) Existing FDM transmission in a network is not designed for outband signalling.

Outband is the preferred AC line-signalling method owing to its simplicity, particularly when continuous signalling is used. The few line signals are acceptable when inter-register signalling applies. Its use, however, tends to be limited in many networks owing to the above application constraints. This accounts for VF being the most widely-applied long-haul line-signalling system.

14.10 Inter-register MF signalling

14.10.1 General

Inband on-speech-path inter-register MF signalling functions over all types of transmission media. The available voice bandwidth is exploited to obtain fast signalling and to provide a generous signalling capacity. Speech interference problems do not arise, which contributes to the fast signalling. The supervisory line-signalling system may be of an appropriate type to suit the

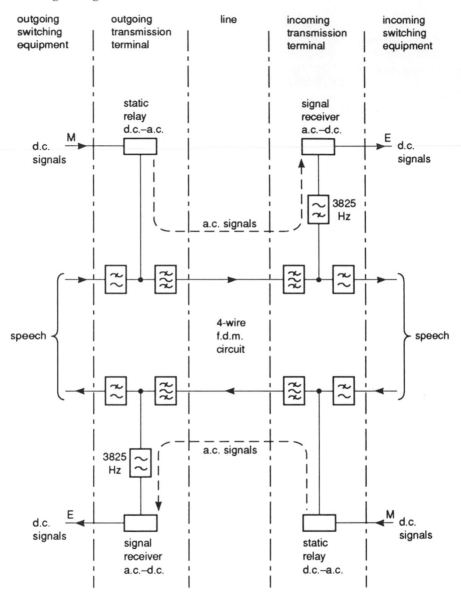

Fig. 14.13 General arrangement for outband signalling

transmission media [10]. Fig. 14.14 shows the various modes of MF signalling. Signalling may be forward and backward (preferred) or forward only. It may also be link-by link or end-to-end.

Combinations of two tones out of six (2/6 MF), giving 15 combinations, are generally used (Fig. 14.15) and the frequencies are 120 or 200 Hz spaced. 200 Hz spacing is used when signalling is forward only, or for signalling in

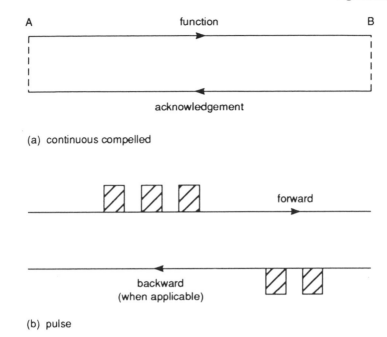

A   function   B

acknowledgement

(a) continuous compelled

forward

backward
(when applicable)

(b) pulse

Fig. 14.14 Types of MF signal

the two directions when the direction of transmission of certain signals changes as the call set-up proceeds, fewer signal frequencies then being used, as in the French Socotel system [11].

The same frequencies, 200 Hz spaced, could be employed in the two directions for wholly-4-wire working (switching and transmission). For 2-wire switching and/or 2-wire transmission, the bandwidth is divided for forward and backward signalling and 120 Hz spacing is adopted. For application flexibility, 120 Hz spacing is usually adopted for both 4-wire and 2-wire working, with forward and backward signalling (Fig. 14.15). The register association splits the speech path at each end, which minimises interference to MF signalling.

14.10.2 Bell R1 system
The Bell R1 system [1, 12] is an unacknowledged pulse link-by-link 2/6 MF system with forward signalling only. It uses frequencies of 700—1700 Hz, 200 Hz spaced. It provides 12 signals only:

KP (start of pulsing)
Digits 1—10 (10)
ST (end of pulsing)

Address information only is transferred. The main objective is to transfer this as fast as possible; otherwise the *en bloc* of the address information at the originating register (Bell standard practice) would result in a long post-dialling delay. The facilities are extremely limited. With the link-by-link

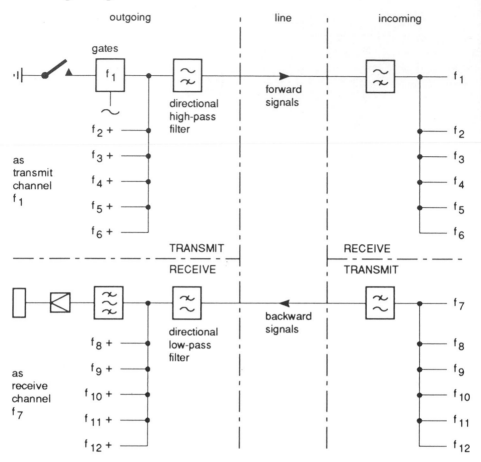

Fig. 14.15 Inter-register MF signalling: general arrangement

signalling adopted, the registers release in turn as the set-up progresses, and originating-register control does not apply.

Modern networks require enhanced facilities. This necessitates backward (in addition to forward) inter-register signalling. The Bell R1 system is not favoured by most administrations for this reason.

14.10.3 CCITT (CEPT) R2 system

The CCITT R2 system [1] is a continuous compelled end-to-end 2/6 MF system with forward and backward signalling. Frequencies (120 Hz spaced) are as follows:

Forward: 1380, 1500, 1620, 1740, 1860, 1980 Hz
Backward: 1140, 1020, 900, 780, 660, 540 Hz

Originating-register control applies. The system has a considerable reper- toire, embracing international, regional and national requirements. A national network adopts relevant signals from this repertoire. A shift technique increases the basic 15-signal repertoire in each direction. Both the forward and backward signals are in two levels, a particular 2/6 MF combination having primary and secondary meanings. A semi-compelled version (continuous functional signal, pulse acknowledgement) is applied to satellite circuits.

R2 is standardised by the CCITT. It is gaining world-wide acceptance for regional and national application in the analogue environment.

14.10.4 UK MF2 system

The British MF2 system [15] is a guarded-pulse end-to-end 2/6 MF system with forward and backward signalling, using the same frequencies as R2. The guarding, a 2/6 combination in each direction, protects the system in conditions of severe interference. The system is faster than R2, and it has more facilities than R1. It has the same potential as R2 for two-level signalling in each direction by shift.

14.11 Link-by-link and end-to-end signalling

Link-by-link signalling is implicit with DC outband PCM and with common-channel signalling. Link-by-link or end-to-end may apply with inband signalling, VF and MF. Link-by-link signalling allows signalling evolution in networks, since new signalling systems may be applied to individual circuit groups without undue disturbance to existing signalling-system provision.

For this reason, VF signalling is usually preferred link-by-link, to be consistent with other types of line signalling. While the same point of signalling evolution may be made for inter-register MF signalling, end-to-end signalling is usually employed for reasons of reduced holding time (thus needing fewer expensive registers) and to achieve originating-register control.

14.12 Channel associated (built-in) PCM signalling

14.12.1 General

PCM systems generally have a built-in signalling system, as described in Section 6.12. This is equivalent to the provision of outband signalling in FDM systems. This is used when common-channel signalling (CCS) is not used, i.e. when PCM systems are applied point-to-point. All PCM systems have a 64 kbit/s channel capability, but this is not always available for signalling, channel-associated or CCS.

14.12.2 Bell D1 24-channel system

In the Bell D1 24-channel (1.544 Mbit/s) PCM system [1], the eigth bit in the time-slot of each channel is used for in-slot signalling at the 8 kbit/s rate. The code pattern (1111–– or 0000––) gives the required Bell 2-state continuous signalling in each direction (Fig. 14.16). Multiframing is not used.

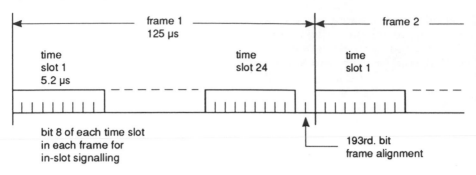

Fig. 14.16 Bell DI 24-channel PCM

Because one bit in each time-slot is used for signalling, only seven bits are used for speech encoding.

14.12.3 UK 24-channel system

The UK 24-channel (1·536 Mbit/s) PCM system [1, 19] had to provide more signals than Bell, so a 4-frame multiframe was adopted to increase the signalling capability. Multiframing allows each frame in the multiframe to be identified individually.

As shown in Fig. 14.17, bit 1 of each channel time-slot in frames 1 and 3 of every 4-frame multiframe is used for in-slot signalling for that speech channel. Speech encoding is 7 bits. Signalling is thus at the 4 kbit/s rate and the two time-displaced signalling bits per speech channel increase the signalling capability. This system is now superseded by the 30-channel system in UK.

14.12.4 CCITT (CEPT) 30-channel system

In the CCITT 30-channel (2·048 Mbit/s) PCM system [20, 21], the 64 kbit/s time-slot 16 is used for out-slot signalling for the 30 speech channels. The signalling time-slot carries 8 'bunched' signalling bits in each of frames 1—15 of the 16-frame multiframe adopted.

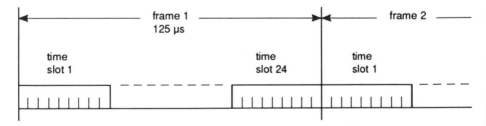

bit 1 time slots frames 1 and 3 in-slot signalling
bit 1 time slots frame 4 and multiframe alignment
bit 1 time slots frame 2 spare
4-frame multiframe

Fig. 14.17 UK 24-channel PCM

The 8 bits are divided into two groups of 4 bits, so each of the 30 speech channels of the system has four time-assigned out-slot channel-associated signalling bits in time-slot 16, as shown in Fig. 14.18. These 4 bits may be regarded as being four independent signalling 'channels' each at 500 bit/s, or as an overall signalling bit rate of 2 kbit/s per speech channel. Speech encoding is 8 bits.

14.12.5 Bell D2 24-channel system

The Bell D1 system is now superseded by the new standard D2 system [1]. The D2 system improves D1 to reduce the quantisation distortion and to obtain the capability of more signals, but it retains 24 channels.

A 12-frame multiframe is adopted. Bit 8 of the 8 bits in each channel time-slot of frames 6 and 12 in each 12-frame multiframe is used for in-slot signalling for that speech channel. Speech encoding is 8-bits in frames 1—5, 7—11 and so on. However, it only has 7 bits in frames 6 and 12, and so on (see Table 14.2). Thus, the signalling per speech channel is two time-displaced eighth bits, each at 0·65 kbit/s, or (as a signalling bit is available every sixth frame), one signalling 'channel' at 1·3 kbit/s.

The 193rd bit per frame is retained for frame alignment. However, in alternate frames it performs the function of multiframe alignment instead.

14.12.6 Comments

Built-in channel-associated PCM signalling deals with line-signalling requirements only. The inter-register MF-signalling coding and decoding

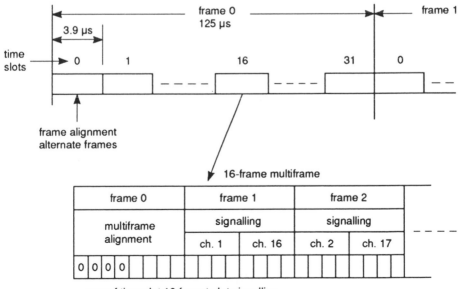

Note: for CCS, time slot 16 is 64 kbit/s bit stream

Fig. 14.18 CCITT 30-channel PCM

Table 14.2   Bell D2 24-channel PCM

Frame number	193rd bit		Bit numbers in each channel time slot	
	Frame alignment	Multiframe alignment	Speech	Signalling
1	1		1—8	
2		0	1—8	
3	0		1—8	
4		0	1—8	
5	1		1—8	
6		1	1—7	8th
7	0		1—8	
8		1	1—8	
9	1		1—8	
10		1	1—8	
11	0		1—8	
12		0	1—7	8th

used as a 4 kbit/s ⎤ 12-frame multiframe
bit stream for Bell
common channel signalling

terminals are located in the switching equipment. For this reason, point-to-point PCM encodes the analogue MF signals as for speech. Built-in PCM signalling is time-assigned, the signalling bits being 'associated' with a speech channel on a time basis, so queueing delays do not arise. There is no requirement to prove speech-path continuity separately.

Should common-channel signalling apply, the multiframing is not required. The signalling bit stream is given by the 64 kbit/s time-slot 16 in the 30-channel system. In Bell D2, the CCS bit stream is given by the 193rd bit in alternate frames at the 4 kbit/s rate (i.e. using the otherwise multiframe alignment bits for CCS signalling). This relatively-slow CCS speed with D2 is a penalty. All D2 time-slots give 8-bit speech encoding when CCS is applied. Common-channel signalling will not be applied to Bell D1 nor to UK 24-channel PCM.

In common-channel signalling the signals are not time-assigned and require speech-circuit labelling, CCS serving the speech channels of many PCM systems, as distinct from one. Signal-queueing delays arise, and speech-path continuity must be proved by a separate arrangement.

14.13 Common-channel signalling

14.13.1 General
With stored-program control of switching (SPC), it is inefficient for the processor to deal with on-speech-path signalling. All information between

processor-controlled exchanges is more efficiently transferred over a bidirectional high-speed data link. Hundreds of speech circuits share a common-channel signalling (CCS) link sequentially for signalling purposes, as shown in Fig. 14.19. CCS replaces all conventional on-speech-path signalling systems (DC, VF, outband, MF etc.).

CCS has the following merits:

(*a*) Fast signalling
(*b*) Potential for a large number of signals
(*c*) Freedom to handle centralised service signalling (management, maintenance etc.) and signalling during speech
(*d*) Flexibility to add or change signals by simple means
(*e*) Economic for large speech-circuit groups
(*f*)   As a CCS link can be routed quite separately from speech-circuit groups; one CCS system may serve a number of speech-circuit groups to make CCS economic for small circuit groups
(*g*) Potential for rationalised signalling in networks

However, CCS gives rise to requirements which do not arise with on-speech-path signalling:

(*a*) High-order error-rate performance

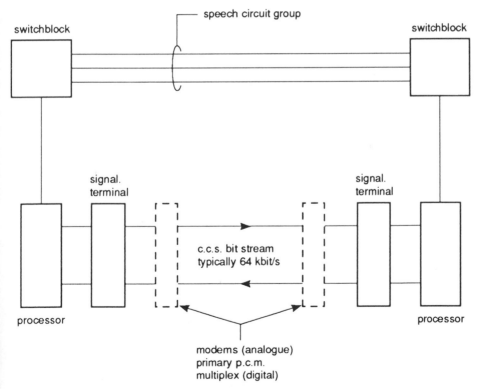

Fig. 14.19 Common-channel signalling: basic arrangement

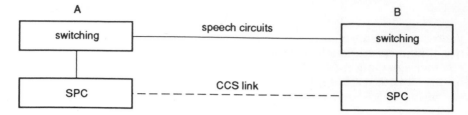

(a) Associated signalling between A and B

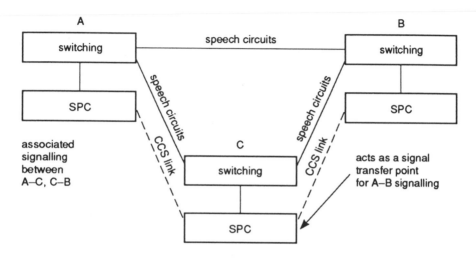

(b) Quasi-associated signalling between A and B

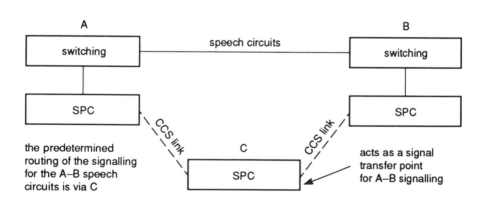

(c) Quasi-associated signalling between A and B

Fig. 14.20 Common-channel signalling modes
a Associated signalling
b Quasi-associated signalling
c Quasi-associated signalling

(*b*)   Signalling operational-security backup
(*c*)   Assurance of speech-path continuity, since CCS does not establish speech-path integrity.

Each message signal on CCS requires indication of the speech circuit, and thus the call, to which it belongs; it therefore includes a circuit-label bit field for this purpose. CCS operates in the synchronous mode, but it is not necessary for the forward and backward signalling paths to be synchronised to each other. Fill-in signal units (SYU-idle etc.) are transmitted when message signals are not.

14.13.2 CCS bit streams

All CCS requires a signalling bit stream, and is thus digital (Fig. 14.19). In the analogue environment, the bit stream is supplied by modems, and it is convenient to call this 'analogue CCS'. Since the line bit-error rate increases with bit speeds, the adopted analogue CCS bit rate is a compromise between an acceptable degree of error correction (which reduces the throughput) and desired high signalling bit rate (to reduce emission time). Rates of 2·4 kbit/s and 4·8 kbit/s are presently adopted for analogue CCS.

A signalling bit rate 64 kbit/s is preferred in the digital environment. This is obtained from the digital primary multiplex and modems are not required. Other bit rates (4, 8 kbit/s etc.) are permissible.

14.13.3 Signalling modes

CCS may be associated with, or dissociated from, the speech circuits served. When associated, the CCS link terminates at the same two exchanges as the speech circuits, as shown in Fig. 14.20*a*.

When CCS is dissociated, the signal messages are transferred between the two exchanges over two or more CCS links in tandem via signal-transfer points, and thus on a different routing. In the quasi-associated mode (a form of dissociated) the signals are transferred over predetermined paths only, and through predetermined signal-transfer points (Fig. 14.20*b* and *c*). This predetermined routing permits the same method of circuit labelling as in the associated mode.

14.14 CCITT No. 6 CCS system

Main features of the CCITT no. 6 CCS system [1, 13] are:

(*a*)   Bit rates of 2·4 and 4·8 kbit/s in the analogue version and 4 and 56 kbit/s in the digital version.
(*b*)   Fixed-size signal units of 28 bits (20 information, 8 parity check). Four padding bits (not carrying signal information) are added for the digital versions for compatibility with the 8-bit time slot.
(*c*)   Lone signal units (LSU) and multi-unit messages (MUM) are used.
(*d*)   11 (or more) of the 20 information bits are required for circuit labelling, leaving few bits for functional purposes. Provision is therefore made for MUMs in which the initial signal unit (ISU) carries the circuit label, but the subsequent signal units (SSU) do not.

(*e*)   Error control by error detection (by redundant coding) and error correction by retransmission.

(*f*)   SUs transmitted in blocks of 12. The 12th unit is the error-control acknowledgment unit (ACU), which indicates whether or not each of the 11 SUs in the block just received contains an error. Each corrupted LSU is retransmitted in another block. If a corrupted SU is part of a MUM, the whole MUM is retransmitted. If an ACU is corrupted, all the message SUs in the block are retransmitted unrequested, SUs are not retransmitted.

(*g*)   Signalling automatically transferred to a backup on link failure or on excessive error rate.

(*h*)   Speech-path continuity check.

The pioneer No. 6 system was produced primarily for analogue networks. The digital versions were produced subsequently, but retained the basic features of the analogue system. As a result, No. 6 has limitations in the digital environment and will not be used by most administrations. The limitations arise from:

(*a*)   The error-control method
(*b*)   The involvement of the processor in the error control. (On retransmission, the processor can receive SUs out of correct sequence or duplicated, necessitating complex processor 'reasonableness' checks)
(*c*)   The small fixed size and non-byte structure of the SU
(*d*)   The SU not being a multiple of 8 bits.

There are other undesirable features and the new No. 7 system is much superior in all respects.

14.15 CCITT No. 7 CCS system

14.15.1 General
A digital version of the CCITT No. 6 analogue system does not give optimum signalling in the digital environment. With the experience gained in the production of the pioneer No. 6 system and with recognition of the evolution of digital networks to Integrated Services Digital Networks (ISDN), the CCITT specified the No. 7 CCS system [1, 14]. The following main signalling areas are identified in the ISDN environment:

(*a*)   Digital Access Signalling System (DASS) single-line
(*b*)   Digital Access Signalling System (DASS) multi-line
(*c*)   Inter-exchange CCITT No. 7 CCS (and national variants of No. 7).

Areas (*a*) and (*b*) are discussed in Section 14.16.
   The No. 7 CCS system is optimised for the 64 kbit/s digital-signalling environment, typically using time-slot 16 in 30-channel PCM systems. It is based on the OSI (Open Systems Interconnection reference) concept of the International Standards Organisation (ISO) [22] and on High Level Data

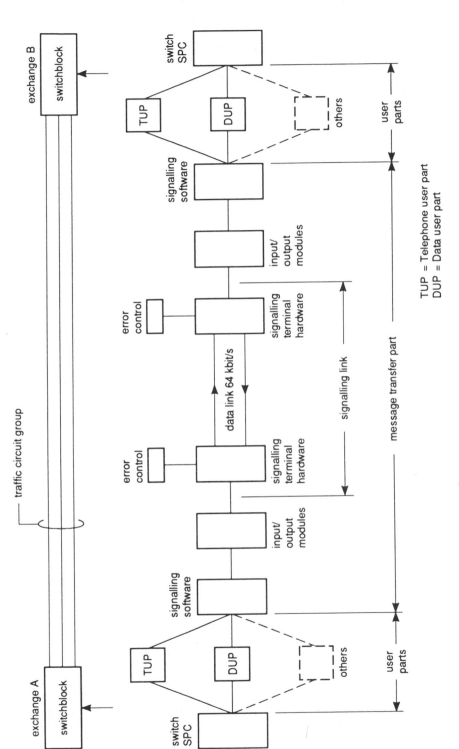

Fig. 14.21 CCITT No. 7 CCS system: general arrangement

Link Control (HDLC) [23]. Various signalling bit rates are admissible in No. 7, ranging from a 4·8 kbit/s analogue version to 64 kbit/s digital. No. 7 is much superior to No. 6 in both the analogue and digital environments, reflecting further knowledge of the CCS art since No. 6 was produced.

Fig. 14.21 shows the general structure of No. 7. The basic principle of the structure is the division of functions into a common *message transfer part* on the one hand and separate *user parts* for different users on the other. User parts are functional entities that utilise the common transport capability provided by the message-transfer part and they may be regarded as being the service–dependent parts of No. 7. The CCITT has so far specified a *telephone user part* (TUP), *data user part* (DUP) and *ISDN user part* (ISDN-UP), with potential for additional user parts as may arise in the future [14].

The message-transfer part allows corresponding user parts at different exchanges to communicate with each other by means of addressed messages, the addressing identifying the service (and thus the user part) for which the message is intended. Should corresponding user parts communicate with each other for the interchange of information not related to call processing, such messages are also conveyed by the message-transfer part.

The functional concept adopted allows change to one function without significant impact on others. It involves, for example, flexible arrangements in the message-transfer control without involving the switching SPC. This implies that in No. 7, and unlike No. 6, the switching SPC has no function in the error-control process, this being completely attended to at the message-transfer signalling terminals.

14.15.2 Error control

In the No. 7 error control [24], it is required that signal messages (MSUs) should not get out of correct sequence, nor be duplicated, to the user part or to the switching processor. As a contribution to this all types of SU are sequence numbered cyclically and MSUs are accepted in correct sequential order only.

The CCITT specifies two optional error-control modes for international No. 7: (*a*) basic and (*b*) preventive, mode (*b*) being a variant of (*a*).

Error detection is by redundant coding (user generator polynomial $X^{16} + X^{12} + X^5 + 1$) and error correction is by retransmission. As shown in Fig. 14.22, all types of SU in the two signalling directions carry a message-transfer control part consisting of:

(*a*)   A 7-bit forward sequence number (FSN)
(*b*)   A 7-bit backward sequence number (BSN)
(*c*)   A backward indicator bit (BIB)
(*d*)   A forward indicator bit (FIB)
(*e*)   A 16-bit check field

All non-message SUs (fill-in, link status) transmitted between message SUs carry the FSN as the last message SU transmitted. Non-message SUs are not retransmitted when an error is detected and they are not deposited in the retransmission store.

The BSN and BIB in a signalling direction refer to MSUs transmitted in the other direction. The BSN is the positive acknowledgment. When sent back from an end, say B, it has the same binary-code value as the FSN of the message SU just accepted by B. Receipt of this BSN at A clears the message SU with that FSN binary value from the retransmission store at A.

The BIB is the negative acknowledgement. The basic error-control method uses it to initiate a retransmission. When a corrupted message SU(s) is detected at B, the polarity of the BIB sent back to A is reversed. Detection of this BIB code reversal at A causes retransmission of all the message SUs from the retransmission store at A (cyclic retransmission), the SUs retransmitted are then redeposited in the retransmission store.

The FIB sent from A informs B by inversion of the FIB bits that the retransmission requested by the BIB reversal is taking place, the code polarity of the BIB and FIB bits now being the same. Thus, the combination of FIB and BIB is a 'handshaking' process.

Preventive error control is a variant of the basic mode (*a*) above and may be applied to improve the throughput on long-propagation-time signalling links (satellite, some terrestrial). It is a positive-acknowledgement cyclic-retransmission forward-error-correction method. Cyclic retransmission of all the MSUs in the retransmission store takes place whenever there are no new MSUs or link-status SUs available to be sent. It thus eliminates error correction by retransmission under the control of negative acknowledgments. The error-control format of the basic mode is used. However, the BIB does not apply; the preventive mode uses FSN and BSN indications only. Arrangements are adopted for forced retransmissions should the retransmission store become filled with MSUs or with MSU octets.

14.15.3 Operational security
Signalling is diverted from a regular link to an alternative (backup) when the regular link is recognised as having failed due to:

(*a*) Loss of the bit stream
(*b*) Continuous consecutive failure of any type of SU to check correctly to a degree depending upon the signalling bit rate, typically 64 SUs at 64 kbit/s
(*c*) Intermittent failure of any type of SU to check correctly to a degree depending upon the signalling bit rate, typically a SU error rate of 1 in 256 at 64 kbit/s.

There is additional monitoring for an acceptable number of octets in SUs.

14.15.4 Loading
The maximum loading of a CCS link in terms of the number of traffic circuits served is greatly influenced by the signalling bit rate; the higher the rate, the greater the number of circuits. In practice, the loading reflects acceptable queueing delay and the size of circuit groups in networks. Assuming traffic circuits allocated to specific signalling links, a maximum of 1000 or 1500 traffic circuits per CCS link is usually adopted in normal operation;

more in abnormal (backup) operation when the load of a failed link is transferred to the existing load on a working backup link.

14.15.5 Continuity check of traffic circuit path

Unlike on-speech-path signalling, CCS does not monitor the speech path of the call (signalling integrity) and other arrangements must be made to assure continuity of the traffic path (line plus switching). Otherwise, there would be the possibility of a connection being set up by the signalling path, but without the traffic path. The CCITT recommends separate line and cross-office checks and that the line check be link-by-link. The cross-office check may be on a per-call basis, or by a statistical method of routine testing of idle switching paths. The latter is usually preferred.

In circumstances where the line transmission system itself does not assure traffic-path continuity in normal course, a loop link-by-link per-call check is made prior to connection to the called party. A 2000 Hz transceiver is connected to line at the outgoing end and a check loop is applied at the incoming end. The link check is successful when the check tone sent on the forward path is received on the backward path.

14.15.6 No. 7 signal unit formatting principles

14.15.6.1 High level data link control (HDLC)
Signal messages consist of a number of constituent bit fields, each having its own function. Thus, a protocol must be adopted to enable individual fields to be identified. The formatting principle and protocol of No. 7 signal messages are based upon the *high level data link control* (HDLC) procedures [23] as follows:

(a)   The signal messages are of variable length and organised in a byte (octet) structure.
(b)   Opening and closing flags of the unique binary-code pattern 01111110 act as delimiters to serve as reference points. Since this code pattern must not occur elsewhere in the signal message, the HDLC procedures require that a 'O' be inserted ('stuffed') after any succession of five continuous '1's in the signalling information. The receive end removes a '0' that follows a received succession of five '1's. Inserted and removed '0's are not included in the error check.
(c)   The error-check field is immediately before the closing flag. This flag thus indicates to the receive end that the bit field (16 bits) just received constitutes the error-check field.
(d)   Header bit fields of the signalling information follow the opening flag.
(e)   The remaining part of the signalling information is located between the header and check bit fields.

14.15.6.2 Basic signal unit format
No. 7 identifies three types of SU, differentiated by means of a length indicator carried in all types of SU [14], the types being message SU,

link-status SU and fill-in SU. Fig. 14.22 shows the format principle. Consideration will be limited here to the message SU and for the case of a typical basic *initial address message* (IAM).

For a typical IAM (Fig. 14.22a) the opening and closing flags and the error-control bit fields (check, BSN, BIB, FSN, FIB) have been discussed above. Other fields are as follows:

(*a*) Length indicator (6 bits)
This is used to indicate the number of octets following the length-indicator octet and preceding the check bits. It is a number in binary code in the

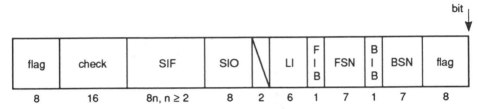

(a) Basic format of message SU (MSU)

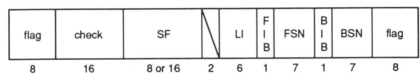

(b) Format of link status SU (LSSU)

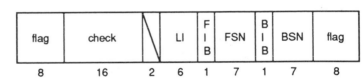

(c) Format of fill-in SU (FISU)

SIF = Signalling information field
SIO = Service information octet
SF = Status field
LI   = Length indicator

Fig. 14.22 CCITT No. 7 signal-unit formats

a Message signalling unit (MSU)
b Link-status unit (LSSU)
c Fill-in unit (FISU)
 SIF = Signalling information field
SIO = Service information octet
 SF = Status field
 LI = Length indicator

range 0–63. It discriminates between three types of SU in international application as follows:

$$\text{Length indicator} = 0 \qquad \text{fill-in SU}$$
$$= 1 \text{ or } 2 \qquad \text{link status SU}$$
$$> 2 \qquad \text{message SU}$$

In national application of No. 7, should a signalling-information field in a message SU be 62 octets or more, the length indicator is set to 63. The present interest is the message SU.

(*b*) *Service information octet (SIO)*
In message SUs, the service information octet (SIO) is divided into the service indicator (4 bits) and the subservice field (4 bits). The service indicator is used to associate message-SU information with the appropriate user part, e.g. telephone user part or data user part. The sub-service field indicates national or international call, or any other which may arise (e.g. local in national).

(*c*) *Signalling information field (SIF)*
In message SUs, the signalling-information field (SIF) is the total message information required by, and passed by the signalling terminal to, the appropriate user part dealing with that message SU. In international working, it consists of at least two octets and up to 62 octets. In national working, it may consist of up to 272 octets.

In CCS, the IAM is the first signal in call processing, since a separate seizure signal is not given. The SIF of the IAM carries information additional to the address. For the basic IAM it has the following constituent bit fields in order of transmission to line:
(i)   *International label*: This 40-bit label is divided into:
Destination-point code (14 bits) indicating the exchange for which the MSU is intended.
Originating-point code (14 bits) indicating the exchange which is the source of the MSU.
Circuit-identification code (12 bits) indicating the traffic circuit taken for the call from those directly interconnecting the originating and destination exchanges.
The purpose of including the originating and destination exchange codes is to ease the real-time processing. However, this 40 bit label may be reduced and varied in national application.
(ii)   *H0/H1 octet*: The H0 (4-bit) code identifies message groups (e.g. forward address messages, supervisory messages etc.). The H1 (4-bit) code identifies a particular type of message in the relevant H0 group. This H0/H1 arrangement, with potential for 16 H0 groups and 16 different MSUs in each group, is more than adequate for MSUs presently defined, and it has ample spare for future requirements. In the present IAM example, H0 is coded 'forward address message group' and H1 is coded 'IAM variable length'.
(iii)   *Calling party category* (6 bits): This indicates ordinary subscriber, priority, language operator etc.

(iv) *Message indicator* (12 bits): This indicates such points as control of satellite routing, control of echo-suppressor provision, national call, international call, signalling path used etc.

(v) *Number of address signals* (4 bits): This indicates the number of address digits in the IAM and thus gives the length of the address-digit part of the IAM.

(vi) *Address signals*: The address-digit signals conveyed by the IAM are in successive 4-bit fields, an unused 4-bit field (the last) being coded 0000 (filler).

14.16 Integrated digital access

14.16.1 Digital access signalling (DASS) single line [25, 26]

In the ultimate ISDN, the connection between the user's terminal equipment and the local exchange may well be 4-wire, thus enabling a wide variety of facilities and services to be offered to users.

The 2-wire customer line plant in present networks is a significant proportion of total network cost. Two such pairs for the 4-wire facility would increase this cost considerably. This means that, for a considerable introductory period, the digital services that can be offered to the majority of users will be restricted to those that can be provided using the existing 2-wire customer line to carry digital signals. Thus, for the foreseeable future, integrated digital access (IDA) to the ISDN will be provided on the 2-wire access line that caters for voice, data etc., control and signalling. These information components are processed in a *network terminating equipment* (NT) at the user's location (Fig. 14.23).

The transmission limitations of customers' 2-wire lines require that the aggregate bit rate should be as low as possible. The CCITT has defined a basic-rate single-line IDA operating at some 192 kbit/s. It provides two 64 kbit/s traffic B-channels and a 16 kbit/s D-channel. The B-channel can be used for voice or data. The D-channel is used for signalling. The remaining bits are used for housekeeping purposes, such as maintenance etc.

Since duplex operation is required over a 2-wire line, it is necessary to avoid components of the transmitted signal being received on the receive

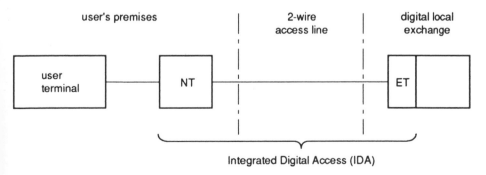

Fig. 14.23 Single-line integrated digital access (IDA)

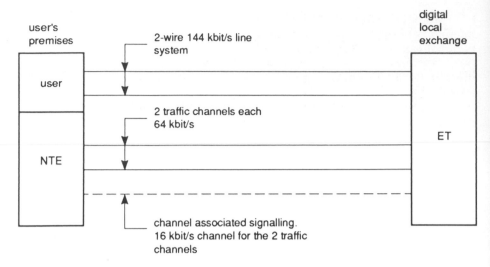

user's premises

digital local exchange

2-wire 144 kbit/s line system

user

2 traffic channels each 64 kbit/s

ET

NTE

channel associated signalling. 16 kbit/s channel for the 2 traffic channels

Fig. 14.24   Single-line IDA signalling

path of the user's 4-wire terminal. A number of different methods have been proposed to solve this problem, e.g.:

(*a*)   Frequency separation
(*b*)   Time separation ('ping pong' or 'burst mode')
(*c*)   Hybrid separation

These are explained in Chapter 15. Hybrid separation is now favoured.

A digital access (DASS) single-line system provides channel-associated signalling, as shown in Fig. 14.24. For explanation, it is convenient to describe the BT DASS2 single-line system. DASS is based on OSI and the format of HDLC (Fig. 14.25). The signalling interchange is based on a 'command' and 'response' concept. The response is an acknowledgment to a command signal and carries no message information.

The fields convey the following broad functions:

(*a*)   *Opening and closing flags*: These have been discussed previously.
(*b*)   *Address*: This indicates whether the message is a command or a response and indicates to which of the two traffic channels the message relates. Also,

first bit

flag	check	information	control	address	flag
1 octet	2 octets	0–45 octets	1 octet	1 or 2 octets	1 octet

Fig. 14.25   DASS frame format

a bit coding in the first Address octet indicates whether the Address field is one or two octets.

(*c*)   *Control*: This indicates the control of the interchange and acknowledgment of the message.

(*d*)   *Information*: This is the message (e.g. digits, requests for service, call connected etc.).

(*e*)   *Check*: This provides the error check in the error control (see Section 14.16.2).

When there are no messages to be sent, continuous flags (code pattern 01111110) are transmitted.

14.16.2 DASS multi-line

For explanation, it is convenient to describe the BT DASS2 multi-line system [25, 26]. Typically, the multi-line situation concerns a digital PBX accessing the public digital local exchange. This PBX circuit group is realised as 2 Mbit/s PCM and thus all the traffic circuits are 4-wire, as shown in Fig. 14.26.

SPC applies to the switching at each end, so the logical approach to signalling is common-channel. However, PBX circuit groups are relatively 'thin' and do not carry high concentrations of traffic. Also, the propagation time is short. Advantage can be taken of these conditions to achieve a relatively-simple system. The main features of DASS multi line are as follows:

(*a*)   *Loading*: The DASS CCS system serves the signalling requirements for the 30 traffic channels in one PCM systems only (Fig. 14.26). The signalling

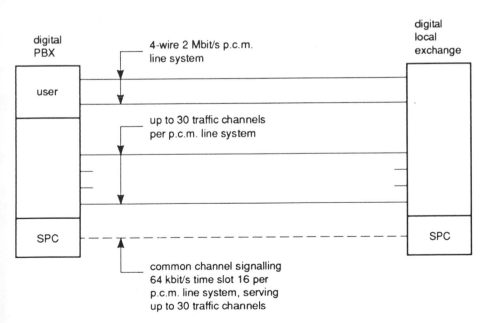

Fig. 14.26   Multi-line IDA signalling

is contained within each 2 Mbit/s system, using time-slot 16 for the 64 kbit/s signalling bit stream.

(*b*)   *Circuit labelling*: The circuit-label capacity is limited to 30. The Address field (Fig. 14.25) of each frame indicates the traffic channel to which the message relates.

(*c*)   *Format*: This based on HDLC (Fig. 14.25). The Address, Control, Information and Check field have functions much the same as in DASS single-line systems.

As in DASS single line, continuous flags are transmitted when there are no messages to be sent.

Since DASS CCS is per-PCM-system and uses time-slot 16 of that system, the signalling 'proves' the continuity of the traffic channel. Thus, a continuity check is not required to ensure signalling integrity.

For both DASS single and multi-line systems, error detection is the same as for the No. 7 system. However, the No. 7 error control is noncompelled. DASS single and multi-line systems adopt the compelled mode. Here, one signal frame only is transmitted at a time, the frame being repeatedly transmitted (some 60 times in the limit) until acknowledged by a responding frame in the return direction. This is a form of positive acknowledgment, which informs the transmit end that the signal frame has been received error free and instructs it to cease sending that frame and to send the next signal frame (if available).

Corrupted signal frames are discarded. The receive end waits until it checks that a frame is error free and then returns the acknowledgment response. The compelled mode is thus a form a forward error correction.

This compelled mode is a considerable simplification compared with the noncompelled mode, because:

(*a*)   No retransmission store is required.
(*b*)   No negative acknowledgments are required.
(*c*)   Sequence numbering of signal frames is not required for error-control purposes. (Sequence numbering of both the command and response signals are included for other purposes.)

14.17 References

1 WELCH, S.: 'Signalling in telecommunications networks' (Peter Peregrinus, 1979)
2 FLOOD, J.E. (Ed.): 'Telecommunication Networks', (Peter Peregrinus, 1975) pp. 66–67
3 FLOOD, J.E. (Ed.): 'Telecommunication Networks', (Peter Peregrinus, 1975) pp. 40–45, 91, 190
4 BULLINGTON, K., and FRASER, J.M.: 'Engineering aspects of TASI', *Bell System Tech. J.*, 1959, **38**, pp. 353–364
5 WELCH, S.: 'Signalling arrangements for dialling over transatlantic telephone cables', *PO Elect. Eng. J.*, 1960, **53**, pp. 201–205
6 WELCH, S.: 'The fundamentals of DC pulsing in multiexchange areas', Printed Paper No. 184, Inst. Post Office Elect. Engrs., 1944

7 BREEN, C., and DAHLBOM, C.A.: 'Signalling systems for control of telephone switching', *Bell System Tech. J.*, 1960, **39**, pp. 1381–1444

8 WELCH, S.: 'The influence of signal imitation on the design of VF signalling systems', Printed Paper No. 206, Inst. Post Office Elect. Engrs., 1953

9 NEWELL, N.A., and WEAVER, A.: 'Single-frequency signalling system for supervision and dialling over long-distance telephone trunks', *AIEE Trans.* 1951, **70**, pp. 489–495

10 WELCH, S.: 'The signalling problems associated with register-controlled automatic telephone exchanges', *Proc. IEE*, 1962, **109B**, pp. 465–475

11 GAZANION, H., and LEGARE, R.: 'Systèmes de signalisation Socotel', *Commutation et Electron.*, 1963, **4**, p. 32

12 DAHLBOM, C.A., HORTON, A.W., and MOODY, D.L.: 'Multi-frequency pulsing in switching', *Electr. Engg. (USA)*, 1949, **68**, pp. 505–510

13 DAHLBOM, C.A.: 'Common channel signalling — a new interoffice signalling technique'. International Switching Symposium Record, 1972, IEEE, Boston, USA

14 CCITT: Red Book VI, Fascicle VI.7 'Specifications of signalling system No. 7'. (ITU, Geneva, 1985)

15 MILLER, C.B., and MURRAY, W.J.: 'Transit-trunk-network signalling systems: Part 1: Multi-frequency inter-register signalling', *PO Elec. Engrs. J.*, 1970, **63**, pp. 43–48 'Part 3: Line-signalling systems', ibid, pp. 159–163

16 HILL, R.A., and Gee, G.J.H.: 'A miniaturized version of signalling system AC No. 9', *ibid.*, 1973, **65**, pp. 216–227

17 HORSFIELD, B.R., and GIBSON, R.W.: 'Signalling over carrier channels that provide a built-in out-of-speech-band signalling path', *ibid.*, 1957, **50**, pp. 76–80

18 GIBSON, R.W., and MILLER, C.B.: 'Signalling over carrier channels that provide a built-in out-of-speech-band signalling path', *ibid.*, 1957, **50**, pp. 165–171

19 BOLTON, L.J., and BENNETT, G.H.: 'Design features and application of the Post Office 24-channel pulse-code modulation system', *ibid.* 1968, **61**, pp. 95–101

20 POSPISCHIL, R., and SCHWEIZER, L.: '30-kanel PCM system fur den nakverkehr', *Tech. Rundsch.*, 1971, (9)

21 VOGEL, E.C., and MCLINTOCK, R.W.: '30-channel pulse-code modulation system, Part 1: multiplex equipment', *PO Elect. Eng. J.* 1978, **71**, pp. 5–11

22 International Organisation for Standardisation (ISO): 'Open Systems Interconnection (OSI)', ISO TC/97/SC/537

23 ISO: 'High level data link control (HDLC)'. ISO 4335

24 CCITT: 'No. 7 error control'. Red Book VI, Fascicle VI.7, Recommendation Q703 (ITU, Geneva, 1985)

25 CCITT: 'ISDN user network interfaces'. Red Book III, Fascicle III.5, Recommendations 1412, 1420 and 1421 (ITU, Geneva, 1985)

26 BIMPSON, A.D., RUMSEY, D.C., and HIETT, A.E.: 'Customer signalling in the ISDN', *J. Inst. British Telecomm. Engrs.*, 1986, **5**, pp. 2–10

Chapter 15
Customer access
P F Adams, P A Rosher, P Cochrane

BT Laboratories

15.1 Introduction

The local access network, which provides the means of connection between customers and their primary switch or concentrator, is currently dominated by analogue telephony. The metallic twisted pair cables used are, however, capable of supporting signals of much greater bandwidths. This capability has been exploited in the past by pair-gain systems [1], supervisory services, baseband modems and data services at 9·6 and 64 kbit/s. Digital transmission in the long and shorthaul networks, coupled with the introduction of an Integrated Digital Network, has stimulated interest in the provision of full digital customer access. Access technologies are required to be inexpensive as the average utilisation is very low compared with other parts of a network. Also, duplex transmission over all types and lengths of local pair connection has to be possible, despite the variety of cable types, and lengths ranging from a few tens of metres to several kilometres. This is a challenging objective. However, advances in large scale integrated circuit technology have made possible the implementation of complex systems at low cost [2] which support the 144 kbit/s required for ISDN basic rate access over most connections.

In addition, higher rate systems operating at a few hundred kbit/s and even 2 Mbit/s are envisaged. Existing local network cables can, in general, only support these higher rates over limited distances owing to crosstalk and impulsive noise interference [3].

Radio can provide particular advantages in certain situations and it has been deployed for fixed access where the use of cables is difficult or expensive. It can also offer the rapid provision of access for broadband communications. However, its main use is in access for mobile telecommunications and in cordless telephones, but both of these are beyond the scope of this chapter which is focussed on fixed access technologies.

To some extent, these developments have forestalled the introduction of fibre into the local network, which has been compounded by the relatively high cost of optical devices and components. Recently, however, cost competitive [2] solutions have emerged that do allow optical systems to address a significant, and fast expanding, range of applications. Specifically, the provision of wideband services at >0·5 Mbit/s can now be provided at lower cost (on new installations) using optical fibre. For services at >2 Mbit/s and >2 km there is no contest — existing twisted pair cables are generally crosstalk limited and will not support transmission and optical fibre cables offer a lower cost solution than new metallic cable.

15.2 Local access strategy

The telecommunications operating companies of the developed world divide neatly into two strategic groups; those allowed to combine telecommunications services with CATV and those who are not. This has produced two different sets of operating/economic criteria. The combining of CATV with telecommunications services considerably relaxes the economic constraints on telecommunications provision which, in effect, can ride on the back of entertainment services.

It is also interesting to reflect on the penetration of CATV systems. In many parts of the USA and Europe where TV reception is poor, and/or the networking of programmes across international borders is popular, CATV is a success. However, in countries with good national coverage and the economic means to produce good quality programmes, (and perhaps no international boundary) then CATV has so far failed to make any significant impression. In the UK, for example, there are few CATV systems, and an apparent low level of interest from the viewing public. However, over 70% of homes in the UK do own video recorders — which are mainly used for watching hired films! In most first world countries, the problem of local line optical access to the business customer thus remains the key challenge. If CATV is provided to the home, then data and telephone access come almost for free. If it is not provided, then the revenue from telecommunication services is usually insufficient to warrant a dedicated fibre pair per customer access unless it is the lowest cost technology where a universal service obligation is imposed on the service provider.

For the above reasons, a preoccupation for many operators is the economic realisation of optical digital telephone access to the home with wideband systems (>2 Mbit/s) for business customers. For the residential customer, outside the CATV domain, the emphasis is currently on the use of optical fibres linked to remote switches or concentrators with conventional copper distribution.

15.3 Copper access

Fixed local access networks have a tree-like topology with cables in ducts radiating from the primary switch site to branch at permanently connected joints and at cross connection flexibility points [4]. Cables containing different numbers of twisted pairs of wires of different gauges are used in different parts of the network with the larger size, lighter gauge, cables near the switch and smaller, heavier gauge cables at the extremity of the network. Each connection from a customer to the primary switch is a concatenation of lengths of twisted pairs of different gauges. In some countries permanently connected bridged taps occur which have a marked effect on the transmission characteristics. Some typical connections occuring in the UK and the USA are shown if Fig. 15.1. The average length of a connection in the BT's network is about 2 kms; Table 15.1 shows the percentages of each conductor gauge used in the network, with 0·4 and 0·5 mm conductors predominating. The 0·32 mm gauge cable is used in short lengths for the first section out from the switch and larger gauges are used on longer routes to meet planning limits. The planning limits in force when the majority of BT's access networks were installed dictated a maximum loss at 1600 Hz of 10 dB, and a maximum loop resistance (the resistance measured with the end of the connection short circuited) of 1000 ohms. Although these have progressively increased to 15 dB and 2000 ohms over 95% of connections have a loop resistance of 1000 ohms or less. Loop resistance is not of primary interest in determining the transmission characteristics of a cable, although it does correlate well with pair attenuation. However, customer loops are often required to be line powered and under maximum voltage limitations the loop resistance affects the amount of power available at the customer end.

The extreme variability of local connections in length and cable make-up means that the electrical characteristics which affect digital transmission also vary dramatically from connection to connection.

15.3.1 Copper pair characteristics

The electrical characteristics of copper pairs are mostly frequency dependent and some are dependent on the gauge of the conductors. Of interest in the design of transmission systems are: attenuation, phase, characteristic impedance, crosstalk, impulsive noise and random noise [3, 4]. The variation of attenuation with frequency follows a square root of frequency law except that between approximately 10 kHz and 100 kHz the attenuation slope is shallower. Pairs terminated in impedances which are not identical to the pair characteristic impedance have frequency responses that deviate from the pair response at low frequencies where the response is dominated by the terminations. Fig. 15.2 shows some typical responses for pairs terminated by 140 ohm resistors coupled to them by transformers. The phase characteristics for all conductor gauges exhibit linear phase with a constant phase

Table 15.1 Percentage of each conductor used in BT's local networks

Conductor Diameter, mm	%
0·32	2
0·4	33
0·5	48
0·63	14
0·8	1
0·9 + 1·27	2

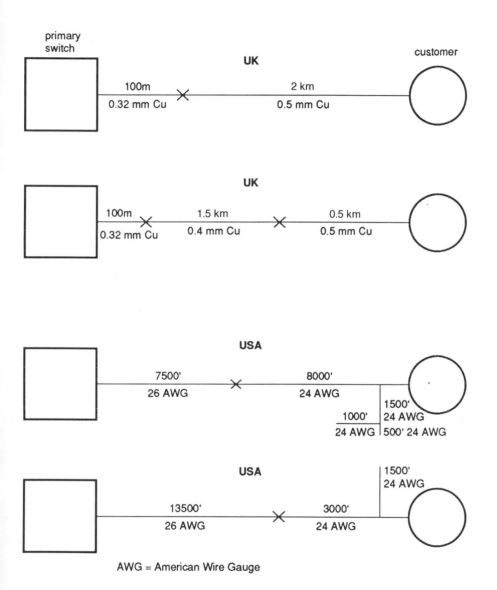

Fig. 15.1 Typical loop make-ups

offset above 10 kHz, but are non-linear at lower frequencies. Fig. 15.3 shows typical characteristics for four copper pairs. The extremes of the modulus of the characteristic impedance of pairs used in BT's network is shown in Fig. 15.4; other gauges lie between the extremes. At higher frequencies it approaches a constant resistive value but at lower frequencies varies dramatically with frequency and with conductor gauge. The characteristic impedance phase angle is zero at high frequencies, but it varies with frequency at low frequencies.

The crosstalk characteristics vary markedly between pairs in the same cable and with frequency. For the situation where many similar systems are in use on a multi-pair cable, the crosstalk can be averaged and it is found that the near end crosstalk (NEXT) attenuation follows a $f^{1.5}$ law [5]. Measurements have shown that 98% of pairs have NEXT power sum attenuations of better than 55 dB. Far end crosstalk (FEXT) is less of a problem and can be ignored where NEXT limits system performance.

Impulsive noise, arising from electromagnetic interference from a variety of sources is the most serious kind of noise found in the local network. It is not easily characterised, but extensive measurements in the UK local network have been made. Fig. 15.5 shows a typical result obtained from one copper pair after filtering. Random interference, e.g. thermal noise is not significant, but radio interference can be in locations close to transmitters.

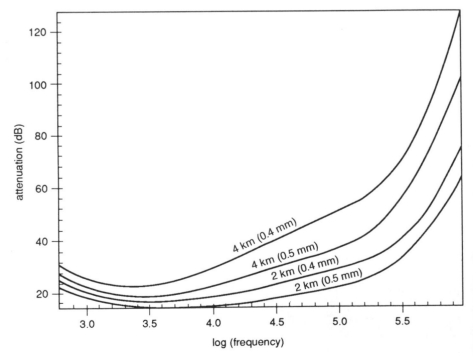

Fig. 15.2   Attenuation characteristics

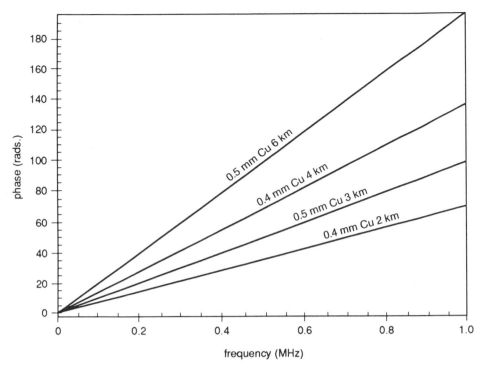

Fig. 15.3 Phase characteristics

15.4 Pulse transmission

Digital transmission requires the modulation of an electrical signal source by the digital information; for baseband transmission the most common way of doing this is to use Pulse Amplitude Modulation (PAM). Amplitude modulated pulses are transmitted over the copper pair and, ideally, the sent amplitude is detected without error at the far end. The pair attenuation and phase characteristics introduce pulse attenuation and pulse distortion which causes inter-symbol interference (ISI), and crosstalk and noise which cause errors if the received amplitude is too small. The degree to which these factors affect transmission is determined by the line code used and the shaping of the pulse by both the transmit and receive filtering. Baseband line codes with no redundancy introduce no spectral shaping. Redundancy, introduced by signalling faster than necessary, or by extra amplitude levels in the line code, can be used to lessen the degree of ISI by reducing the line code spectrum at low frequencies. However, this can only be achieved by enhancing crosstalk because reducing the spectrum at low frequencies will boost it at high frequencies. This is illustrated in Fig. 15.6, where the power spectra for a common transmit power of five line codes (with rectangular pulses) of increasing redundancy (2B1Q, SU32, MS43, B8ZS and

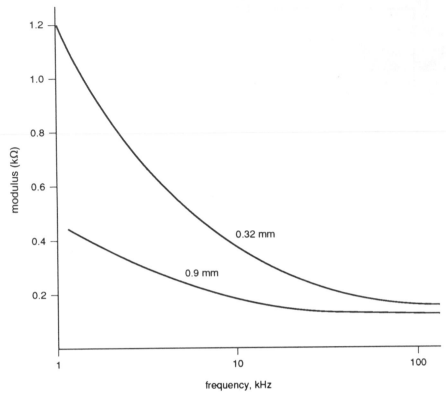

Fig. 15.4   Characteristic impedance of copper pairs

WAL2) are shown. 2B1Q has no redundancy. SU32, MS43 and B8ZS are all 3-level codes whose first null would be at 0·63 if they had zero redundancy. WAL2 is a binary code with a redundancy of 0·75. Pulse shaping has a similar effect; indeed linear line codes are directly equivalent to pulse shaping and the same trade-off between distortion and spectral shaping occurs.

Although line coding and pulse shaping can give a reduction in ISI, the more crosstalk tolerant pulse shapes/line codes do not suppress ISI sufficiently. Consequently, when the received signal after filtering is sampled, the sample at time iT (where T is the baud interval) is given by:

$$s(iT) = D^T(iT) \cdot H + n(iT) \qquad (15.1)$$

where $D(iT)$ is the column vector of $p+q+1$ data elements modulated on to pulses from time $(i-p)T$ to $(i+q)T$, H is the column vector of samples of the overall transceiver and pair time response from p samples before the largest sample to q samples after, and $n(iT)$ is the noise term caused by crosstalk and other interference. The ISI can be reduced by a linear equaliser which effectively multiplies $S(iT)$, where $S(iT)$ is the vector of samples of the received signal, by the vector E which approximates the spectral inverse of H. Because of the variability of the pair characteristics

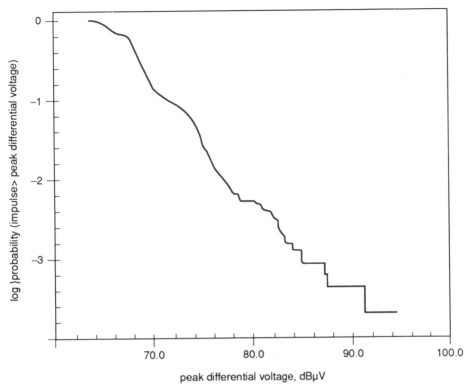

Fig 15.5 Peak voltage cumulative distribution for impulse noise

a fixed equaliser is inadequate for all copper pairs. It is, therefore, necessary to use some form of adaptive equalisation [7].

Linear equalisation can be provided in two ways: by equalising over the whole bandwidth of the line signal or, after sampling at the baud rate. The former is best achieved by an adaptive fractional-tap (FT) discrete time equaliser. By using an appropriate adaptation algorithm the FT equaliser is also capable of optimally filtering the interference to jointly minimise the ISI and the noise [8]. For such an equaliser the incoming received pulses can be sampled at any constant phase and timing recovery is reduced to tracking any frequency differences between the receiver and the far end transmitter.

For equalisation after sampling at the baud rate a T-spaced equaliser is used. This is easier to implement and provided a good sampling phase is used gives almost as good performance. A good sampling phase is that which produces maximisation of the band-edge components [9], thus ensuring that the equaliser in boosting those components does not enhance noise unduely. However, linear equalisation usually introduces noise enhancement; for white noise the noise power is increased by a factor $E^T E$. If the sampling of the received signal can be arranged such that the first p terms of H are negligible, then a decision feedback equaliser (DFE) can be employed

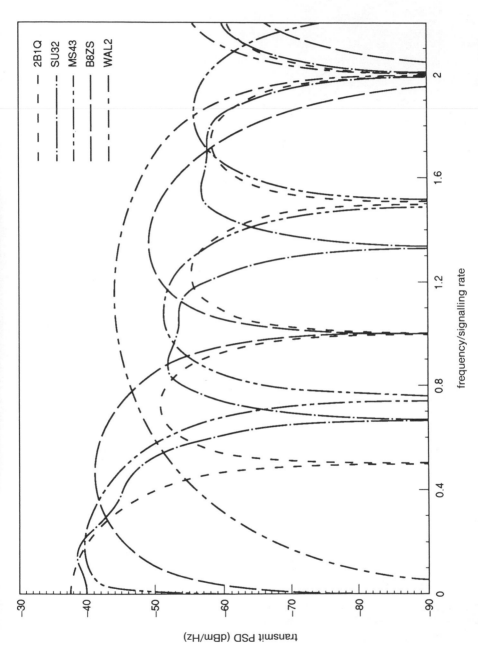

Fig 15.6 Line code frequency spectra

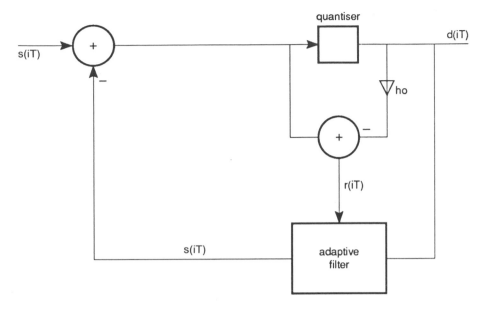

Fig 15.7   Decision feedback equaliser

(Fig. 15.7). The DFE coefficients model the first M significant post–cursor samples of the pulse response, i.e. the first M terms but one of **H**.

The DFE does not enhance interference because the ISI is cancelled and the noise is not modified by the equaliser. However, because detected data is fed back, any errors cause magnification rather than cancellation of the ISI which can create further errors; a phenomenon known as error propagation.

To adjust DFE coefficients to their optimum values the error signal at the output of the subtractor is minimised by an appropriate adaptive algorithm. The most commonly used is the least-mean-squared (LMS) algorithm which seeks to minimise the mean square error signal. The error signal is

$$r(iT) = s(iT) - s'(iT) - h_0 \cdot d(iT) \tag{15.2}$$

where s' is the output of the DFE adaptive filter and h_0 is the first significant term of **H**, and the LMS algorithm performs the simple recursion

$$\boldsymbol{H}'((i+1)T) = \boldsymbol{H}'(iT) + v \cdot r(iT) \cdot \boldsymbol{D}(iT) \tag{15.3}$$

where $\boldsymbol{H}'(iT)$ is the vector of DFE coefficients at time iT.

Sampling at the required phase to allow only a DFE to be used can be achieved by using a decision directed timing recovery scheme [10] to force the precursor ISI to a minimum. Alternatively the DFE can be preceded by a linear equaliser designed to reduce pre-cursor ISI. This allows a wider choice of timing phase to be used but does introduce some noise enhancement.

15.5 Transmission performance

So far it has been indicated that the choice of line code, pulse shaping and technique for equalisation all affect the transmission performance. Ultimately the choices made will depend on the target performance required and the complexity of the signal processing that results. Before comparing specific systems it is worth evaluating the fundamental limitations to transmission capacity which is imposed by crosstalk. Shannon's theory can be applied to compute the capacity as a function of pair length [3]. Using a value of NEXT power sum attenuation of 55 dB at 100 kHz and typical measurements of pair gain frequency response, the graphs of Fig. 15.8 were obtained for two different bandwidths. The capacity is clearly pair-length dependent but even over long lengths it is sufficient to provide useful data rates. The graphs also suggest that on longer connections there is little to gain from using wider bandwidth signals.

A detailed comparison of different systems can be obtained based on measured pair responses and a simple but realistic model of a general transmission system. The model is shown in Fig. 15.9 which allows us to analyse the effects of crosstalk and noise by assuming appropriate model for the various forms of interference and, where necessary, idealised system components. The multi-level encoder transforms binary data into an L-level code at rate $1/T$ Baud. Amplitude modulated impulses from the coder are applied to a transmit filter $S(f)$ which bandlimits the signal spectrum. The response $C(f)$ is the gain/frequency characteristic of the terminated twisted pair. Crosstalk and impulsive noise are introduced at the input to the

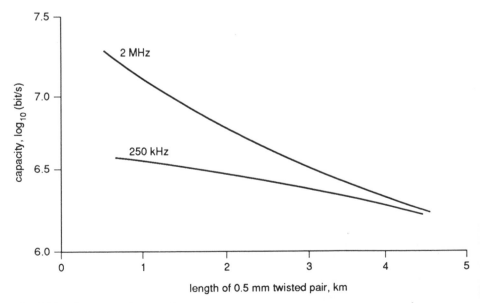

Fig 15.8   Copper pair capacity

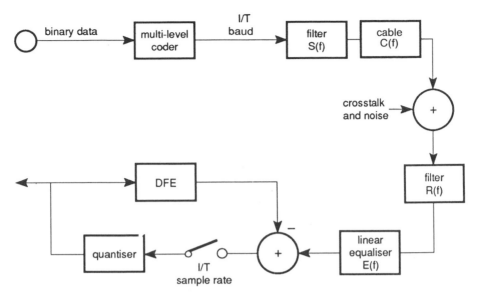

Fig 15.9 Transmission system model

receiver. A receive filter $R(f)$ bandlimits the input. The equaliser $E(f)$ is a linear equaliser that either inverts the channel or, in the case of the DFE equaliser being used, is a fixed compromise equaliser. The filters are constrained so that the overall transfer function $S(f) \cdot R(f) \cdot C(f)$ introduces zero ISI and unity pulse gain for a zero length pair.

15.5.1 Crosstalk analysis

The crosstalk power at the output to the quantiser is obtained by evaluating

$$P = \int_{-\infty}^{\infty} \frac{(L^2-1)}{3T} \cdot |S(f)|^2 \cdot X(f) \cdot |R(f)|^2 \cdot |E(f)|^2 \cdot df \qquad (15.4)$$

where $X(f)$ is the crosstalk power gain. If we assume that common designs of filters $S(f)$ and $R(f)$, with appropriately scaled time responses, are used for all transmission rates, then it can be shown [3] that the receiver quantiser signal to noise ratio Q is given by:

$$Q = 3a^{2Z} y T^{1\cdot5+2bZ}/(I \cdot (L^2-1)) \qquad (15.5)$$

assuming that $X(f) = f^{1\cdot5}/y$, where y is the crosstalk attenuation at 1 Hz, Z is loop resistance, I is an integral which is rate independent but is a function of the system filters and the type of equalisation used, and a and b are pair parameters which are dependent on the type of equalisation. In Fig 15.10 this equation is used to show the maximum data rate for $Q = 20$ dB as a function of loop resistance for various types of transmission system. The results show that decision feedback equalisation is better than linear equalisation and that low redundancy multi-level coding allows greater data rates and/or greater reach.

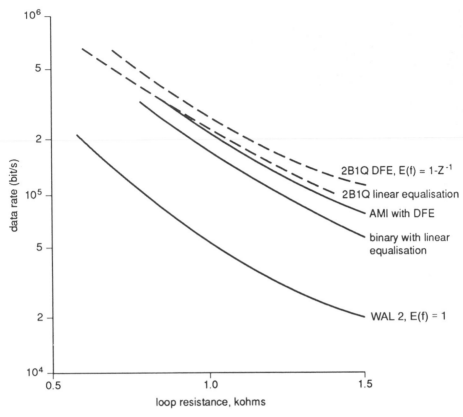

Fig 15.10   Crosstalk limited performance

15.5.2 Impulse noise performance

If a system is designed to operate in a crosstalk limited environment, the main source of errors will be from impulsive noise. Although there are techniques for ameliorating the effects of impulsive noise by, for example, smearing the impulses in time to make them of lower amplitude, generally impulse noise errors cannot be completely avoided.

To design a transmission system to have the best tolerance to impulse noise requires a knowledge of the time waveforms of the noise. Measurements [11] of impulsive noise have revealed that waveforms are very variable. In the absence of specific information the best that can be done is to compare systems by their tolerance to various carefully chosen impulse shapes.

If all that is known about an impulse is its peak magnitude (V) then the worst possible impulse has a time response that is equal to the sign of the time reverse of the response of $R(f) \cdot E(f)$, i.e., $V \cdot \text{signum} \, [r(-t)^*e(-t)]$. This, however, is a rather unrealistic impulse shape as well as having infinite energy! A better choice is the appropriately scaled time-reverse response $r(-t)^*e(-t)$, which is the pulse that for a given energy gives the maximum

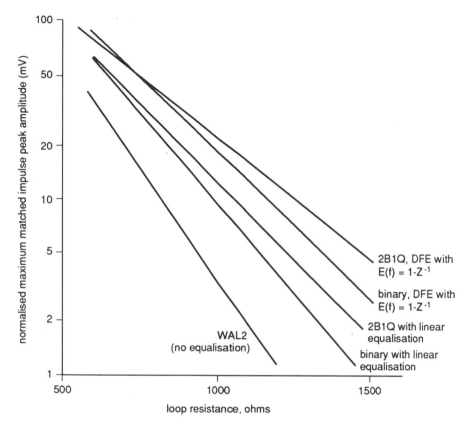

Fig 15.11 Normalised maximum error-free impulse amplitude for basic rate transmission

peak output from the input filter and equaliser combination. Fig. 15.11 shows the calculated tolerance of the various 144 kbit/s transmission systems to this kind of impulse when the transmit power is 10 mW into a 140 ohm load. The 4-level systems have their transmit power increased by a factor of two to account for the fact that the impulses required to cause errors in the 4-level case have twice the energy of the impulses required in a binary system. Low-redundancy line codes with decision feedback equalised systems have a clear advantage, as does 4-level coding.

15.6 Duplex transmission

Digital customer access transmission systems are required to operate in duplex mode over a single pair of wires. Theoretically, this is possible, by the use of a perfect hybrid balancing network. However, as illustrated in Fig. 15.4, the characteristic impedance of a twisted pair is a complicated function of frequency and is difficult to match exactly. Typically, a balance return loss of about 10 dB for PAM signals is all that can be achieved reliably with a simple compromise balance impedance. As the signal from the far end will be attenuated by anything up to 40 dB, this is totally inadequate. Other methods are required to remove the local echo (and any others from impedance mismatches on a mixed pair gauge connection) from the received signal. Early attempts at this were either to separate the signals in the two directions in frequency or time, the so-called frequency-division and burst-mode techniques. Both these, however, result in reduced crosstalk limited reach because they both lead to greater signal attenuation and increased bandwidth.

Burst mode [12] separates the two directions by transmitting and receiving at different times. The transmitted signal is removed from the receive signal path by gating it out with an appropriate timing waveform. Burst mode is very simple to realise and, by using a well-balanced line code/pulse shape, no adaptive equalisation is necessary. It is argued that by synchronising the bursts at the local exchange end, the effects of NEXT can be avoided, but it is doubtful whether such a constraint is acceptable from an operational point of view, and the degradation in tolerance to noise and FEXT still remains. On shorter connections burst mode gives simple and effective duplex transmission. For this reason it has found application for providing digital access to switches within customer's premises.

The only method that does not degrade performance by forcing the use of wider bandwidth signals is echo cancellation.

15.6.1 Echo cancellation

Echo cancellation involves modelling the echo path response which modifies the locally transmitted signal before it is sampled by the receiver. The model is driven by the transmitted signal and its output subtracted from the combined echo and received signal, as illustrated in Fig. 15.12. The signals sent in each direction can occupy the same frequency band and be continuous; therefore the disadvantages of frequency division and burst mode are avoided. The penalty is that an echo canceller requires a complex hardware implementation. However, modern technology enables implementation at acceptable cost.

The echo canceller shown in Fig. 15.12 is quite general. The echo replica it produces is a function of the current transmitted data and past data, i.e. the echo canceller output is

$$e(iT) = F(D(iT)). \tag{15.6}$$

The echo response it has to model normally decays with time and so the echo canceller can be constrained to provide a function of a finite number (N) of past data elements. In a binary system the elements of $D(iT)$ are binary valued; in a multi-level system of L levels the elements are represented by B bits where B is equal to or greater than $\log_2 L$. A straightforward echo canceller implementation is to use the data elements as the address lines to a look-up table which contains the 2^{NB} distinct values of $e(iT)$. An echo canceller realised in this way is known as a table look-up echo canceller [13] and has the great advantage of simplicity and no constraints on the function

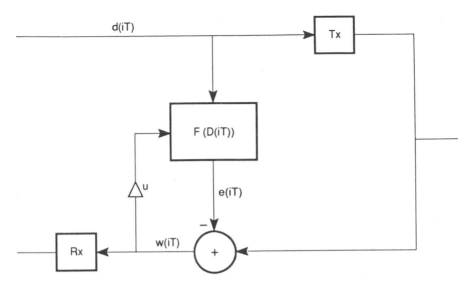

Fig 15.12 Echo cancellation

$F(\cdot)$. The coefficients in the table look-up echo canceller can be adapted by the simple algorithm:

$$e((i+1)T) = e(iT) + u \cdot w(iT) \qquad (15.7)$$

where $w(iT)$ is the error signal which is the difference between the received signal and $e(iT)$. $e((i+1)T)$ then replaces $e(iT)$ in the appropriate look-up table location. The gain constant u determines the rate of convergence. For random data each location is addressed on average only 2^{-NB} of the time so convergence can be very slow.

The implementation is not only slow to converge it has the disadvantage that the memory size increases exponentially with N. Both these disadvantages are diminished if it is assumed that the function $F(\cdot)$ is linear. The echo canceller [14] then becomes identical to the DFE structure previously described. The amount of memory required has been reduced to the minimum of N locations, but the processing is increased to N multiplications and additions. Adaptation is again by the LMS algorithm.

The table look-up and linear echo canceller structures have markedly different convergence properties for a given final residual echo power. It can be shown for example that for binary signals, u and v much less than 1 and the same final residual echo power to received signal ratio that the linear echo canceller is $2^N/N$ times faster to converge than the table look-up echo canceller.

The echo canceller has to work in the presence of the signal from the far-end that will be superimposed on the error signal, causing fluctuation of the echo canceller coefficients about their desired mean values. The mean value of the coefficients will be unaffected provided the received signal and

the local transmitted signal are uncorrelated. This is normally assured by
the use of different scramblers in each transmission direction. It will,
however, result in a significant residual echo power unless u (or v) is small.
A small gain constant results in slow convergence of the adaptive echo
canceller, but if a half-duplex training procedure is used to converge the
filter initially then a high gain constant can be used followed by switching
to a low value for tracking during duplex transmission. An alternative is to
use an update algorithm that has the received signal removed from its error
signal. As the received signal is determined by the transmitted data from
the far end and the pair pulse response, an adaptive channel estimator fed
by the received data is required, in order to produce an estimate of the
received signal that can be removed from the error signal. Such an echo
canceller is known as an adaptive reference echo canceller [15]. In fact, if
the transmission system employs a DFE, then such a reference is already
available, i.e. $r(iT)$ and a joint canceller decision feedback equaliser system
results.

Some timing recovery schemes require processing of the whole of the
received signal spectrum before sampling. For such schemes to work correctly the echo must be removed from the whole of the received signal
spectrum. Therefore, the echo canceller must operate at a sample rate that
is at least twice the highest signal frequency in the received signal. Such
fractional-tap spaced echo cancellers are easily obtained by multiplexing m
(m greater than or equal to 2) echo cancellers such that their outputs are
obtained every T/m seconds.

15.7 Interfaces and standards

In parallel with the advances in digital transmission system technology and
now that most countries are installing or planning to introduce some form
of ISDN, international standards organisations such as the CCITT have set
the standards for the ISDN and access to it [5]. CCITT has formed the I
series of recommendations for this purpose. However, so far an international
standard for the transmission system for the local loop has not emerged.
Instead a new interface has been defined (CCITT Rec. 1420) which gives
the subscriber two 64 kbit/s circuit switched channels and access to a 16 kbit/s
signalling channel. This overall bit rate of 144 kbit/s must be supported by
the local loop transmission system. In addition the need for framing information and an auxiliary maintenance channel will increase the overall bit rate
of the local loop to, typically, 160 kbit/s. The recommendation also impacts
on the transmission system design in other ways. For example, the jitter on
the clock provided by the transmission interface to the customer is constrained to have a peak-to-peak jitter of less than 5% relative to the network
clock at the primary switch.

The 1420 interface is a possible boundary between the telecommunications network and customer terminals. As we have seen the variation and
complexity of transmission systems and the differences between the local
access networks of different countries make this is a sensible boundary in

engineering terms. In addition the small amount of electronics at the end of the local loop allows a demarcation for the purposes of maintenance, so that it is possible to distinguish between network and terminal faults. However, the regulatory position in some countries, for example the USA, forces the boundary to be the wires entering the customer's premises. In the USA, a standard transmission interface [16] using 2B1Q line code has been defined for ISDN basic rate access that will allow interworking between the network end of the loop and competitively provided terminal equipment.

15.8 Radio local access

Radio is, in theory, the ideal method for local access in that it requires no cables to be laid in the ground, avoiding expensive capital expenditure and giving rapid provision of access. However, in practice, the restricted bandwidth allocations for fixed communications applications and the cost of equipment have severely limited its use in the UK. In other countries, especially those with large rural areas or geographical features that make cabled access difficult, radio has been used more extensively.

Radio access systems can provide specific advantages such as: fast and/or cost effective provision where there is no available cable plant; flexibility for re-configuration of the network topology; easy equipment recovery and redeployment if access is no longer required or cable plant is installed later; an alternative technology and routing for network security. These advantages coupled with modern technology that potentially will reduce equipment costs suggest that the application may grow [17].

There are two basic system types for fixed radio access: point-to-point and point-to-multipoint. There are many proprietary radio systems in each of these categories. Here only the general attributes of typical systems and the role they may play are discussed.

15.8.1 Point-to-point systems

A single-channel point-to-point system is essentially an alternative to the copper pair line. The cost of current systems means that they are only applicable in situations where the provision of a copper pair connection is high. VHF systems are available which operate up to about 50 km range and can tolerate some degree of path obscuring. 2 GHz microwave systems can offer a similar range.

Multi-channel point-to-point systems share the high cost of the terminal equipment over many customer channels. Other ways in which point-to-point radio could be effectively applied to provide local access are to link remote concentrator units (RCUs) to the primary switch. Point-to-point radio links could be used in some places to provide alternative routing between RCUs and the primary switch and to close rings. Alternatively, pole top multiplexers could be developed to link say 30 customers back to the primary switch by 18 or 2 GHz radio.

18 GHz systems require line of sight operation and achieve up to about a 20 km range. The restriction to line of sight is not too serious a problem in practice and the range gives a useful coverage.

15.8.2 Point-to-multipoint systems

Microwave radio systems are also available for point-to-multipoint operation. Time Division Multiple Access (TDMA) techniques are used to give a number of out-stations access to a central station. However, the costs of out-stations are too high for one per customer circuits and so a number of customers are usually connected to an out-station by conventional copper pairs. Cost comparisons suggest that the use of point-to-multipoint systems in urban areas is very limited. Rural areas may well be a more suitable application where point-to-multipoint radio is used to replace much of the cable route back to the primary switch. However, the multiplexing required to reduce out-station costs makes the application less attractive. If customer density is low then a lot of cabling is required from the out-station. Ideally a single customer out-station is required. Currently this is too expensive. What is required is a technology somewhere between the technologies of the emerging new generation of cordless telephones and the current multi-point systems.

Another problem with the use of current point-to-multipoint TDMA systems in the UK is that they can introduce an unacceptable time delay of typically 8–12 ms into the local loop. The overall time delay in the UK is limited to 23 ms in order to avoid the use of echo control devices on national calls, with 8 ms allocated to each local access network.

15.9 Optical access

The next decade is expected to see a major growth in optical technology in the access network [18]. At present, the relatively few systems deployed have been targetted at large business customers, and have been based on point to point optical systems developed for trunk applications. Indeed, in most cases the very same optical systems have been used! Optical transmission has been so widely utilised for trunk applications because it can offer long distance, point to point transmission at many hundreds of megabits per second at very low cost. The transmission requirements in the access network, even to large business customers, are, however, very different. The access network provides many different services, e.g. PSTN, ISDN, private circuits, CATV, etc to a range of customers extending from the residential to the corporate user. These have to be carried in an economic manner and must also be responsive to the introduction of new services. The access network is the most cost sensitive and the deployment of new technology has to be thoroughly researched. For these reasons, operating companies worldwide are actively investigating differing network topologies that exhibit the characteristics described above with the objective of significant roll-out programmes spanning the next decade. It is, however, already apparent that an all-fibre solution to the home is unlikely in the short to medium term unless CATV is included, and for this reason hybrid solutions of fibre, radio and copper technologies are also being considered.

15.10 Optical system options

The dominance of Single Mode Optical Fibre (SMOF) for all telecommunications applications [19, 20] has prompted studies into LED, ELED and SLED options for bit rates up to 560 Mbit/s and distances of 5–30 km. In Table 15.2 results reported by various organisations following this approach are presented with 'champion results' at 107 km—16 Mbit/s and 25 km—560 Mbit/s. A critical feature of these types of system is the fibre coupling arrangement, which tends to require a moderately complex lens between the chip and SMOF. Typical coupling losses are generally <12 dB, with −24 dBm being a representative launch power coupled into a fibre tail. None of the reported work so far has been supported by cost comparisons and it is therefore difficult to make comment on applicability. However, these results do support the notion that there are many different technical solutions to the cost equation.

Table 15.2   Reported LED — single mode fibre results

Laboratory	Mb/s	km	LED	DET
Bellcore	560	25	SL	APD
Bellcore	560	15	EE	APD
NEC	560	5	SE	APD
Bellcore	560	4·5	SE	APD
Sumitomo	400	10	SE	APD
Bellcore	280	7·5	EE	APD
AT&T	180	35	EE	APD
Bellcore	140	50	SL	APD
Bellcore	140	35	EE	APD
GTE	140	30	SL	pin
ANT	140	30	EE	APD
NEC	140	25	SE	APD
Bellcore	140	22·5	EE	pin
Bellcore	140	15*	EE	pin
AEG	140	12	EE	pin
Bellcore	140	4·5	SE	pin
BTRL	34	11*	EE	pin
BTRL	34	16	EE	pin
CNET	27 MHz	12	SE	APD
Plessey	16	107	SL	pin
BTRL	8	20	EE	pin

* Duplex Working

In many cases the ELED structures appear to be as, or more, complex than lasers, but with the claimed dual advantages of being less temperature sensitive and simpler to operate. In a number of the reported cases to date, however, the ELED/SLED devices are actually lasers operated in an LED mode! This situation is further confused by statements regarding the device cost trends which apparently favour LED solutions as their price has been falling faster than that of lasers. What can be stated with some certainty is that; the cost of producing the LED chips is about the same as that for lasers and cost savings must therefore be sought from the alignment, packaging and testing processes. In the long run, however, lasers are likely to be the most favoured solution.

15.10.1 A 2 Mbit/s low cost SMOF system example

The dominant cost in terminal equipment for $1 \cdot 3$ μm operation is currently the optical devices which approximately equal the cost of cabled fibre in a 3 km link. For this reason, a system operating in the 820–950 nm region exploiting lower cost optical devices has been developed [20]. Although the fibre, which is single mode at $1 \cdot 3$ μm, is not specially intended for 850 nm operation, the loss is typically only $2 \cdot 2$ dB/km, which is sufficiently low for a local line access system. The bandwidth, however, is limited due to the propagation of $2 - 5$ modes (depending on precise fibre type) giving a distance bit rate product~60 km M/bits. Laser systems at these wavelengths are prohibited because of interference effects between the small number of modes. Fig. 15.13 shows a schematic of a typical system and power budget; other key features include:

- HDB3 Code Transparency: Using the CCITT interface code [21] as the optical line code avoids the need for code conversion at the transmitter

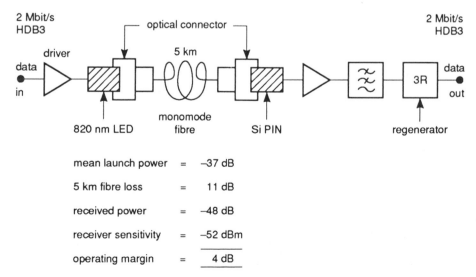

Fig. 15.13   2 Mbit/s optical access-system configuration and power budget

and receiver. The optical penalty for employing three, instead of two, level signalling at 2 and 8 Mbit/s is small (<4 dB).

- Silicon PIN: A simple low cost PIN-FET receiver combination gives a sensitivity of <-52 dBm (BER $<10^{-9}$) at 2 Mbit/s.
- Fibre Alignment: Because the LED and PIN are relatively large area devices (1 mm diameter for the active area), coupling into both SMOF and MMOF does not involve high precision alignment-connectors — see Fig. 15.14.

This approach can also be used at 8 Mbit/s although a further 7 dB of optical power is required to maintain the range. For 34 Mbit/s and above traffic density allows the use of more expensive devices such as $1\cdot3 \, \mu$m ELEDs and lasers.

15.10.2 $1\cdot3 \, \mu$m LED Systems

For greater reach and/or higher digit rates than the above system, $1\cdot3 \, \mu$m operation is necessary. The cheapest sources currently available are LEDs which are capable of a typical launch power into SMOF of -24 dBm. Since SM tailed devices are usually at a considerable premium over multimode devices, MM tailed devices have also been used to launch directly into SMOF. The main disadvantage of this method is the variability of the loss at the MM–SM fibre interface. This has been calculated to be <19 dB, but it is

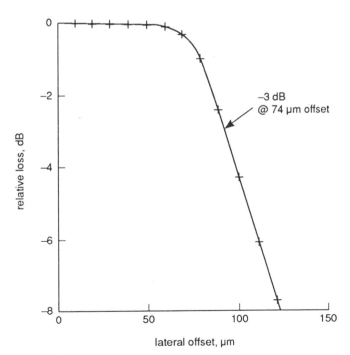

Fig. 15.14   2 Mbit/s access system fibre connector tolerance

strongly dependent upon the mode fill and has been measured between 9 and 18 dB [22].

Employing these types of device and conventional two-fibre operation, the systems shown in group 'A' of Fig. 15.15 have been engineered. State-of-the-art devices are capable of launching considerably more power, but it may turn out that highly optimised LEDs are ultimately more expensive than low performance lasers and the latter are thus favoured by some system designers.

15.10.3 Duplex transmission by mux and directional couplers

With 1·3 μm and 1·5 μm systems it is possible to reduce costs further by using a single fibre for both directions of transmission. This may either be accomplished by using separate wavelengths or directional couplers [18, 23]. At present, directional couplers are very attractive since they are relatively inexpensive and are the only additional system component required. With wavelength multiplexing not only are the devices currently expensive, but the cost of optical sources is increased by the need to control the wavelength more tightly.

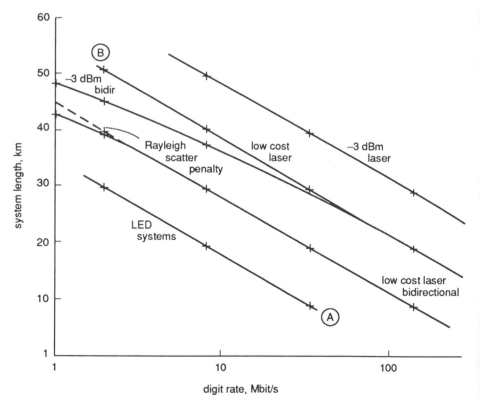

Fig. 15.15   Practical operating limits for 1·3 μm local line systems

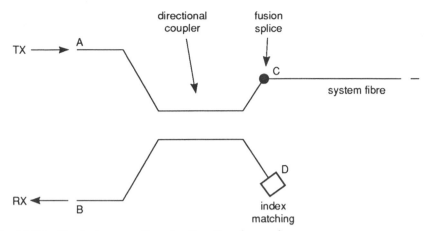

Fig. 15.16 Typical connections to directional compler

As indicated in Fig. 15.16, it is necessary to fuse the couplers onto the system fibre to minimise reflections which could create a near end crosstalk path. Each coupler introduces an inherent 3 dB loss which is also wavelength dependent, resulting in a total end-to-end penalty of about 7 dB which must therefore be included in the system budget [22]. Provided that discrete reflections are eliminated from the fibre path, this loss represents about a 10 km reduction in reach. There is also a penalty on lower bit rate systems caused by increased shot noise due to light reflected into the receiver by connectors, splices and Rayleigh scattering. A typical power budget for an ELED based duplex system is given in Table 15.3.

Table 15.3 Typical duplex power budget

	Transmission path	
	A	B
ELED centre wavelength nm	1260	1280
TX power (mean), dBm	−19·5	−20
MM-MM connectors	2	2
MM-SM splice	17	17
All other splices	0·5	0·5
Coupler loss	2·2	3·6
Fibre loss	3·5	3·5
Excess fibre loss	1·4	0·8
Coupler loss	4·0	2·5
RX power, dBm	−50·1	−49·9
RX sensitivity, dBm (BER $= 10^{-9}$)	−50·5	−50·5
Margin, dB (BER $= 10^{-9}$)	0·4	0·6

15.10.4 Duplex transmission by burst mode

An alternative method of avoiding crosstalk and echoes in single fibre working is to use burst mode transmission. This is potentially quite effective if a single optical device can be adopted as both receiver and transmitter [23]. However, compromises have to be made in device and system design. In particular:

(a) An optical source is normally optimised for low photon absorption, which implies poor performance as a receiver.

(b) The peak receiver sensitivity under reverse bias does not coincide with the peak of the emission spectrum under forward bias.

(c) The fibre alignment and device area giving optimum receiver sensitivity do not generally provide ideal coupling of source modes and vice versa.

(d) The system needs to operate at slightly more than twice the mean digit rate which implies a penalty of 3·5–5 dB.

The combination of these factors can result in a 15–20 dB penalty relative to conventional two fibre system and they are therefore only suitable for low digit rates and/or very short spans. A further significant penalty, especially for high speed operation, is associated with the amount of additional electronics required which leads to a significant reduction in system reliability and increased power consumption.

15.11 Problems of optical implementation

The cost of implementing SMOF in the local network is not solely concerned with the choice of optoelectronic devices but also includes costs associated with fibre installation and optical connectors. One technique which has demonstrated considerable advantage for the local network is 'Blown Fibre' [24]. In this method, low cost empty plastic tubes are initially buried or drawn into existing earthenware or metal ducts and fibre bundles are subsequently inserted with the aid of compressed air. The key advantages are that: fibres may be provided as required rather than to anticipate future expansion; the inherent low cost; a dramatic reduction in splices necessary; low fibre fibre strain; rapid installation rate.

For SMOF, connectors are significantly more expensive than for MMOF, simply because of the tighter mechanical tolerances required [21]. In the case of the low-cost 2 Mbit/s system this has been avoided by the use of relatively large area sources and detectors. Fig. 15.14 shows the coupling loss of a micro-lensed SLED source as a function of the lateral displacement of monomode fibre. An offset of several tens of microns is required at the source before any significant system degradation occurs; the receiver diode is similarly insensitive. Whilst some systems currently use low cost connectors, these involve both assembly and polishing operations and future versions may be coupled directly to a cleaved fibre in order to further reduce installation overheads.

15.12 Future fibre local access

Many of the early installations of local optical system for the business community are configured in a dedicated star. This is the direct analogy to the existing copper network and unlikely to be viable for fibre to the home as single star networks require the installation of very high fibre count cables. However, single stars do offer significant advantages since each customer has his own physically independent fibre link to the local switch. It is also a future proof architecture with no limit on future service provision. It has significant maintenance advantages, since one customer's service cannot be directly affected by the faulty operation of another customer's equipment. The actual optical technology required is also very simple since the transmission distance is just a few kilometers, and, in principle, as previously discussed, almost any type of optical source, fibre and detector combination can be used. Despite this, replicating the copper star network is unlikely to be economically viable for anything other than the very early systems for business access [4].

15.12.1 Active double star

This topology is analogous to the carrier serving principle that is common in the USA [25]. It is based upon the deployment of remotely sited multiplexing and switching electronics. These are fed by common optical feeder links from the local exchange. Hybrid arrangements with different technologies deployed on either side of the active node can also be supported, Fig. 15.17. The network cost is reduced by the use of these common feeders, but maintains the advantage of the single star architecture in the final link to the customer. The major disadvantage is the need for remote provision of powering and siting of electronic equipment, the associated environmental control and the relatively high initial costs.

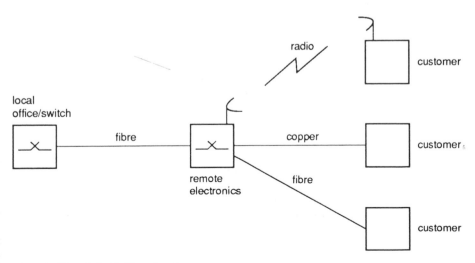

Fig. 15.17   Hybrid fibre local access

One development of the architecture is termed the Broadband Integrated Distributed Star (BIDS), Fig. 15.18. It uses remote switching for broadband, and multiplexing for narrowband at the remote access point [4]. Narrowband services, derived from 64 kbit/s switches, are synchronously multiplexed into a 140 Mbit/s stream for transmission to the remote access point. Broadband video material is fed to the remote access point in blocks of 16 channels on each fibre. Each block of 16 video channels is carried in an analogue, FM/FDM transmission format (27 MHz/channel). This necessitates an optical system design that provides a linear analogue transfer function. Parameters such as laser relative intensity noise (RIN) and laser linearity now become important in determining system performance. Each customer is simultaneously able to select from the remote access point any two video channels, plus radio channels, plus PSTN services. The final drop uses low cost burst mode transmission for the PSTN service, and a low cost laser scheme for the broadband services. These are combined over a single fibre using passive optical couplers. Burst mode transmission is realised using a single optical device biased to act as a transmitter and receiver at different times. For such a short link (<1 km), a semiconductor laser with its forward bias removed can be used as a PIN detector.

15.12.2 Passive star topology

This topology retains the shared capacity of the fibre feeder but passively splits the signals, in one or more locations, using optical couplers rather than active electronics. All signal processing is performed either in the local switch or in the customer's premises. The fibre utilisation is good but higher bandwidth is required together with a method to ensure privacy and integrity since the signal is distributed to all locations. Initial network costs are low with majority costs being incurred when customers are connected. It is also well suited to supporting asymmetric traffic such as telephony and broadcast television distribution.

One development of this architecture is termed Telephony on a Passive Optical Network (TPON). In this system, a single fibre emerging from the

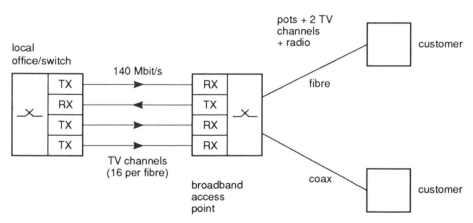

Fig. 15.18   Broadband integrated star (BIDS)

local switch site is fanned out via passive optical splitters at suitable points in the external network to feed a number of customers (Fig. 15.19). The system can use either WDMA or TDMA protocol similar to many point to multipoint radio systems. The signal is broadcast to all customer terminals on multiple wavelengths for WDMA and on a single optical wavelength for TDMA. Each customer terminal accesses the multiplex and selects only the channels intended for that destination. In the case of TDMA the return direction data from each customer's terminal is transmitted at a pre-determined time into a time division multiple access frame such that it arrives back at the exchange in its assigned timeslot. A dynamic ranging protocol ensures that customer transmissions never overlap at the local exchange. Current TDMA development systems use either bit or byte interleaving and operate between 20 and 40 Mbauds with WDMA systems demonstrators using up to 100 wavelengths demonstrated in the 1·3 and 1·5 μm windows. Deployed digital systems have, however, been far more modest with only 4–8 wavelengths. Typically, a single TPON system can provide up to 128 customer terminals with 144 kbit/s access or equivalent with modest fibre split ratios at each step.

Such an optical split from a single fibre emerging from the local exchange implies a relatively large end to end optical loss. The optical 'building block' is currently a 2:1 splitter of the form shown in Fig. 15.16 (although 1×7 has been shown to be viable). When light is injected into port A it is split equally between ports C and D, with typically 60 dB isolation to port B. Each coupler is inherently bidirectional and can thus be used to construct a bidirectional passive optical network (PON). Each customer is therefore served by a single optical fibre. The loss of each coupler is approximately 3 dB. A 128 way duplex PON requires 9 such couplers in each customer's signal path, implying a minimum optical loss of 27 dB. Once allowance is made for coupler tolerances, fibre and connector losses, plus operating system merging, the required optical budget is extended beyond 40 dB.

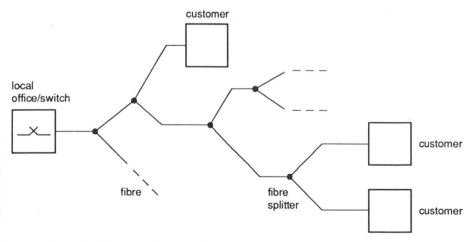

Fig. 15.19 Passive optical network

Clearly, relatively expensive optoelectronic devices are required to satisfy this optical budget, but it is anticipated that the sharing of local switch and plant resources inherent in this architecture will more than compensate.

15.12.3 Some realism

So far, it has been assumed that fibre cables will be connected to all homes. This is unlikely in the short to medium time frame. A more likely deployment of the PON architecture is in a hybrid scheme with the fibre terminating in the street, the so-called fibre to the kerb (FTTK) scenario. The final customer drop will then be on existing copper pair or via short range microwave radio. The stimulus for extending fibre to the home will come from additional broadband services. The PON infrastructure is ideally suited to carrying additional services using wavelength division multiplexing (WDM) techniques. One example is the provision of a broadcast cable television service. The signal to be transmitted is applied to a second laser at the local exchange at, say, 1550 nm, which is coupled into the PON using a wavelength division multiplexer. At the customer terminal a wavelength division demultiplexer is used to separate the narrowband and broadband services. The broadband system may be injected into the PON at a lower split level if it has a significantly lower optical budget than the narrowband system. The choice of optical multiplexer will depend on the level of isolation required. They can be based on either fused fibre, interference filter or diffraction grating technology. Fused fibre and interference filter technologies can be used to separate systems operating in different fibre windows, usually 1300 and 1550 nm. Fused fibre devices can offer isolations up to 20 dB, whilst interference filter devices offer isolation up to 40 dB. Diffraction grating technology can separate ten or more wavelengths operating in the same fibre window. The use of grating devices is expected to become widespread in the longer term when multiple services are provided over each PON using dense WDM.

15.13 CATV

The competing technologies for the broadcast cable television service have renewed interest in analogue optical transmission. Much of North America, for example, is served by extensive coaxial tree and branch analogue CATV networks with the most common format being amplitude modulation/vestigial sideband (AM/VSB) in line with domestic television. The transport of this analogue format over PONs is a possibility. Another alternative is the frequency modulation (FM) format as used in direct broadcast satellite (DBS) service provision which can also be supported by analogue transmission over fibre. The system requirements for the passage of digital, AM/VSB and FM formats are very different, see Table 15.4. The differing optical budgets and suitability for transmission over PONs are very apparent. However, in the short to medium term it is likely that the AM/VSB format will remain dominant. At present, cable television operators, who have

Table 15.4 Comparison of differing broadcast cable television formats

	Digital	AM/VSB	FM
Channel Bandwidth	70 Mbit/s	8 MHz	27 MHz
Number of Channels	32	40	16
Multiplex – type	TDM	FDM	FDM
– bandwidth	2·2 Gbit/s	50—550 MHz	950—1750 MHz
Optical Transmitter	Fabry Perot laser*	isolated DFB laser	isolated Fabry Perot laser*
Linearity	N/A	high	medium
RIN	N/A	very low <−155 dB/Hz	low <−125 dB/Hz
Optical Receiver	APD transimpedance	resonant PIN 75 ohm amplifier	APD/50 ohm amplifier
Sensitivity† (typical)	−34 dBm	−6 dBm‡	−32 dBm
Optical Budget§	37 dB	9 dB	35 dB

* DFB if operating in 1550 nm (dispersive) fibre window
† For a video signal to noise ratio of 50 dB
‡ Shot noise limited
§ For +3 dBm optical launch power

traditionally been different from telecommunications operators, are installing point to point AM/VSB optical systems over a few kilometres to set up pseudo headends in existing networks. This reduces the number of electrical line amplifiers through which a given customer's signal passes and hence improves the received picture quality. The next step will be the introduction of low level optical splits to share transmitter costs. In the medium term the use of erbium doped fibre amplifiers [27] is expected to permit optical splitting levels that are not compatible with those for narrowband services. In the longer term, once a fibre infrastructure has been developed, a transition to a digital format with all the benefits of digital transmission may be realisable.

15.14 Final comments

The last decade has seen a revolution in long haul networks with the widescale replacement of coaxial cables, microwave radio and satellite transmission with single mode optical fibre. This revolution has also largely penetrated the medium and short haul arena. The next stage will see the widespread introduction of fibre into the access network. It is unlikely to be a straightforward transition from copper to fibre as, unlike long haul, there is no single solution to the complex access requirements. We are therefore likely to witness the contiguous introduction of point to point star systems, PONs, joint fibre-copper-radio systems, remote concentrators and many more. The coexistence of PSTN, wideband services and CATV will be influenced by both the technology and (probably more drastically) by politics through governments and regulatory bodies.

In the long term (>20 years) we can reasonably expect to see the eradication of copper and a gradual spread of fibre to all telephones with a retrenchment to fewer and fewer switching centres with longer reach access networks carrying a full range of integrated services. Until then, there is likely to be a wide global diversification of the approach to fibre in the access network.

15.15 References

1 VOGEL, E.C.: 'Digital moves into the local loop' (ISSLS, Toronto, Canada, 1982)
2 Special issue on the local network, *BTTJ*, **17**, No. 2, April 1989
3 COX, S.A., and ADAMS, P.F.: 'An analysis of digital transmission techniques for the local network' *BTTJ*, **3**, No. 3, 1985
4 GRIFFITHS, J.M. (Ed.): 'Local telecommunications' (Peter Peregrinus Ltd., Stevenage, United Kingdom, 1983)
5 GRIFFITHS, J.M.: 'ISDN explained', John Wiley & Son, 1990
6 BYLANSKI, P., and INGRAM, D.G.W.: 'Digital transmission systems' (Peter Peregrinus Ltd., Stevenage, United Kingdom, 1976)
7 COWAN, C.N.F., and GRANT, P.J.: 'Adaptive filters' (Prentice-Hall, New York, USA, 1985)
8 QURESHI, S.: 'Adaptive equalisation', *IEEE Comm. Soc. Mag.*, March 1982, pp. 9–16
9 LYON, D.L.: 'Timing recovery in synchronous equalized data communications, *IEEE Trans.*, **COM-23**, 1975, pp. 269–274
10 MUELLER, K.H., and MULLER, M.: 'Timing recovery in digital synchronous data receivers': *IEEE Trans.*, **COM-24**, 1976, pp. 516-531
11 COOK, J.W.: 'Wideband impulsive noise survey of the access network', submitted to Globecom '91
12 BYLANSKI, P., and TRITTON, J.A.: 'Extending the capabilities of burst-mode transmission in the local loop' (ISSLS, Toronto, Canada, 1982)
13 HOLTE, J., and STUEFLOTTEN, S.: 'A new digital echo canceller for two-wire subscriber lines', *IEEE Trans.*, **COM-29**, 1981, pp. 1573–1581
14 MUELLER, K.H.: 'A new digital echo canceller for two-wire full duplex data transmission', Sept. 1976, *IEEE Trans.*, **COM-24**, pp. 956–962
15 FALCONER, D.D.: 'Adaptive reference echo canceller', *IEEE Trans.*, **COM-30**, No. 9, 1982, pp. 2083–2094
16 American National Standard for Telecommunications, 'Integrated Services Digital Network (ISDN)—Basic Rate Interface for Use on Metallic Loops for Application on the Network Side of the NT (Layer 1 Specification)', ANSI T1.601-1988
17 HARRISON, F. G.: 'Microwave Radio in the British Telecom Access Network', 2nd IEE National Conf. on Telecommunications, York, April 1989
18 'The 21st century subscriber loop', *IEEE Com-Soc Mag*, 1991, **29**, No. 3
19 SHUMATE, P.W.: 'Applications of LEDs with single mode fibre for broadband loop distribution', *IEEE*, **CLEO-86**, p. 252

20 BETTS, R.A., MOSS, J.P., and HALL, R.D.: 'A low cost 2 Mbit/s optical local line transmission system using single mode fibre', *IEE Elec. Lett.*, 1986, **22**, No. 3, pp. 143–144

21 YOUNG, W.C.: 'Design considerations for single-mode fibre connectors', *SPIE*, 1984, **479**, Fibre Optic Couplers, Connectors and Splice Technology

22 HALL, R.D., BETTS, R.A., and MOSS, J.F.: 'Bidirectional transmission of 11 km of single-mode optical fibre at 34 Mbit/s using 1·3 m LEDs and directional couplers', *Elec. Lett.*, 1985, **21**, No. 14, pp. 628–629

23 HUBBARD, W.M., and KEHLENBECK, H.E.: 'Light emitting diodes as receivers with avalanche gain', *Elec. Lett.*, 1978, **14**, No. 17, pp. 553–554

24 HORNUNG, S., CASSIDY, S.A., YENNADHIOU, P., and REEVE, M.H.: 'The blown fibre cable', *IEEE Trans. Selected Areas in Comms.*, to be published

25 AMICONE, M.R.: 'Fibre optic subscriber feeder systems', *Telephony*, 1985, pp. 70–74

26 REEVE, M.H. *et al.*: 'Design of passive optical networks', ibid, **4**, pp. 89–99

27 COCHRANE, P.: 'Future directions in long haul fibre optic systems', *BTTJ*, 1990, **8**, NO. 2, pp. 5–17

Chapter 16
Transmission planning
A.W. Muir

BT

16.1 Introduction

The objective of any planner can be basically stated as anticipating and assessing the needs of his customers and putting plans into place to meet those needs in the most straightforward and economic way possible. Within a stable market, and with a modest rate of technological advancement and change, this apparently simple objective can only be achieved in practice through a very complex planning process. When the market exhibits rapid change, along with the technology and political operating framework, then the planning function becomes extremely challenging. For many of the developed nations, the past five years have seen moves towards the de-regulation and control of telecommunications [1—4] with the subsequent rapid growth in services and demand [5, 6]. The planning departments of the operating companies have therefore had to cope with the arrival of new forms of competition, unprecedented growth in new services and capacity demands which were hardly conceived of ten years ago [7, 8]. During the same period the maturing of optical fibre transmission technology, plus the complementary advances in terrestrial and satellite radio systems, have served both to fuel and satisfy this exceptional user demand. In addition, fibre optics has also realised a transmission cost reduction of about tenfold over its copper predecessors during the last decade [8].

In this chapter, we therefore have to consider the planning process from two basic standpoints: a well-esablished stable exponential growth of the dominant telephony services with fifty years of significant history, and the new rapidly-growing areas, such as cellular radio, data transport and corporate networks, with a history of less than five years. A measure of the difficulty that we might assume from the outset can be gauged from the cellular radio network in the UK which has grown from zero to more than 1% of all the telephones within a three-year period of its inception, and has realised growth rates [9] of between 25 and 50% across most European countries. This type of rapid growth is also reflected in the deployment of optical fibre technology, which now supports about 50% of all the global long-lines traffic (Fig. 16.1) and gives a further indication of the scale of the problem facing today's network planner.

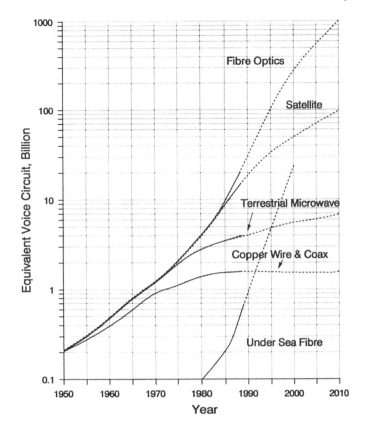

Fig. 16.1 Global transmission capacity

16.2 The planning process

Most modern operating companies have, to some degree, automated their planning, procurement, installation and operating processes in a manner similar to that shown in Fig. 16.2. This augmentation cycle starts with the forecasting of demand, based on marketing intelligence which is generally expressed as quantities of individual circuits required between switching nodes. The ultimate objective of the planning process is to combine these circuits to form hierarchical blocks [10] in the transmission plan, then complete system blocks (for example of 140 or 565 Mbit/s in Europe), and then to ensure that line plant and radio systems are available to interconnect switching units as required.

In practice, transmission planning has often been divided into two basic areas in order to optimise the balance between the complexity and efficiency of the process. The two areas are known as network capacity and system planning, both of which are described by this chapter.

The output of the planning process is a request to the procurement and works organisations to implement the plans in the most economic way. It

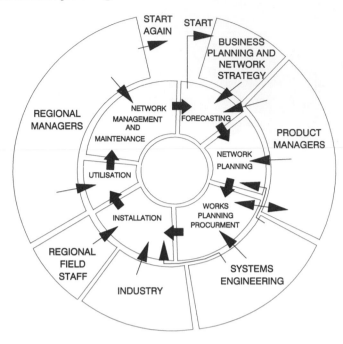

Fig. 16.2   Network augmentation cycle

is perhaps worth noting at this stage that the whole process, from identifying the need for a new circuit to commissioning the transmission plant, usually takes up to 2—3 years, but can require anything up to 5—7 years if new ducts or buildings are required. Inevitably, there will be some mismatch between the actual utilisation and the plans; thus the final stage in the network augmentation cycle is the feedback of this information to ensure correction in subsequent planning cycles [11].

16.3 Network structure

Before discussion of the planning process in detail, it is useful to examine the structure of the telecommunications network, and to remember that it is an evolutionary and not a static entity. For example, during the last decade there has been a rapid move from analogue to digital operation, from thermionic valves to transistors and latterly to integrated circuits, and from copper to glass. In the UK network, 62% of all long-lines traffic is (at the time of writing) now carried by single-mode fibre, 14% by terrestrial radio and 24% by coaxial and pair cables. For the purposes of this discussion, we will assume a four-level digital network [12] of the form shown in Fig. 16.3. The four levels of switching used are: the remote concentrator unit (RCU), the digital cell centre exchange (DCCE), the digital main switching unit

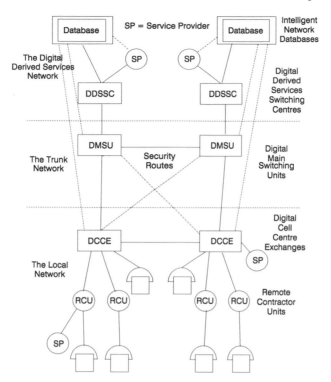

Fig. 16.3 The digital network

(DMSU) and the digital derived switching unit (DDSU). The routing principles for interconnecting these switches ensure that a high level of network security can be achieved through traffic and route diversity and, in particular, the complete failure of any individual transmission system or switching unit in the network will not result in the isolation of groups of customers. For example, Fig. 16.3 shows that each DCCE has routes to at least two DMSUs, each of which is routed on diverse line plant to ensure a high level of security.

Although networks of the form shown in Fig. 16.3 have been designed around the specific needs of the public switched telephony network (PSTN), a significant proportion of any operator's traffic is now generally devoted to other services, such as private circuits, international circuits, broadcast, mobile radio, point-of-sale, and corporate networks. Wherever possible, these non-PSTN circuits follow the same general routing plan, but departures are necessary to meet the individual requirements of particular services. Such is the present level of demand for circuits falling outside the 64 kbit/s PSTN or the ISDN capacity frame [13], new transmission and switching technology is being developed to provide T1 (in the USA) and 2 Mbit/s (in Europe) customer-to-customer services in the near future [14]. Whilst this will provide an interesting bias to the future planning process, it will not significantly change its basic character.

16.4 Capacity planning

The capacity planning process is illustrated in Fig. 16.4 and its main objective is to aggregate forecast circuit requirements into 'capacity' modules (typically blocks of 34 Mbit/s capacity) and to route the capacity modules through to the evolving transmission network in the most efficient manner without compromising network security principles [15]. Circuit forecasts have to be revised on a regular basis, and each revision compared with the previously

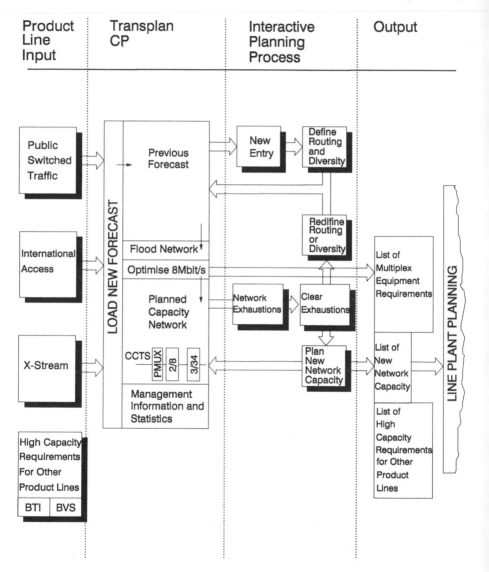

Fig. 16.4 Capacity planning

planned network in order to highlight exhausted sections of capacity. There are various ways in which these exhaustions can be relieved, such as providing more capacity modules over the same section, or alternatively by offloading traffic via a different routing. It may, for example, be more efficient to provide a new longer section to replace or relieve two shorter sections.

Under stable conditions the above process is relatively straightforward, requiring little more than an extrapolation of previous network growth history [10, 16]. However, even in this mode stability is only relative and it is generally necessary to call upon local knowledge to account for developments such as new industry, housing estates and building programmes. Use can also be made of government statistics and projections to account for population movements and economic changes across a country region by region [17]. Correlations have also been established between national economic indicators and the basic telephone service, as well as the individual income per head [18].

Since the demand for telecommunications services, equipment and call rate is correlated with a country's economic activity, there are a number of metrics that can be used to forecast growth, market size and revenue. For example, the telephone density with Gross Domestic Product [16, 17] for a developed nation is generally of the form:

$$\log D_T = -4 \cdot 16 + 1 \cdot 68 \log x \qquad (16.1)$$

$$\uparrow \qquad\qquad\qquad \uparrow$$

$$\text{telephone} \qquad\qquad \text{GDP}$$

$$\text{density}$$

There are many examples of growth starting slowly, increasing rapidly and finally decaying as saturation is reached [18]. Over many decades telecommunication systems have been found to behave in this manner [7, 19], which can be modelled by the logistic growth curve [16, 19]:

$$y = \frac{1}{1 - \exp(a - bt)} \qquad (16.2)$$

By these means relatively accurate forecasts can be produced.

When a new service is demanded or is to be offered, or it has not long been provided, the process is far more difficult and is almost always full of surprise. There are basically four approaches [17] to the problem:

(i)   *Total neglect*:   This is the monopolist's approach, with the non-provision of a service; it is often the adopted solution in a bureaucratic environment.

(ii)   *Provision oriented*:   This is the opportunist's approach: 'it looks like a good idea so someone is bound to buy it'. This ideal has been dramatically borne out by the growth of the basic telephone service, trans-Atlantic cables, satellites, mobile radio, etc. But it is not always true, as demonstrated by the British Prestel system [20] or Germany's 'BIGFON' [21].

(iii)   *Demand oriented*:   This is the marketing man's approach: 'ask the user or potential user what he would like and then supply it!' This is quite valid for the things the customer knows about and can visualise, and which can be realised in a short time. However, it is not of much benefit for new concepts that may take years to develop.

(iv)   *Forecast oriented*:   This is the idealist's approach: 'if only we could get all the available information then we could predict everything'. This is probably as difficult an objective as trying to predict the weather; a laudable ideal but probably impossible. In practice, it seldom seems to give the right answer.

On the face of it, the outlook is bleak for the planner trying to efficiently accommodate some new service; as indicated above, the traditional and the new forecasting techniques are of uncertain use. However, there are a number of sensible, if unorthodox, and to some extent reliable, techniques that can be employed as follows:

(i)   A review of other previously-established services of a similar nature if they already exist
(ii)   A review of other similarly-developed countries if they have introduced the same or similar services
(iii)   A review of similar consumer products, price thresholds, rate of take-up/consumption etc.
(iv)   Courage and prayer!

In short, the planner can be reduced to a heuristic approach based on any useful information he can glean. Some services, like the fully-mobile cellular radio system, appear to be an obvious choice for an instant success. But no one anticipated a growth rate of up to 50% per annum when the initial price per mobile unit was about £1500. Only three years after the original launch Europe had in excess of 500 000 units deployed, bandwidth was almost universally exhausted, new bands were being opened and the price per unit had fallen below £500 [9]. Moreover, it was originally anticipated that a mobile telephone in the car would be dominant, but in the outturn it was mobile in the pocket that was destined to become the main use.

The first impact of the cellular network as far as the planners were concerned was the required rapid deployment of 2 Mbit/s line systems to the cell transmitter sites. This was rapidly followed by an increasing telephone call traffic, which exhibited a new (longer) holding time pattern.

To put the capacity planning problem into perspective, it is interesting to note that the UK national network has over 6500 telephone exchanges interconnected by line and radio systems and services a population of 55 million with some 24 million telephone lines and over 500 000 cellular mobiles. In the long-lines network alone there are over 2000 different and technologically-diverse routes with multiple cables and fibres providing a full digital transport and switched trunk system [22]. Moving down the hierarchy towards the customer level, this network complexity increases by orders of magnitude in terms of routings and plant diversity. Clearly, the problem requires the capability and sophistication of the computer to realise economic solutions.

16.5 System planning

The system planning operation is illustrated in Fig. 16.5. The first stage is to route the capacity modules onto any existing line and radio systems which are not already full loaded. The aim is to make maximum use of all existing

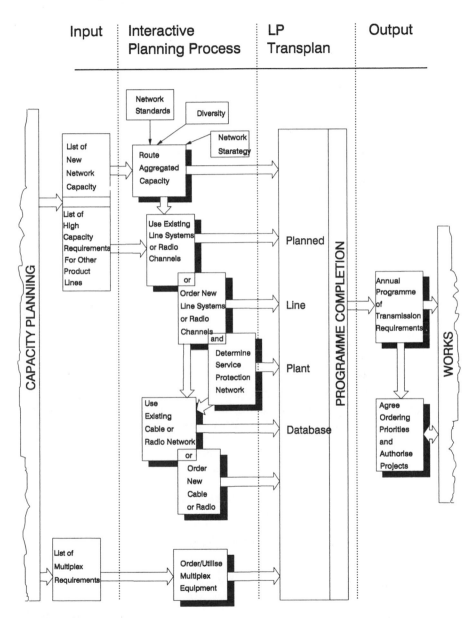

Fig. 16.5 System planning

assets, but with due regard to diversity and performance requirements and safeguarding the network reliability and resilience. The planner may need to consider re-arrangements to the existing network in order to maximise the use of plant already provided or in the course of provision. Usually, the planner will find that there are a number of capacity requirements which cannot be met on existing plant and will therefore need new plant provision. The decision as which plant to provide is often a complex one and it is necessary to find a satisfactory balance between the following conflicting objectives:

(i)   Minimum network cost
(ii)   Maximum network security
(iii)   Maximum network flexibility
(iv)   Maximum network performance
(v)   Optimum deployment of new technology
(vi)   Optimum replacement of old technology
(vii)   Optimum deployment of resources
(viii) Acceptable lead time

The decision on which plant to provide is generally taken in a layered or hierarchical fashion. At the easiest level, it may be possible to provide new plant by just equipping spare line or radio system capacity with additional multiplexers. At the next level, new systems may have to be added to routes which have spare bearer capacity, e.g. where there are spare fibres in the cable. Beyond that level, new bearers may have to be established; for example, it may be necessary to open up a new radio band between existing radio stations. Finally, completely new radio or cable may have to be considered. The number and complexity of the factors which have to be taken into account tends to increase at the lower layers. However, the planner cannot afford to ignore forecasts for future years, as these will generally have a very strong influence on the decisions made in the current planning year.

In order to tackle these complexities for a large transmission network, the planner requires access to a very effective automated planning database for his network. Such a system has to allow on-line access to comprehensive records of the network including topological information, digital block routings, cable, line equipment and multiplex deployment and designation. A relatively high level of human interaction is still required with these computer models, although a number of research organisations are developing tools to fully automate the process. However, optimisation will not be easy; the objectives are generally ambitious and include life-cycle costing, maintenance minimisation, optimum flexibility, minimum spares holding and further new semi-operational/planning parameters.

16.6 Technology

Telecommunications transmission technology has seen more change in the last decade than it saw in the previous ten. Planners are now simultaneously having to contend with network migration from analogue to digital technology whilst at the same time accommodating the move from copper to glass as the dominant bearer [8]. The mix of copper, glass and radio varies from country to country and the rate of deployment of the new optical fibre systems has also been dependent on many economic and sociological factors. A number of clear decisions have, however, been taken by the majority of operating companies. The first of these is to use single-mode fibre at all levels in the telecommunication network. The USA, Canada, Japan and the UK have opted for 1·3 and 1·5 μm operation for all services from business access through to long lines. Bit rates generally span 1·544/2·048 Mbit/s at the customer-access level through to 90/140/565/1700 Mbit/s at the long-lines level. As an engineering rule of thumb it has been shown [7] that each time the bit rate is quadrupled, the cost per bit transported falls by a factor of 3. Even more impressive savings have been made compared with the earlier twisted-pair and coaxial systems and transport costs have fallen by approximately an order of magnitude over the last decade [8]. A comparison of system operating costs with distance is given in Fig. 16.6, whilst Fig. 16.7 shows the progression of technology deployment for the specific case of the UK network.

At the time of writing the *synchronous digital hierarchy* (SDH) was at a planning stage and is the next logical step in the technological progression [14, 23]. With the rapid demise of repeater sites as fibre systems are deployed, attention becomes focused on the terminal equipment as the key source of unreliability. Moreover, with the associated deployment of digital switches and their subsequent synchronisation, it rapidly becomes apparent that the need for a 2–8–34–140 Mbit/s structure with justification at each stage is unnecessary and wasteful in hardware terms [24]. Moving to a synchronous operating strategy allows a single step 2 to 155 Mbit/s hierarchy with a 5:1 reduction in electronic hardware, and facilitates simple 'add-drop' and 'routing/switching' functions [23, 25].

The introduction of a fibre-based synchronous digital hierarchy thus offers the next significant advance in transmission technology and will result in further cost reductions, but with far greater network flexibility, utility and reliability. Digital radio systems are still being retained by most operators; they are particularly attractive where the terrain makes provision of cable expensive or where temporary provision is required until cable can be laid. However, the effects of propagation are significant and optical fibre is now favoured even in countries such as the USA, where radio has long enjoyed a dominant position carrying over 80% of long-lines traffic until recently.

Looking to the future, optical fibre offers bi-directional working on single fibres and wavelength-division multiplexing [26, 27]. Later, photonic amplifiers will allow a degree of transparency, leading to the likely emergence of 'optical ethers' which will operate in a radio fashion [25]. It is already clear that the need for a major switching centre in every town can be averted by 'back-hauling' traffic to major nodes and thereby reducing the number of buildings and operating costs for companies in the future. Clearly these possibilities will present the network planner with many new options over the coming decades. If he is to remain effective he must not only keep abreast of the new options, but be prepared to champion the new topologies and operating modes made possible.

Whilst a planner may be perceived to be predominantly concerned with the future and be looking forward to the new technology and what it can offer, he must also continually consider in detail the optimum time for the recovery of the old technology. An over-cautious approach to new technology can result in a commitment to the old technology when improved cheaper new technology is available. Conversely, an over-optimistic commitment to new technology can result in the equipment not being fully available when required.

Large-scale changes of network technology and topology, such as the recent and ongoing conversion from analogue to digital working and from copper to glass bearers, generate complex interworking requirements between the different technologies. This is a very high-cost area and the planner must keep the level of these interworking requirements to an absolute minimum or the economy and quality benefits of the new technology can easily be lost in the amount and the expense of the interworking equipment necessary. In particular, the manner in which new technology is deployed is critical and overlay versus growth strategies need careful resolution.

16.7 Economic considerations

The transmission planner is essentially looking for the lowest life cycle cost which will meet all his requirements. The first stage is to identify fully the various transmission schemes which meet the capacity and diversity requirements. Costs have to be identified for all elements in the scheme and ideally the dates when they will be incurred throughout the life of the system. Fig. 16.6 shows the distance-related costs for 140 and 565 Mbit/s optical fibre and radio systems averaged over a complete network. The step nature of the radio system curve is a function of the need for periodically-spaced radio towers, whilst the linear curves for the fibre systems indicate their almost repeaterless point-to-point nature and the dominance of cable and fibre costs relative to terminal and repeater equipment.

As most transmission schemes incur costs at different times over a number of years, it is necessary to bring all expenditure to a common base date. It is also necessary to allow for the investment potential of the money; i.e. the money associated with the scheme can be invested and show a rate of return until the expenditure is incurred. Investment-appraisal techniques such as discounted cash flow are thus required to determine the true present value of the various schemes under consideration. Inflation can generally be ignored for cost comparison exercises, except where differential inflation is involved over a period of years. One effective form of differential inflation is the real reduction in most transmission costs compared with the retail price index. In the past this has shown a most dramatic effect; as with most consumer electronics, the cost in pounds has stayed relatively constant or

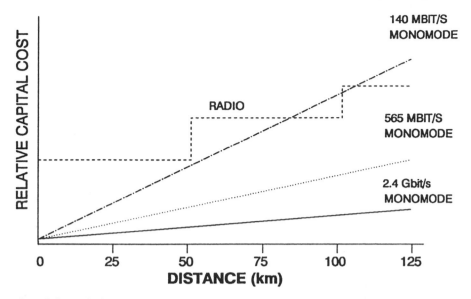

Fig. 16.6 Relative transmission costs of a 140 Mbit/s path

has even fallen as inflation has increased and monetary value has fallen. For example, Fig. 16.7 shows the rapid reduction in the cost of an average 140 Mbit/s path on optical fibre over the period 1980–86.

The planner is thus faced with the need to achieve a fine balance between fully meeting the network demand on time and yet still getting the maximum advantage of falling costs by leaving capital investment as late as possible.

16.8 Network security and performance

In most developed countries both business and residential customers are relying more and more on telecommunications networks as a basic utility

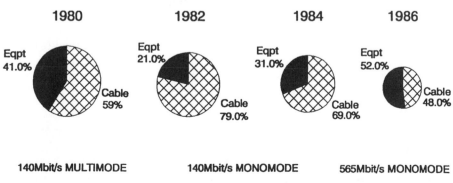

Fig. 16.7 Line transmission cost trends for a 140 Mbit/s path

in their everyday life. This dependence has increased the quality and performance expectations of the customer, and thus the attention that the planner must now give to these factors when designing the network. There are two principal parameters by which quality and performance are determined. The first is the quality of the transmission path; for example, background error rates and error-free seconds, etc. [28]. The second is the availability of the transmission path, which is the percentage of the time that a circuit is available for use by the customer as a specified level of performance [29].

A high level of availability can only be achieved by incorporating security into the network in a number of ways. Firstly, traffic routes between switching units are designed to have minimum diversity of 2 and are generally split such that no more than two-thirds of the traffic is carried on an individual transmission path. The diverse paths are provided on physically-separate transmission plant such as two separate cable routes or a cable and a radio system [30].

Whilst the provision of diversity and standby circuits reduces the impact of circuit system and bearer failures, it can lead to a high level of equipment over-provision. To combat this and provide not only a high level of security for the future, but also for the future network switching requirements, new systems are being developed using a synchronous TDM hierarchy with both time and space switching built into the basic transmission network [23]. It is expected that these new facilities will allow both manual and automatic restoration and switching. Such facilities will, however, introduce yet another layer of complexity for the network by necessitating the generation of 'make good plans' for each working line system. These will require modification every time the network is augmented by new line systems or by switched-traffic changes. The problem is currently being studied by a number of organisations and will undoubtedly require a high level of computer modelling.

16.9 Resources and lead times

Development, manufacturing, procurement, installation and commissioning lead times are among the most difficult non-network items the planner has to consider. They are often made more difficult to quantify by the political environment, interaction of several diverse programmes, the need for staff training and the inevitable 'teething troubles' associated with new equipment and systems. All these factors have to be allowed for and can, in practice, outweigh the minimum life-cycle cost estimates embodied in the main planning cycle.

16.10 Conclusions

Telecommunication services are very much one of the principal building blocks for any modern society. The services provided are largely dependent upon the transmission network infrastructure; thus, it is up to the network planner to ensure that the quality, performance and cost objectives of the transmission network are fully realised. In this chapter we have considered the general transmission-network planning process and given an indication of its very complex nature. As new technology emerges, network planners will have to continue to adapt to the new (and exciting) options offered, which will serve to not only change the nature of the network, but also the quality and type of services provided.

16.11 References

1 DTI Report: 'Deregulation of the radio spectrum in the UK'. HMSO, March 1987
2 'Evolution of the UK communcations infrastructure'. HMSO, Summer 1988
3 Various papers: 'Telecommunication deregulation', *IEEE COMSOC Mag.* 1989, **27**, (1)
4 VIETOR, R.H.K., and DYER, D.: 'Telecommunications in transition — managing business and regulatory change'. (Harvard Business School, 1986)
5 'World communications—New power—New hope'. (ITU, 1986)
6 ROBINSON, O.G.: 'Communications for tomorrow—Policy perspectives for the 1980s'. (Praeger, 1978)
7 COCHRANE, P.: 'Future trends in telecommunication transmission', *Proc. IEE.*, 1989, **131F**, pp. 669–683
8 MONGOMERY, J.D.: 'Broadband intelligence for the 21st century'. Monterey Conference on Fibre Optics to the Year 2000 (Electronicast Corp, 1988)
9 DAVIES, J.: 'Retuning for growth', *Commun. Int.*, 1988, **15**, pp. 50–53
10 FLOOD, J.E.: 'Telecommunications networks' (Peter Peregrinus, 1975)
11 LEE, J.C.: 'Nested Rotterdam model: Applications to marketing research with special reference to telecommunications demand', *Int. J. Forecasting (Netherlands)*, 1988, pp. 193–206
12 MUIR, A., and HART, A.: 'The conversion of a telecommunication network from analogue to digital operation'. First IEE National Conference on UK Telecommunications Networks—Present and Future, 1987
13 Various papers: Special Issue on 'ISDN — Recommendations and field trials', *IEEE J.*, 1986, **SAC-4**, (3)
14 BALLART, R., and CHING, Y.C.: 'SONET—Now it's the standard optical network', *IEEE COMSOC Mag.*, 1989, **29**, pp. 8–15
15 CRAVIS, H.: 'Communications network analysis' (Lexington, 1981)
16 SAUNDERS, R.J., WARFORD, J.J., and WELLENIUS, B.: 'Telecommunications and economic development'. World Bank Publication, 1983
17 CATTERMOLE, K.W.: 'Communications Services'. University of Essex Lecture Notes, 1975
18 LITTLECHILD, S.C.: 'Elements of telecommunications economics' (Peter Peregrinus, 1979)
19 BEWLEY, R., and FIEBIG, D.G.: 'A flexible logistic Gravitt model with applications in telecommunications', ibid Ref. 11 pp. 177–192
20 'Prestel sets out to seek its fortune in the business world', *Computer Magazine (GB)*, Feb. 1985, p. 28

21 OHNSORGE, H.: 'The potentialities of BIGFON'. Telematica 84, Vol. 2, pp. 61–78
22 CROOKS, K.R.: 'Programming the digital modernisation of a national local exchange network', Second International Network Planning Symposium
23 HAWKER, I., WHITT, S., and BENNETT, G.: 'The British Telecom transmission core network'. IEE Second National Conference on Telecommunications, York, April 1989, pp. 364–368
24 ROWBOTHAM, T.R. *et al.*: 'Transwitching — The next step in optical networks', *IEEE Trans.*, **ICC–87**, pp. 1527–1528
25 COCHRANE, P., and BRAIN, M.C.: 'Future optical fibre transmission technology and networks', *IEEE COMSOC Mag.*, Nov. 1988, pp. 45–60
26 HILL, G.R.: 'A wavelength routing approach to optical communication networks', *BTTJ*, 1988, **6**, pp. 24–31
27 WAGNER, S.W., and KOBRINSKI, H.: 'WDM applications in broadband telecommunication networks', *IEEE COMSOC Mag.*, 1989, **29**, pp. 22–30
28 YAMAMOTO, Y., and WRIGHT, T.: 'Error performance in evolving digital networks including ISDNs', *IEEE COMSOC Mag.*, 1989, **27**, pp. 12–23
29 Various papers: 4th European Planning Workshop, Chantel, France, March 1988
30 SCHICKNER, M.J.: 'Service protection in the trunk network', *BT J.*, 1988, **6**, pp. 89–109

Chapter 17
Future trends
P. Cochrane

BT Laboratories

17.1 Introduction

Ask any transmission engineer 'where do we go from here — what happens next?' and the reply will almost certainly contain predictions for the expansion of services, higher bit rates and more bandwidth. For the past 100 years this has been an accurate and convincing prognosis [1, 2]. However, recent events in other areas of human activity such as transport and those consuming large amounts of raw materials have highlighted that there are finite limits to growth [3]. It is therefore expected that many of these activities will have experienced a reduction in the rate of growth, leading to a levelling off and ultimately to a decline [4] by the early part of the next century.

In this chapter we take the view that no such downturn in demand for telecommunication transmission is yet evident; nor will it be so, on a global basis, even in the far future. Furthermore, it is proposed that the exponential growth experienced so far is merely the tip of the iceberg, and can be expected to continue unabated. Unlike other activities that are energy and/or material intensive, and therefore strongly linked to economic factors such as the gross national product, telecommunications technology is viewed as being largely decoupled from such influences and devoid of many of the natural limits to growth. Moreover, the fact that constraints are being placed on other physical processes that can, in many respects, be replaced by enhanced communication, is seen as a prime catalyst for an increase in the global network and an expansion of the service offered [5, 6].

In order to set the scene for our discussion on transmission system trends, the growth of telecommunication is first reviewed drawing on the global distribution of telephone density to highlight latent demand. Problems of interconnection and the basic types of service required are discussed within the constraints of present technology development forecasts. After postulating a number of service objectives the implications of new service requirements on transmission are considered. Some forecasts for the likely advances in technology are then made, and used to justify our claim for continued exponential growth into the far future.

17.2 Motivating technology

The widespread use of the transistor and integrated circuit from about 1965 [7] created a rapid increase in circuit demand and provision capability. By the early 1970s the nature of telecommunications began to change significantly, and it became increasingly apparent to both operators and users that the provision of a 'Plain Old Telephone Service' (POTS) was not enough [8, 9]. This was principally brought about by the influx of computer technology and a desire to not only pass messages, but also to access and exchange information on an increasing scale. The manner in which this was brought about is perhaps best exemplified by Moore's law [7], shown in Fig. 17.1,

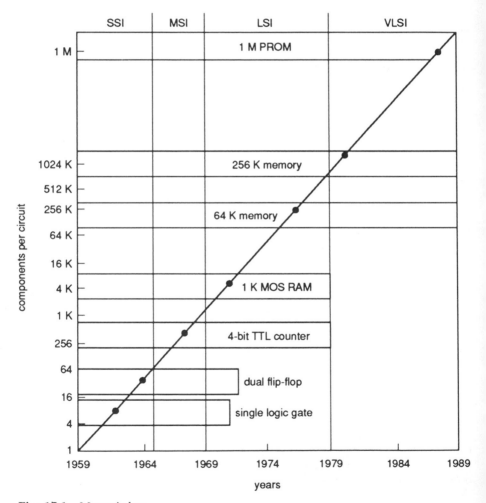

Fig. 17.1   Moore's law

which predicted a doubling in the complexity of integrated circuits each year from about 1964 onwards.

A consequence of this development has been the widespread introduction of low-cost high-complexity integrated and microelectronic circuits into all aspects of the communications network, allowing the deployment of digital transmission and switching systems associated with sophisticated terminal equipments [10]. In turn, this led to significant increases in the traffic capacity of pair and coaxial line plant, which have also been augmented by similar improvements in microwave radio and satellite system performance. As a result, transmission costs have continued their steady downward trend (Fig. 17.2) and, the demand-capability gap has rapidly closed (Fig. 17.3).

Within the last ten years these developments have led to the following key realisations:

Global network: A fully interconnected world (and national) telephone network of well defined and controlled standards could be most economically provided using digital transmission [10—11].
Switching: Combined digital switching and transmission gives the most economic network realisation, by a factor of about 50% [10] (Fig. 17.4).
Services: A wholly digital network can provide much more than a POTS. Within the confines of the switched bit rate (64 kbit/s), it is capable of

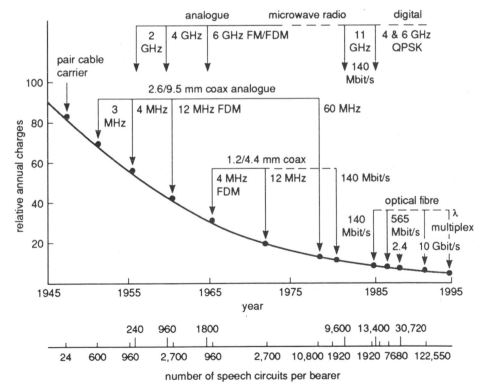

Fig. 17.2   Transmission plant relative annual charges for a 100 km circuit

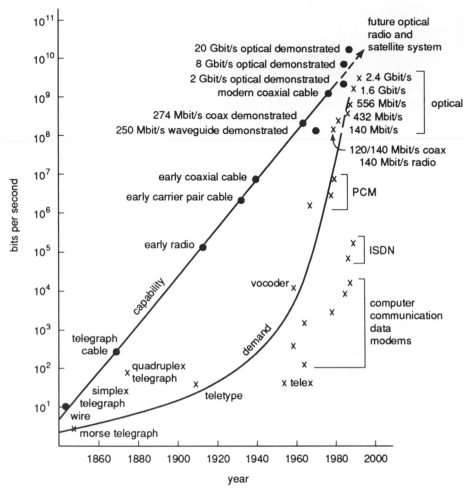

Fig. 17.3   Telecommunication capacity and demand

economically providing many more services and facilities [6, 12, 13] at an exchange and customer terminal level (see Fig. 17.5). When augmented by the addition of high-bit-rate (wideband carrier) circuits, the possibilities become even broader, allowing the provision of high quality audio, video and superfast computer communication.

To a large extent this has led to a situation where the 'tail is now wagging the dog'. No longer are operators striving to find solutions to the basic problem of providing a POTS; they can now see limitless possibilities and are seeking ways of economically providing new and untried facilities — in effect, creating new markets. A major snag with this new situation is that many of the proposed services have to be introduced wholesale, or at least

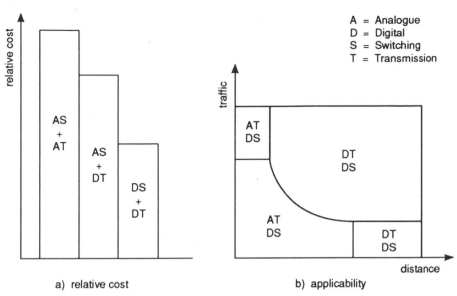

A = Analogue
D = Digital
S = Switching
T = Transmission

a) relative cost   b) applicability

Fig. 17.4 Network costs and applicability for analogue and digital switching and transmission

on a large scale, to prove the market exists. Obviously, this involves a high degree of risk and creates great difficulties in forecasting and planning.

To illustrate the uncertainty inherent in the present situation, it is interesting to consider two analogues market situations from the past, one a resounding success and the other a dramatic failure:

(i) *First Transatlantic Telephone Cable* (TAT1):   When this project was first proposed, it was without economic justification; HF radio circuits, although crowded, were apparently proving adequate for UK–USA traffic. Shortly after its installation in 1956, with 36 circuits, TAT1 had to be enhanced with channel bandwidth reductions to give 48 circuits, and then to 85 via TASI. Within three years of offering this new service a second, almost identical, TAT2 had to be installed, thereby starting an exponential growth in transatlantic traffic which continues today (see Fig. 17.6). Clearly, there had been a high latent demand that had largely gone unrecognised. The reason for this was the suppression of traffic by the relatively high cost and poor quality of the HF radio circuits.

(ii) *Picturephone*:   During the early and mid 1960s Bell Laboratories developed a picturephone system with an effective video bandwidth of 0·5 MHz for operation over existing pair cables. By 1968 extensive trials had been completed and confident predictions of vast penetration into the US network by the late 1970s were being made. As we now know, this did not happen. The reasons for failure were too high a price, an immature technology, and a service ahead of its time; as potential customers were not all that concerned about 'seeing when speaking'.

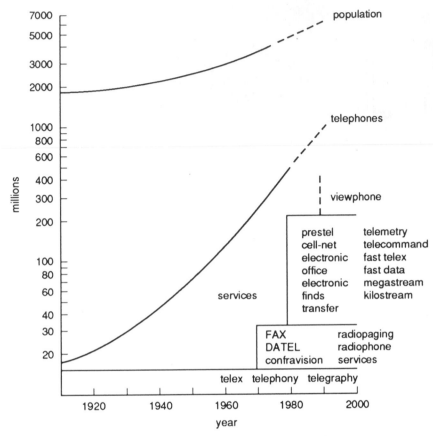

Fig. 17.5   Global growth of population, telephones and services

17.3 World perspective

To date the POTS has reached 100% household and workstation penetration (or very near) in many of the First World countries, and by definition, at a price the majority of people can afford. This has been achieved with a predominantly analogue transmission and switching network, which is now undergoing major modernisation and conversion to digital operation [14]. Within these First World countries, it is not unusual to find a fully automatic interconnection on all internal calls and a high percentage of automatic access to the rest of the world. In the UK, for example, it is possible to dial direct from any telephone some 90% of the world's estimated 720 million telephones in 200 different countries.

Sadly, the above figures hide a disproportionate distribution between the rich and poor nations. In fact, the First World, with only 15% of the world's population, has 85% of all the telephones. Clearly a modest provision of telephones (per capita) coupled with increases in industrialisation of the

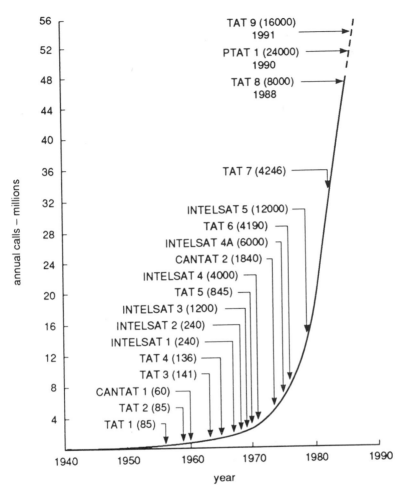

Fig. 17.6 Telephone calls between the UK and North America

Second and Third World countries, could significantly swell the size of the total network and greatly influence global traffic demand.

17.4 Present (1991) status of transmission

Today, telecommunications stands at a technological crossroads, with the need to merge telephone, computer and entertainment services [15—17] to create what has generically been termed Information Technology (IT). It seems likely that over the next decade this will create at least a tenfold increase in information flow [9, 13, 18, 19] that cannot easily be accommodated by existing copper and microwave-radio based media.

At national levels, digital hierarchies have been implemented to meet local needs, which have often pre-empted any agreed international standards. Consequently, there are in existence at least five hierarchies, which are dominated by the North American, Japanese and European (CCITT) versions. Obviously, this situation is a significant impediment to efficient global unification, which is likely to be required to meet future demands, and is an echo of the similar problems previously encountered within the old FDM hierarchies [20]. The introduction of the recently proposed global SONET/SDH standard should largely overcome this problem within the next decade [21].

Fortunately, developments in microelectronics, and the associated manifestation of the above problems, have been coincident with the development of waveguide transmission. First developed for microwaves [22], and then more economically and conveniently for optical signals [23, 24], the transmission of information over glass fibres offers the key to the future. This fact is evidenced by the qualitative comparison of information capacity given in Fig. 17.7 and the loss with frequency curves of Fig. 17.8. Moreover, the added advantage of very small size (about 1—10 mm diameter) compared with copper pair and coaxial structures (about 100 mm diameter), coupled with a reduction in the need for cable cladding and duct space per bit carried, gives an even more dramatic advantage than indicated by a bandwidth comparison alone.

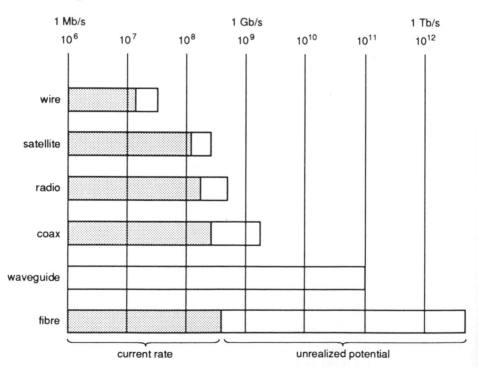

Fig. 17.7 Current data rates and ultimate limits for transmission

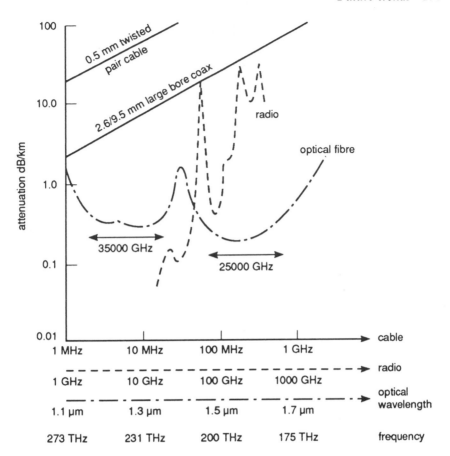

Fig. 17.8 Transmission media losses

17.5 Future systems facilities

Before we go on to consider specific transmission technology trends, it is both useful and instructive to consider some projected targets for both facilities and global provision. It will then be possible to conceive a network capable of satisfying these postulated requirements.

17.5.1 The three 'Captain Kirk problems'

The projected aims of the telecommunications engineer, that encompass most of the projected IT possibilities for the merging of computer and communication technology outside the sphere of entertainment, have, previously [25] been classified as the 'three Captain Kirk problems':

(i)   Instant mobile man–man voice and data communication at any time and place on the planet

(ii) Instant man–machine voice communication

(iii)   Instant machine–machine communication from remote locations

Let us now consider each of these categories in turn, and see how they may be realised.

17.5.2 Mobile man–man communication

The current evolution of cellular mobile radio and paging networks throughout the world has demonstrated an increasing demand for instant mobile communication. It is not unreasonable therefore to suppose that the business man and private user of the future will expect to have voice and data communication via a pocket or wrist-size device.

With the present world population of about five billion we could envisage allocating numbers to people rather than telephones. Only ten digits would be required today — say 11 tomorrow — and perhaps 12 with check bits. Using computer control it should be possible to keep track of each person on a global scale, and to identify the cell of occupation at any time their number is called. Of course, we would probably wish to include an answerback facility that says 'the person you are calling does not wish to receive calls at present, and what is more, does not wish to reveal his location' [26].

How could these voice and data objectives be achieved? The solution to this problem falls neatly into the following three categories:

(i)   *Wilderness locations*:   Here, a communication cell could be vast. A mid-ocean or Australian outback region, for example, has a very low population density, and can be most economically served by satellite technology [27].

(ii)   *Rural and urban locations*:   Cells of varying size are required to match the low-to-modest population densities and can be achieved by existing satellite spot beams, in addition to new 'pin-point' beams, made possible with the giant space antennas [28]. Low orbit asynchronous satellites could also be used, at perhaps an even lower cost.

(iii)   *Dense city locations*:   Existing cellular systems based on low-power terrestrial transmitters would probably be required with very small area coverage for the most densely populated business areas such as London and New York [29]. Again, these could be augmented by low cost, low orbit asynchronous satellites.

All of this presumes there is enough radio spectrum available, but for the more densely populated regions there is not, within present frequency allocations, as current estimates for user density suggest that a 40 MHz slot could service between 100—200 thousand mobile stations. More of the microwave spectrum could be made available by placing many of the fixed point-to-point and broadcast services on cable, a medium to which they are better suited. Also, frequencies in the 24, 60, 120 and 180 GHz absorption bands could be allocated, the high attenuation in these bands being a 'plus' in terms of frequency re-use on a cell-to-cell basis [30]. The larger satellites now being planned may also have sufficient power available and antennas large enough to utilise higher microwave frequencies than hitherto possible [31].

17.5.3 Man–machine interface

A prime limitation in the relationship of man and machine (computers) has been the need to operate a keyboard and very often to write programs also [33, 34]. The advent of low cost speech synthesis and recognition circuits holds out a solution. At the trivial end of the possibilities we can see the eradication of the telephone dial and pushbuttons: products are already on the market that accept spoken numbers and names, under both static and mobile operation. At a more sophisticated level, the eradication of telephone operator services and language translation seem a not unreasonable target within the next ten years.

The now demonstrated possibility of real time language translation is very interesting from the standpoint of dramatically increasing transmission demand. Perhaps the most far reaching device would be an on-line real-time translator. For example, international traffic from Western nations to Japan consists of 80% fax, telex and data; only 20% or so is telephony. The reason why is clear. Not many people outside Japan speak the language, nor do many Japanese speak any of the European tongues with sufficient skill to permit efficient telephonic communication. If this situation is true for Japan, then it is also true for many other countries like China, Russia and India. Clearly, a modest growth in the industrialisation of such countries, with a subsequent increase in telephone densities, could, with the introduction of a translation device, create a vast amount of telephone traffic. Even in the West we could expect to see a significant increase in international traffic with the provision of such a service.

17.5.4 Machine–Machine Interface

So far, interconnection has mainly been at the behest of man for the transfer of data and accessing of information. The evolution of expert systems dedicated to specific disciplines such as medicine, communication, geophysics etc. will change this situation. A 'reasoning' machine will wish to access data automatically from files scattered across the globe, and it may in addition wish to enter into discussion with other expert systems on its own as well as other topics.

From a transmission viewpoint it is clear that such activities will increase traffic demand, but to what degree depends on the evolution and deployment of the fifth and sixth generation machines [34]. At present, expert systems are operating in isolation throughout the West and are highly individualistic, making any interworking extremely difficult. Despite this limitation, they are, in many instances, able to outstrip their human counterparts within the confines of their particular expertise; and their 'intelligence' is estimated to be increasing by an order of magnitude every seven years. This growth is startling; in just 40 years man has created intelligence of a magnitude it took evolution about 40 million years to create! Within the next decade it is estimated that they will be on a par with the human brain across a broad range of descriptions.

It would thus seem reasonable to anticipate that computer-prompted and controlled machine–machine communication will be a significant source of telecommunication traffic.

17.5.5 Other service predictions

Beyond the broad predictions expounded so far there are many specific services described by a number of authors, that are summarised in Fig. 17.5. It is beyond the scope of this chapter to detail each of these, but most are self-evident, and the reader is referred to the publications of Martin [6, 12, 13] in particular, who offers a list of over 100 possible services for the home and office!

A further set of services also exist for the future which we may place in the category of 'not invented yet'. These are often considered as analogous to the relatively recent home computer and video recorder boom, both of which did not exist ten years ago. In future we can expect to see similar opportunities arise as new (and as yet unthought of) technology emerges.

17.5.6 CATV wideband services

These systems have traditionally developed outside the mainstream of the PSTN and other telecommunication activities, but now with the proposals of interactive channels they fall firmly under the IT umbrella [35, 36]. Linking such systems with mainstream telecommunication and computer services offers a significant expansion of service capability, particularly in the areas of video conferencing and home meetings. The reader is referred to the literature for the detail and description of future possibilities; suffice it to say that the expansion of video services eats up transmission capacity faster than any other facility provided so far.

17.6 Future transmission trends

Previous sections have indicated that the demand for transmission will increase at an accelerating rate. The only major uncertainty that now arises is precisely which of the projected services will be commercially successful. Some relative growth projections for services in the developed world are shown in Fig. 17.9. What is clear is that a wholly digital network will ultimately yield the most economic means for continuing global expansion of the telephony network, while also allowing the provision of major new services.

In most developed countries the decision to digitise switching and transmission has been made, with completion targets typically spanning the years of 1988–2010. Within these plans optical fibre transmission systems are destined to play a major role, providing the backbone of the trunk network, low cost junction circuits and ultimately the final connection between exchange and customer. A consequence of these trends, coupled with the introduction of new services, will be the complete restructuring of both national and global networks.

In the sections that follow we consider current network and technology trends and postulate extensions into the future.

17.6.1 Divergence–Convergence of Provision

Under the present climate of deregulation and liberalisation of telecommunications, favoured in a number of countries, the commercial competition

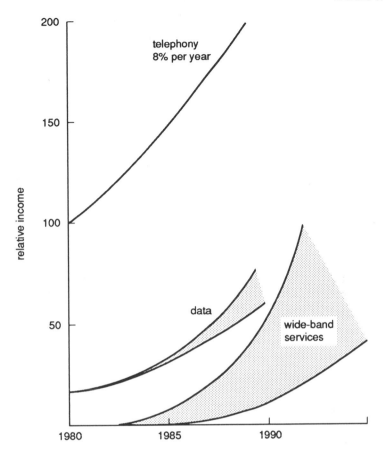

Fig. 17.9 Traffic growth projections in terms of operating company income

engendered is leading to a plethora of expedient solutions to service provision. This is also being fuelled by disparities in the transmission rates of the different services and a dominant trend at present is therefore to introduce overlay networks for speech, high speed data, video, CATV, private circuits and networks etc. Moreover, the provision can be via pair, coaxial and fibre cables, as well as microwave radio and satellite links.

In the short term this trend of dedicated circuit and network provision is likely to continue. However, in the longer term it should logically converge to a single network. A first step in this direction has been taken with the introduction of the 'integrated services digital network' (ISDN), which accommodates both speech and low and medium bit-rate data over the PSTN [35] as depicted in Fig. 17.10

The provision of higher bit rates at the local level, which so far has mainly been associated with entertainment, but in the future could also be prompted by new generations of computers, is likely to still require a separate network. If wideband networks, based initially on CATV, are provided on a large

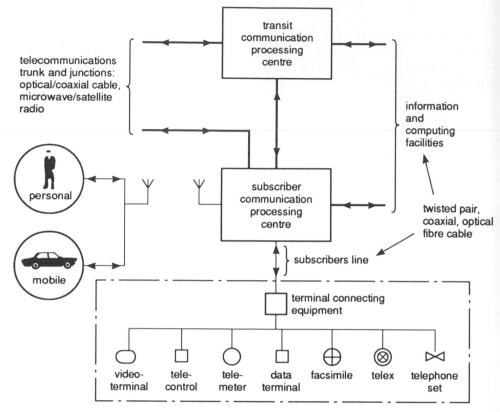

Fig. 17.10   The ISDN of the future

scale, it would seem reasonable to assume that both ISDN and other services could merge on a common carrier. Some commentators do, however, express reservations in that CATV can be seen, to some extent, to have been eclipsed by video recorders, especially in Europe where the broadcast TV networks are generally of good quality over most of the regions served. Even in North America, the home of CATV, numerous operators are in severe difficulties and not making sufficient financial return on investment. Direct broadcast satellite systems [37] also pose a serious threat to the provision of such networks, although a number of these services are also finding it difficult to survive commercially. In the longer term however, it would seem almost inevitable that CATV or wideband services will be provided on a large scale: the only question is when?

17.6.2 Network strategy

Existing networks are based on an ever growing mix of cable, radio and satellite carriers that must, in future, conform to a more ordered assembly to economically (in all respects) provide the many new services. A likely scenario is as follows:

Optical fibres: Now the dominant carriers of fixed point-to-point information on national and international routes (via undersea cables) with monomode systems used throughout.

Radio and satellite: Adopted mainly for the provision of mobile, portable and rapid-response specialised services. They will also be increasingly used for the provision of fixed point-to-point links where it is uneconomic to put in cables; e.g. sparsely populated remote locations, areas separated by difficult terrain and congested locations, where expensive civil engineering costs would otherwise ensue. They will also be retained to provide radio and TV broadcast facilities.

17.6.3 Advances and trends in optical systems

The development of optical fibre transmission systems has so far followed a path similar to their radio counterparts. At present, commercially available or development systems may be likened to the early spark transmitter and receiver combinations; lasers are multimode and unstable, receivers use direct detection. Despite their relative crudeness, the performance of such systems is remarkable when compared with any other cable medium. For example, the very high capacity optical systems currently in production span 90—2400 Mbit/s with repeater spacings between 50 and 170 km [38]. This represents a transmission capacity and repeater span increase of some 2—16 and 50—200 respectively, over their coaxial predecessors.

In the past four years research into more sophisticated optical systems has introduced a number of significant steps forward:

(i) Dispersion shifted monomode fibres giving 0.2 dB/km at 1.55 mm have been produced [39]

(ii) Single mode laser sources have been realised [40]

(iii) Heterodyne and homodyne detection realised giving increased receiver sensitivities [41]

(iv) Demonstration of wavelength multiplexing for uni- and bidirectional working [38]

(v) Optical amplifiers in semiconductor and doped fibre formats have become commercially available.

This has led to a number of impressive demonstrations, with bit rates up to 0.56—20 Gbit/s over 100—300 km of fibre.

These advances have also made available most of the modulation, signal-processing and coding developments so far used in radio technology. It is conceivable that these could be introduced into optical systems at 1.55 mm, and when coupled with an increase in laser launch power level, optical power amplifiers, distributed fibre (cable) amplifiers and supersensitive receivers, ought to permit repeater spacings in excess of 350 km. Allowing for factors such as cabled fibre loss increase, production spreads, joint losses and operating margins, it would seem reasonable to expect commercial systems with repeater spacings of >250 km to be available within the next decade.

Retrofit increases in systems capacity should also be possible by direct bit rate adjustment or by wavelength multiplexing — an echo of previous trends on pair and coaxial carriers!

For both terrestrial and undersea applications, these developments will have profound implications. In the UK, for example, existing 140 and 565 Mbit/s monomode systems do not require dependent repeaters, as all BT surface buildings are less than 30 km apart. Power feed conductors and other equipment are thus no longer necessary, yielding a significant cost reduction. With longer section spans it is possible to link many major cities without a single repeater [43]. Undersea applications will also benefit from these advances, as many locations, such as the Mediterranean and the North Sea, contain relatively short routes of less than 300 km in length.

Given the low loss, low dispersion and almost infinite bandwidth of monomode fibre, its ease of manufacture and lower cost than multimode, it is not surprising that the market place has become dominated by this product. Its deployment in trunk, junction and local links has resulted in a potentially large transmission capacity being available through the optical network, customer to customer. This in turn, has led to the realisation that bandwidth efficiency may be economically traded against terminal signal processing. That is, as the cost per bit falls on a transmission media of such wide bandwidth, the traditionally hailed bit/Hz efficiency criterion of the communication engineer can be neglected in favour of a significant reduction in equipment complexity.

Turning now to the more distant future; research effort is currently being applied to the study of glasses for longer wavelength operation (2—10 nm), where losses as low as 0·001 dB/km have been postulated, and where 0·01 dB/km might be more reasonably expected. However, the glasses being considered are based on various halides, metal-oxide groups and crystal combinations [43, 44], and so far researchers have had difficulty in achieving low loss; the best reported so far is about 0·4 dB/km at 2·5 mm using fluoride glass. This is in marked contrast to the theoretically derived predictions shown in Fig. 17.11. In addition, there appear to be formidable problems associated with also achieving good mechanical strength and other desirable properties. This is a situation somewhat akin to that of silica fibre development in the 1970s when fibre losses lay in the range 100—1000 dB/km! However, the current rate of progress is much faster as indicated in Fig. 17.12. A realistic scenario is that the progress in this area may well be forestalled by the existence of a silica fibre network in a very similar way to the development of gallium arsenide devices in competition with silicon!

Even with existing silica fibre based networks it is possible to conceive of an optical ether, with terminals connected by passive taps in a star or ring topology. The use of WDM channels through coherent technnology would, in principle, yield a possible 60 000 channels on a 1 GHz spacing and 6000 at a more modest 10 GHz [45]. Alternatively, and perhaps in the interim, TDMA, FDMA, WDMA or CDMA could be used at the terminal stations on this new ether. The most significant and new step would be the eradication of a centralised switching function [45]. This could be dispersed to the terminal stations as indicated in Fig. 17.13. Combine optical switching and

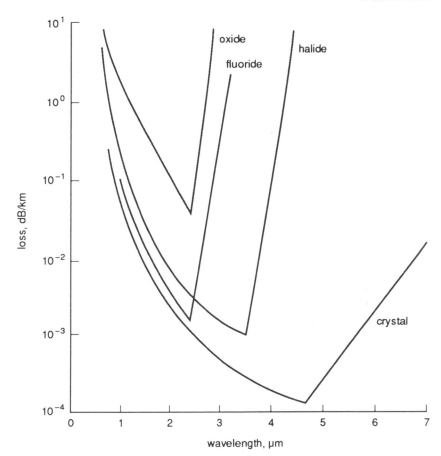

Fig. 17.11 Theoretical predictions for future fibre loss

integrated optics [46] with such thoughts and we have realised a radically new form of network offering unparalled reliability with a uniform bit transport capability throughout. An even more radical possibility also exists with the notion that networks could migrate to an analogue form of operation. If WDM is adopted with repeater spacing of about 100 km, then the economics of networks and service provision could perhaps allow the global transport of carriers without electronic processing at intermediate points. Customers could then be allocated an end-to-end wavelength independent of the service to be transported.

17.6.4 Advances and trends in radio and satellite systems

Microwave radio technology has been in common usage and has undergone significant development from the late 1930s until the present day. It is

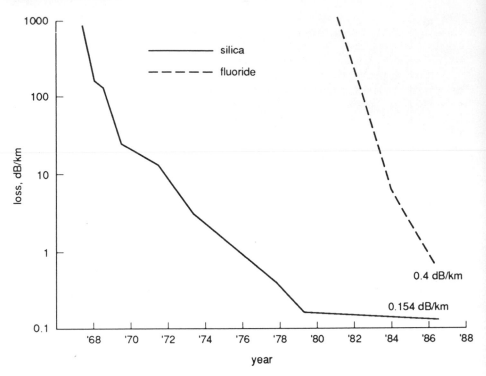

Fig. 17.12   Reduction in fibre loss with time

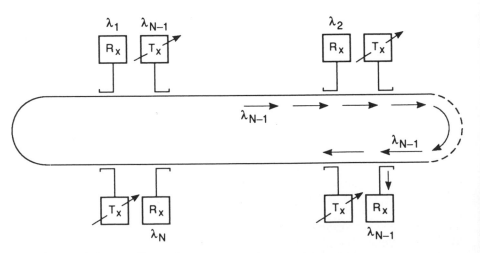

Fig. 17.13   Coherent fibre ether

therefore a mature technology, and unlikely to spawn the great advances currently being made in optics. In the future of radio we can thus expect to see an increasing refinement of what we already have.

Following the main theme of this chapter, the future role of radio is seen as one of broadcast (from satellite), mobile, portable and, because of its relatively low cost and versatility, a rapid-response medium. To be specific, in the future as now, it is likely to be used to support new services which will ultimately be transported by wideband optical systems. However, until the optical network is established, microwave systems will be more economic, less risky and will allow speculative services to be tried and tested in advance.

The major advances in terrestrial microwave radio systems are expected to be realised by the use of microwave integrated circuits (MICs) and more sophisticated signal processing, coupled with the use of higher carrier frequencies.

Commercial systems are available today at 19 and 20 GHz with bit rates spanning 2—140 Mbit/s for trunk and junction applications. QAM, with 4, 16, 32, 64 and 256 states [47], is in common use and higher radix schemes are being considered. However, the non-linear nature of microwave amplifiers poses some limits to this route, but it is thought that new generations of devices and pre/post digital signal processing will accommodate constellations of 1024 and higher states. PFM is also in use for video and other analogue applications. As well as point-to-point links, multipoint systems are becoming available for local area network (LAN) applications within densely populated city areas ('rooftop to rooftop'). Providing ISDN facilities to outlying industrial areas and other remotely located plants is also a major new applications area.

A logical next step will be the use of higher frequencies, including those in the absorption bands (for terrestrial mobile links) at 60, 120 and 180 GHz. The possibilities for personal communication at 60 GHz are currently being investigated by several operators [30], and military 'battlefield' systems are already in use at 90 GHz. Given that technically such systems can be produced, the major development problems are confined to realising systems at the right price.

In the case of satellite communication, many of the above aspects are also applicable, but there are additional degrees of freedom left to explore [31]. These may be categorised as follows:

Size of vehicle: The space shuttle programme has made possible the construction of much larger (space-station or antenna-farm) satellites [48]. Moreover, it has also introduced the demonstrated possibility of 'in-space' maintenance. These two steps, coupled with an expected improvement in orbital stability, will allow the configuration of much more complex units, some aspects of which are now considered. An alternative and interesting possibility also being addressed is for 'cannon launched — football size' satellites placed in low asynchronous orbit to provide low cost mobile (hand held and vehicular) services [49].

Power: Larger (about 50×5 m) solar collectors with >10 kW capability are currently being designed and tested. To overcome solar eclipses as well as peak-load conditions, on-board power storage and generation is being considered to make available 10 kW (or more), compared to the present1 kW or less [50].

Antenna size: Could be expected to increase dramatically (in number also) allowing many more spot and static as well as scanning pinpoint beams. Currently beam widths of 1°, with pointing accuracies of about 0·1°, allow spot beams of about 400 km of less in diameter [51]. It is reckoned that an improvement by a factor of 10 or so is feasible, which will give pinpoint beams of about 40 km diameter. Subsequent increase in link performance; especially when coupled with increased power, will allow smaller earth stations, thereby reducing overall costs. Large earth-bound antennas are very expensive, and are moreover subject to gravitational and wind distortion, which become increasingly important at the higher frequencies (14/11 and 30/50 GHz). Antennas in space can in theory be made very large and elaborate, lightweight and low cost! Present developments are aiming at 50—120 m-diameter parabolic reflectors, and include inflatable structures as well as proposals for constructions of wire mesh only 12 Å thick, but even larger structures — 300 m plus are being contemplated. These features are also a major contributor to portable and personal communication system realisation.

On-board switching: This will permit interconnection between different antenna beams, polarisation, frequency and, in the case of TDMA, time slot. A greater connectivity and flexibility will result, allowing capacity to be allocated on a static as well as a dynamic basis.

Satellite–satellite communication: The limitations of multihop satellite links can, in many respects, be overcome by the use of direct space links, as originally envisaged by Arthur C. Clarke. These are likely to be realised using higher (atmospheric) absorption frequencies (>180 GHz), or by optical links, where a high degree of focus can be achieved with modest antennas/reflectors.

Frequency re-use: The larger antenna size and high power available to future satellites will facilitate frequency re-use on a number of different beams and polarisations. Although this technique is in use today (Intelsat V, for example) the addition of TDMA and onboard switching, will give far greater flexibility and connectivity.

Onboard signal processing: Communication satellites have so far been almost passive 'mirrors in the sky'. Signal amplification and frequency translation have, in most cases, been the only signal processing performed. New generations of satellites can be expected to perform extensive signal processing with full regeneration and forward error correction of digitally modulated carriers. It is probable that even demultiplexing/multiplexing with dynamic allocation of channels will also be introduced.

Broadcasting from Lagrange points: The gravitational null between the earth and the moon provides a suitable site for broadcast satellites where path delay (about 4 s) is not a significant problem. At these points, large vehicles could be expected to assume a higher degree of orbital stability compared to their geostationary counterparts, which tend to suffer from an eastward drift, necessitating periodic adjustments. As the current satellite spacing has now been reduced to 2° this adjustment will become more frequent, leading to the requirement of more fuel for a given satellite lifetime. An amusing phrase used in this context is 'premature orbital retirement', which means the satellite has run out of gas and has slipped out of position — or plain got lost!

New orbits: Satellite services currently provided from the geostationary orbit are satisfactory for static applications covering the majority of the populated globe. However, mobile services in Europe, for example, can be more easily provided using the Molniya orbit (Fig. 17.14) with 3—4 satellites following a highly elliptical path. This gives an effective satellite positioning almost vertically overhead for four hours before and after apogee. This requires a vertically pointing mobile antenna beamwidth of only 15° without recourse to beam steering, which in turn yields less man-made and atmospheric noise/attenuation than steerable antennas pointing to the geostationary orbit 15°—20° above the horizon.

As the geostationary orbit has to accommodate more maritime, aeronautical, broadcast, domestic, military and experimental systems, an ultimate limit will be reached owing to interference noise. It has been estimated that an activity increase by a factor of ten or so over the present day is feasible.

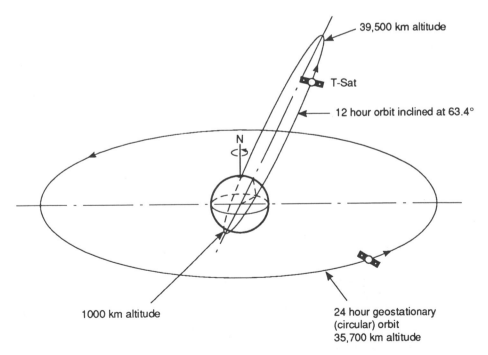

Fig. 17.14 The Molniya orbit

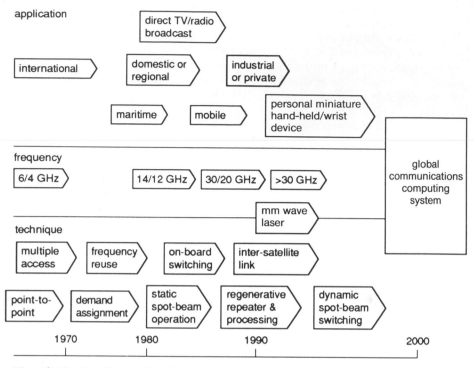

Fig. 17.15   Predictions for future satellite communications

A summary forecast for future satellite applications and techniques is given Fig. 17.15.

17.7 Summary

The progressive development of transmission technology over the past 150 years has resulted in an exponential growth of capacity, services and performance. In reality this has been something of an illusion created by the overlay of a series of logistic curves for each major development, as depicted in Fig. 17.16. Each new innovation has produced a curve that completely swamps its predecessors and dominates our view of the scene. The logistic start up, technical improvement, gathering momentum, leading to eventual decline is thus not visible from the overall characteristic. Moreover, the most recent optical technology offers such an advance that exponential growth looks set to continue well into the next millenium.

Precisely where the optical era will end is impossible to forecast. What is clear is that there is an abundance of silica for glass production and, unlike

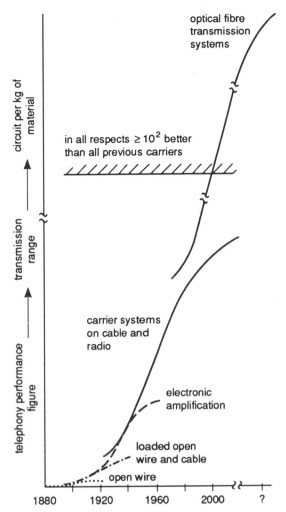

Fig. 17.16   Project logistic curve for further transmission system development

copper, it appears inexhaustible. Our summary prognosis for the future is thus:

(i)   Merging of communication, computing, control, information and entertainment services

(ii)   All digital transmission and switching dominated by optical fibres and integrated optics — but with a distinct possibility of migration back to analogue modes of operation

(iii)   Dominance of (largely repeaterless) monomode terrestrial and undersea global point-to-point backbone network

(iv)   Mobile, personal, portable and difficult-access communication provided by satellite and microwave systems.
(v)   Convergence of digital hierarchies, bit rates of $\geqslant 10$ Gbit/s and versatile (bit-rate-independent) access.
(vi)   Increased bit/Hz efficiency on radio, but deliberate bandwidth wasting on optical bearers in order to reduce processing equipment complexity.

17.8 References

1 'The World's Telephones', AT&T Long Lines, USA, 1982
2 CHERRY, E.C.: 1978, 'World communication — threat or promise?' (John Wiley, 1978)
3 DE MOTBRIAL, T.: 'The energy countdown: A report to the club of Rome', (Pergamon, 1979)
4 ROBERTS, P.C.: 'Limits to growth revisited' (Halstead, 1978)
5 MADDOX, B.: 'Beyond Babel: New directions in communication' (Andre Deutsch, 1972)
6 MARTIN, J.: 'Future developments in telecommunications' (Prentice Hall, 1971)
7 JONES, T.: 'Microelectronics and society', (Open University Press', 1980)
8 'Outcomes from Telecom 2000'. National Telecommunications Planning Report, Australia, 1978
9 'Telecommunications trends and directions', EIA Seminar, June 1982, Hyannis, Mass, USA
10 'Telecoms in the 80s and after', Royal Society, London, 1978
11 *Electron & Power*, 1983
12 MARTIN, J.: 'The wired society', (Prentice Hall, 1978)
13 MARTIN, J.: 'Telematic Society', Prentice Hall, 1981)
14 Proc. ICC, IEEE, June 1983, Boston, CH 1874–7
15 ALVEY, J.: 'A programme for advanced IT' (HMSO, 1982)
16 BLEAZARD, G.B.: 'Telecommunication in transition — A position paper', NCC, Manchester, 1982
17 WILLIAMS, M.B.: 'Pathways to the information society'. Proc Sixth Int. Conf. on Computer Comms. (North-Holland, 1982)
18 *Telephony*, 1983, 200
19 BOTHWELL, J.C.: *Satell. Commun.* 1980, **4**, pp. 32–35
20 FLOOD, J.E.: 'Telecommunications networks' (Peter Peregrinus, 1977)
21 BALLART, R., and CHING, Y.C.: 'SONET Now it's the standard optical network', *IEEE COMSOC Mag*, 1990, **29**, pp. 8–15
22 KARBOWIAK, A.E.: 'Trunk waveguide communications', (Chapman & Hall, 1965)
23 MARTIN-ROYLE, R.D.: 'Optical fibre transmission systems in the British Telecom network — An overview!', *Br. Telecommun. Eng.*, 1983, **1**, pp. 190–199
24 GAMBLING, W.A.: *Electron. & Power*, 1983, **29**, pp. 777–780
25 COCHRANE, P.: 'Future trends in telecommunication transmission', *Proc. IEE*, 1984, **131F**, pp. 669–683
26 MAY, C.A.: 'Opening address', ICC 83, Boston, USA
27 ANDERSON, R.E., FREY, R.L., LEWIS, J.R., and MILTON, R.T.: *IEEE Trans.* 1981, **VT–30**, pp. 54–61

28 NASA: 'Requirements for a mobile communication satellite system'. TRW Space Technology Group Report, 1983
29 'Satellite systems for mobile comms and navigation', 1983, IEE Conf. Publ. 222
30 MCGEEHAN, J.P., and YATES, K.W.: *Telecommunications*, 1986, **20**, pp. 9
31 Satellite Summit Industry Conference, 1983, London
32 GARDINER, J.G.: 'Satellite services for mobile communications', *Telecommunications*, Sept. 1986, pp. 75–88
33 SCHNEIDER, M., and THOMAS, H.C.: *Comms. ACM*, 1983, **26**, pp. 252
34 Special Issue on 'Tomorrow's Computers' *IEEE Spectrum*, 1983
35 Special Issue on 'ISDN', *IEEE J.*, 1986, **SAC–4**(3) and (8)
36 Special Issue on 'Broadband comms', *IEEE J.*, 1986, **SAC–4**,(4)
37 ABRONSON, C.J.: *Microwave Syst. News*, 1983, **13**, pp. 51
38 Special Issue on 'Fibre optic systems for terrestrial applications', *IEEE J.*, 1986, **SAC-4**, (9)
39 Conf. Proc. IOCC 83, Tokyo, June 1983
40 Light Tec', *IEEE J.* March 1983, **LT–1**.
41 BRAIN, M.C.: 'Performance requirements for field application of coherent optical systems'. SPIE Fibre LASER 86, Cambridge, MA, Sept. 1986.
42 O'MAHONY, M.J.: 'Semiconductor laser amplifiers as repeaters'. IOOC–ECOC 85, Venice, Oct. 1985, Vol 2, pp. 39
43 ROWBOTHAM, T.R.: 'Submarine telecommunications', *BTTJ*, 1987, **5**, pp. 5–24
44 Proc. 3rd Int. Symp. on Halide Glasses, 1985, Rennes, France
45 COCHRANE, P., and BRAIN, M.C.: 'Future optical fibre transmission technology and networks', *IEEE COMSOC Mag*, 1988, **26**, pp. 45–60
46 Proc. Int. Conf. Lasers and Electro-optics, 1985, Baltimore
47 Special Issue on 'Microwave radio', *IEEE COM–SOC Mag.*: 1987, **25**(1) and (2)
48 Special Issue on 'Space 25', *IEEE Spectrum*, 1985, **20**
49 BINDER, R., HUFFMAN, D., CRANTE, I., and VENA, P.: 'Cross-link architectures for multiple satellite systems', *Proc. IEEE*, 1987, **75**(1)
50 *Space Communications and Broadcast*, 1983, (1)
51 'Antenna applications', Symposium, Univ of Illinois, 1982

Index